The Fundamentals of Process Intensification

The Fundamentals of Process Intensification

Andrzej Stankiewicz

Tom Van Gerven

Georgios Stefanidis

Authors

Prof. Andrzej Stankiewicz
Delft University of Technology
Process & Energy Department
Leeghwaterstraat 39
2628 CB Delft
The Netherlands

Prof. Tom Van Gerven
KU Leuven
Department of Chemical Engineering
Celestijnenlaan 200F
B-3001 Heverlee (Leuven)
Belgium

Prof. Georgios Stefanidis
KU Leuven
Department of Chemical Engineering
Celestijnenlaan 200F
B-3001 Heverlee (Leuven)
Belgium

Library of Congress Card No.:
applied for

British Library Cataloguing-in-Publication Data
A catalogue record for this book is available from the British Library.

Bibliographic information published by the Deutsche Nationalbibliothek
The Deutsche Nationalbibliothek lists this publication in the Deutsche Nationalbibliografie; detailed bibliographic data are available on the Internet at <http://dnb.d-nb.de>.

© 2019 Wiley-VCH Verlag GmbH & Co. KGaA, Boschstr. 12, 69469 Weinheim, Germany

Print ISBN: 978-3-527-32783-6
ePDF ISBN: 978-3-527-68013-9
ePub ISBN: 978-3-527-68015-3

Cover Design Formgeber, Mannheim, Germany
Typesetting SPi Global, Chennai, India
Printing and Binding Markono Print Media Pte Ltd, Singapore

Printed on acid-free paper

10 9 8 7 6 5 4 3 2 1

To Ewa, Barbara, and Jenny

Contents

Preface

In the coming decades, our society will be confronted with enormous global challenges that include energy transition, climate change, health care, and scarcity of materials, food, and water. The chemical and related process industries present an important element of those grand societal challenges, in particular regarding materials and energy efficiency, environmental burden, and operational safety. Process intensification (PI), aiming at substantial increase in the material and energy efficiency, improved safety, and environment-friendly processing, presents one of the most promising ways of dealing with those challenges and is commonly seen as an important progress area in chemical and process engineering.

This book has come into being as a result of c. 15 years of experience in teaching Process Intensification at MSc level in Delft University of Technology, the Netherlands, and in the University of Leuven, Belgium. It presents a different approach to PI than the ones seen in the numerous books published on that topic so far. Here, we try to explain Process Intensification in a more fundamental way, through four generic principles and four elementary domains: spatial, thermodynamic, functional, and temporal. As the result of such an approach, the vision on PI becomes broader than the traditional one. We go beyond the usual PI methods and equipment aimed to remove the limitations in heat and mass transfer or mixing. We look, among other things, into manipulating the molecules or even atoms, in order to improve the reaction kinetics, or into eliminating the randomness in the systems, in order to even out the processing experience of the molecules. We discuss the mechanisms of the interactions between various energy forms and materials, we look for the synergies at different process levels, and we analyze the ways of manipulating the time characteristics of the events.

The book consists of three parts. In the first part, we present a short history of Process Intensification and discuss the development of its definitions and interpretations (Chapter 1). The generic principles of Process Intensification are introduced in Chapter 2. Part II consists of four chapters, each dedicated to one of the four elementary domains addressed by PI. In the spatial domain (Chapter 3), various forms of structures targeting different phenomena at different scales are discussed. Chapter 4 (thermodynamic domain) focuses on the use of alternative forms and transfer mechanisms of energy in processing systems. In the functional domain (Chapter 5), we examine the possibilities of combining various functions within one processing step or one apparatus, in order to achieve synergistic effects. Finally, in Chapter 6, various aspects of Process Intensification in the temporal domain, such as manipulation of characteristic times of events or introduction of dynamic elements, are discussed. The third, final part of the book (Chapters 7 and 8) focuses on the practical implications of Process Intensification for sustainability. It discusses the methods for ecological assessment of PI technologies and the effects of PI on the inherent process safety. It also presents an approach to the conceptual design of an intensified chemical plant, illustrated with the example of a case study that the MSc students in Delft and in Leuven do every year, namely the PI-based conceptual redesign of the Union Carbide's carbaryl process in Bhopal, once the place of the worst technological disaster in the history of mankind.

Although the present book is primarily aimed as the support in the postgraduate-level teaching of Process Intensification, we believe that a considerable part of the material presented here will also be of interest to engineers working in industry. This includes not only the chemical processing sector, the birthplace of PI, but also other industrial sectors where the word EFFICIENCY is the name of the game.

The preparation of this book has been a long and complex process and we would like to express our sincere gratitude to Keerthivasan Rajamani for the immense editorial work on the text and the images and to Florens Kruik for providing numerous graphics for the book.

March 2019
Delft, Leuven

Andrzej Stankiewicz
Tom Van Gerven
Georgios Stefanidis

About the Authors

Andrzej Stankiewicz: Full Professor and Chair of Process Intensification at Delft University of Technology, the Netherlands, and Director of TU Delft Process Technology Institute. With almost 40 years of industrial and academic research experience, he is the author of numerous scientific publications on Process Intensification, chemical reaction engineering, and industrial catalysis. Prof. Stankiewicz is one of the pioneers of Process Intensification. He is the principal author and co-editor of the world's first book on Process Intensification. The book was translated to Chinese in 2012. Prof. Stankiewicz is also the author of the first full-size academic course on Process Intensification.

Prof. Stankiewicz is the Editor of Chemical Engineering and Processing: Process Intensification (Elsevier) and the Series Editor of the Green Chemistry Books Series (Royal Society of Chemistry). He was the founder and first Chairman of the Working Party on Process Intensification at the European Federation of Chemical Engineering. He currently chairs the Board of the European Process Intensification Centre (EUROPIC).

Current research interests of Prof. Stankiewicz focus on control of molecular interactions and intensification of chemical reactions using electricity-based energy fields (e.g. laser, microwave, and UV). The research in that area has brought him prestigious Advanced Investigator Grant from the European Research Council. More recently, Prof. Stankiewicz has initiated and coordinates the H2020 "ADREM" project on methane valorization using alternative forms of energy in modular catalytic reactors.

Tom Van Gerven: Professor of Process Intensification at the University of Leuven, Belgium, and head of the Process Engineering for Sustainable Systems section in the same university. He is the author of numerous publications on Process Intensification, solid waste treatment, and sustainable processing. He focuses on the use of ultrasound and light for efficient chemical processing. Since 2013, he is the chairman of the Working Party on Process Intensification at the European Federation of Chemical Engineering. In 2019, he chaired the second International Process Intensification Conference (IPIC2) in Leuven, Belgium.

Georgios Stefanidis: Professor at the University of Leuven. He holds a Diploma in Chemical Engineering from the National Technical University of Athens and a PhD degree in the same field from the University of Gent. He has co-authored numerous publications in the broad field of Process Intensification, mostly focusing on alternative energy forms and transfer mechanisms (mainly microwaves and plasma). In the area of plasma processing, he received, among other things, a prestigious grant from the Bill and Melinda Gates Foundation. He is currently the Associate Editor of the Chemical Engineering and Processing: Process Intensification Journal (Elsevier), Vice Chair of the EFCE Working Party on Process Intensification, and serves in the management board of the Association of Microwave Power in Europe for Research and Education (AMPERE).

Part I

Principles

1

Introduction

1.1 Short History of Process Intensification

The timeline of process intensification (PI) (Figure 1.1) is about four decennia long. Although the term "process intensification" started to appear in East European publications on metallurgy already in mid-1960s and early 1970s, it was meant simply as equivalent to "process improvement." In chemical engineering, the first appearance of process intensification as we know it today was marked by the paper on application of centrifugal fields (so-called "HiGee") in distillation processes [1] published in 1983 by Colin Ramshaw from the ICI's New Science Group. The ICI project had been triggered by one of the NASA research projects on producing high transfer rates by using centrifugal fields in the zero gravity environment. Consequently, in the first years after its birth process, intensification was dominated by the rotating equipment, which still presents an important area of PI. Gradually, other technologies such as heat exchanger (HEX) reactors, intensive mixing devices, or microchannel reactors emerged within the PI domain.

Until early 1990s, process intensification was almost exclusively a British discipline. It was also the British BHR Group that organized the first international conference on PI in 1995 [2].

As can be seen in Figure 1.1, the real acceleration came in the last years of the second and, in particular, the first years of the third millennium, when a fast growth of PI-related activities in industry and in academia was observed. National academic–industrial PI networks have been established, first in the United Kingdom, later in the Netherlands, and in Germany. Process intensification has found its way to the university curricula. First books on PI were published [3–5] and the first PI-dedicated journal *Chemical Engineering and Processing: Process Intensification* was launched in 2007. In 2005, the European Federation of Chemical Engineering recognized the importance of PI by establishing the Working Party on Process Intensification (www.efce.eu/wp_pi). Since then, the Working Party organized five European Process Intensification Conferences (EPIC) and two International Process Intensification Conferences (IPIC) held in Barcelona in 2017 and in Leuven in 2019. The *European Roadmap for Process Intensification*, published in 2008 [6] and based on the contributions by the experts from 16 countries, laid down a foundation for short- and mid-term

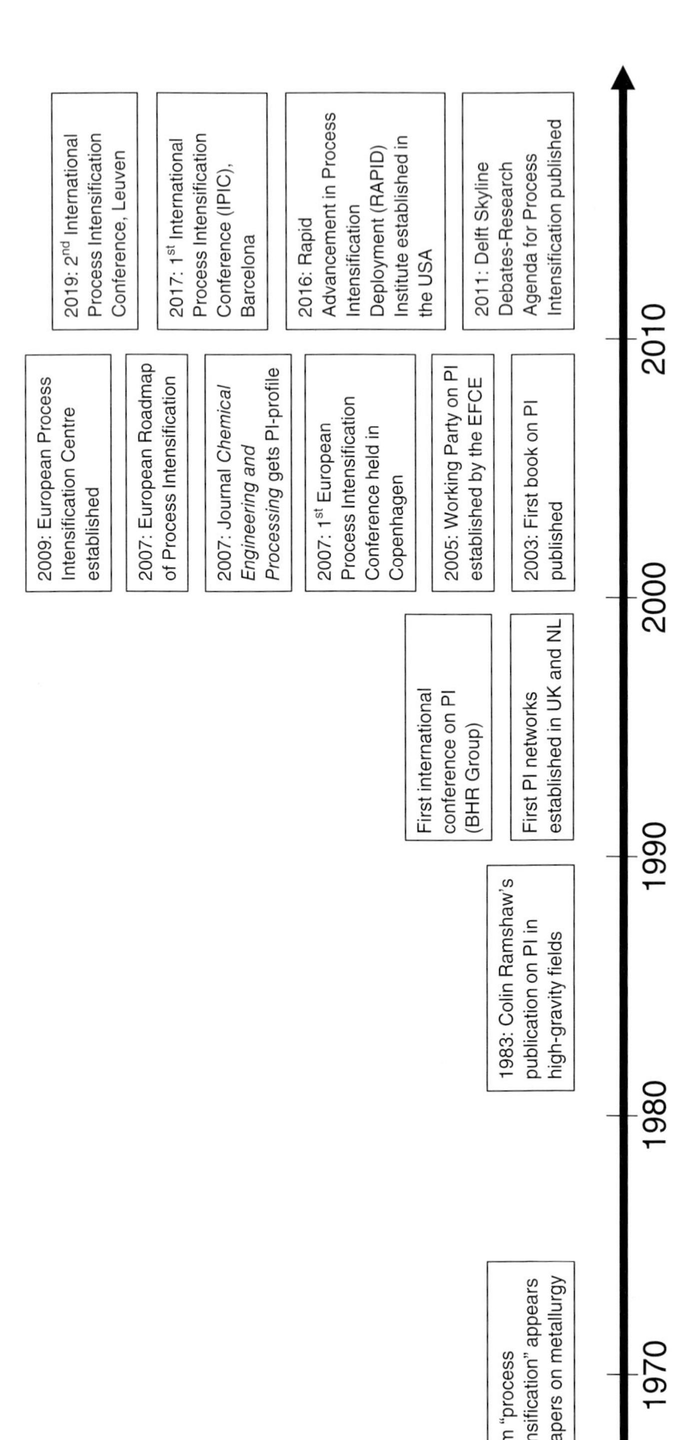

Figure 1.1 Timeline and milestones of process intensification.

research programs in the field. In 2009–2011, the roadmap got a follow-up in the form of the *Delft Skyline Debates* project, during which a multidisciplinary team of 75 leading academics and industrialists from different countries created a scientific vision on long-term developments in the field of process intensification that would reach beyond the horizon of 2050. The vision was published as a series of position papers [7] and also delivered a long-term research agenda for process intensification [8]. It is interesting to note that the above research agenda has gone beyond the traditional application area of PI, i.e. the chemical process industries, and has also addressed other areas including energy, water, and health. Also, in 2009, the European Process Intensification Centre (EUROPIC, www.europic-centre.eu) was established. The center presents an industry-driven platform for knowledge transfer in the field of PI and comprises chemical and pharmaceutical manufacturers, technology providers, equipment vendors, and engineering companies. Last but not least, the recently established *Rapid Advancement in Process Intensification Deployment (RAPID)* Institute (https://www.aiche.org/rapid), with private and governmental (US Department of Energy) funding exceeding 140 million dollars, presents a major development in this field in the United States and a proof of the importance of process intensification for the American economy.

It is interesting to note that after almost 40 years of development process, intensification has been brought again in connection with the space programs. Microchannel reactors for CO methanation are considered a promising technology for the *in-situ* resources utilization (ISRU) on Mars or on the Moon [9]. It appears that process intensification may one day revisit its birthplace – the Space.

1.2 Definitions and Interpretations of Process Intensification

From its very beginning, process intensification has been subject to numerous discussions and interpretations. In particular, various *definitions* of process intensification have been proposed in the literature, as presented in Table 1.1. Here, one can see that a considerable diversity exists in the way the researchers perceive and describe process intensification.

The *interpretations* of process intensification by various authors are quite diverse as well. For some, the *miniaturization* presents the fundamental issue of PI [22, 23], with microreactors being the common example. For others, process intensification is based on *functional integration* [14, 24], with reactive distillation as a prominent illustration. Arizmendi-Sánchez and Sharatt [25] combine different approaches by identifying *synergistic integration of process tasks and phenomena* and *targeted intensification of transport processes*, both on multiple scales, as the main PI principles. Finally, Freund and Sundmacher [26] present the suggestion that PI should and does follow a *function-oriented approach*. Similarly to the previous references, they identify different scales as to which this approach should be undertaken, from phase level via process unit level to plant level (thus not taking the molecular scale into account).

Table 1.1 Definitions of process intensification over the years.

Process intensification ...	Author (year)	References
... [is the] devising exceedingly compact plant which reduces both the "main plant item" and the installations costs.	Ramshaw (1983)	[1]
... [is concerned with] order-of-magnitude reductions in process plant and equipment.	Heggs (1983)	[10]
... [is a] philosophy of plant design and construction whereby a given performance is achieved in very much smaller equipment – typically with a volume reduction of 2–3 orders of magnitude.	Ramshaw (1985)	[11]
... [is the] strategy of reducing the size of chemical plant needed to achieve a given production objective.	Cross and Ramshaw (1986)	[12]
...[is a] novel design approach where fundamental process needs and business considerations are analyzed and innovative process technologies used to meet these optimally.	Green (1998)	[13]
... [is the] development of innovative apparatuses and techniques that offer drastic improvements in chemical manufacturing and processing, substantially decreasing equipment volume, energy consumption, or waste formation, and ultimately leading to cheaper, safer, sustainable technologies.	Stankiewicz and Moulijn (2000)	[14]
... [is the] strategy of making dramatic reductions in the physical size of a chemical plant while achieving a given production objective.	Dautzenberg and Mukherjee (2001)	[15]
... [implies] faster reactions, better conversions, improved or new products and fewer by-products.	Swamy and Narayana (2001)	[16]
... [is the] revolutionary approach to process and plant design, development and implementation. Providing a chemical process with the precise environment it needs to flourish results in better products, and processes which are safer, cleaner, smaller, and cheaper.	BHR Group (2003)	[17]
... refers to technologies that replace large, expensive, energy-intensive equipment or process with ones that are smaller, less costly, more efficient or that combine multiple operations into fewer devices (or a single apparatus).	Tsouris and Porcelli (2003)	[18]
...[is] any chemical engineering development that leads to a substantially smaller, cleaner, safer and more energy-efficient technology.	Costello (2004)	[19]
... [is the] holistic approach starting with an analysis of economic constrains followed by the selection or development of a production process. Process intensification aims at drastic improvements of performance of a process, by rethinking the process as a whole. In particular, it can lead to the manufacture of new products which could not be produced by conventional process technology.	Degussa (Franke 2009)	[20]

(Continued)

Table 1.1 (Continued)

Process intensification …	Author (year)	References
… [provides] radically innovative principles ("paradigm shift") in process and equipment design which can benefit (often with more than a factor 2) process and chain efficiency, capital and operating expenses, quality, wastes, process safety and more.	European Roadmap for Process Intensification (2007)	[6]
… [stands for] an integrated approach for process and product innovation in chemical research and development, and chemical engineering in order to sustain profitability even in the presence of increasing uncertainties.	Becht et al. (2009)	[21]

Next to the diversity in definitions and interpretations of process intensification, the *relation between PI and Process Systems Engineering* (PSE) has also drawn a lot of attention and triggered many discussions within the chemical engineering community. Moulijn et al. [27] in a comparison between PI and PSE argued that PI had "a more creative than integrating character and primarily aims at higher efficiency of individual steps in that chain, for instance, by offering new mechanisms, materials, and structural building blocks for process synthesis." They also saw process intensification acting more on the nanoscale (molecules) to macroscale (reactors) and less on the megascale (plants, sites, and enterprises). We share both these views. Process intensification should indeed be seen as a "supplier" of novel building blocks, from which process systems are synthesized, as shown in Figure 1.2.

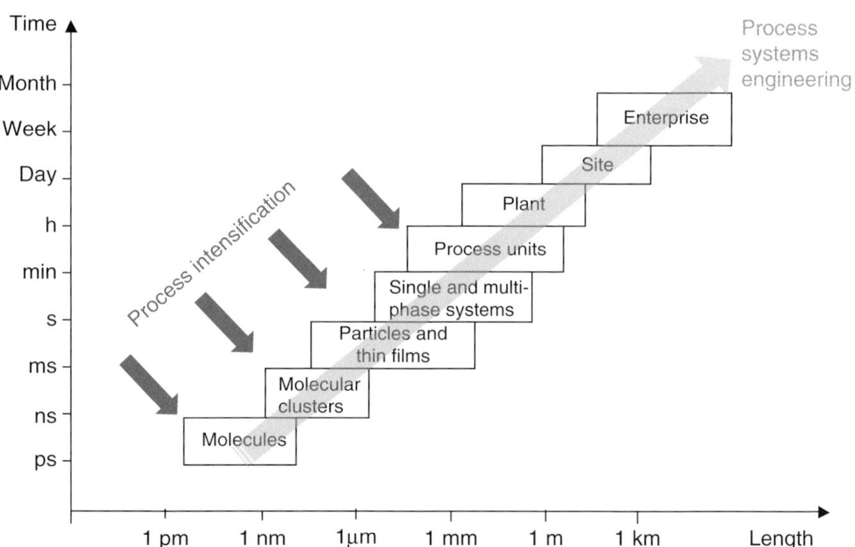

Figure 1.2 Process intensification delivers new concepts of individual steps in Process Systems Engineering chain, at small and medium scales (molecules to process units). Source: Adapted from Grossmann and Westerberg 2000 [28].

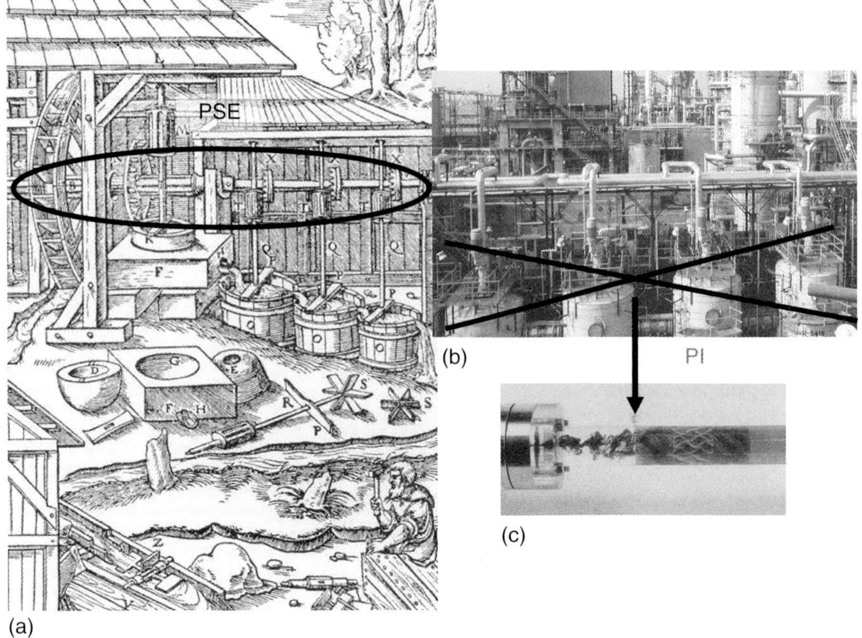

Figure 1.3 The relationship between PSE and PI: PSE integrates the individual steps, whereas PI introduces new concepts of those steps. Source: (a) Agricola 1556 [29]; (b) Stankiewicz 2006 [30]; (c) Sulzer Chemtech Ltd., Switzerland, www.sulzer.com.

Moulijn et al. [27] illustrated the distinction between PI and PSE with old and modern chemical plants shown in Figure 1.3. The PSE thinking in Middle Ages is presented in the left-hand side figure that comes from the sixteenth century book "De Re Metallica" by Georgius Agricola [29]. It illustrates the process of retrieving gold from ore. Ore is crushed by the stamp "C," ground in the mill "F," and mixed with mercury in vessels "O." Remarkably, these three operations are integrated and driven by a water wheel via a shaft and a number of gears. This is an example *par excellence* of the PSE thinking. On the other hand, a motionless mixer as a replacement of the stirred tanks in the modern plant presents a much more efficient, fundamentally different concept of mixing fluids and a classical example of process intensification.

1.3 Fundamentals of Process Intensification – Principles, Approaches, Domains, and Scales

In 2000, Stankiewicz and Moulijn [14] proposed a "toolbox" view of process intensification and divided that "toolbox" into two subdomains (Figure 1.4):

- *Process-intensifying equipment*: such as novel reactors, intensive mixing, heat transfer, and mass transfer devices;

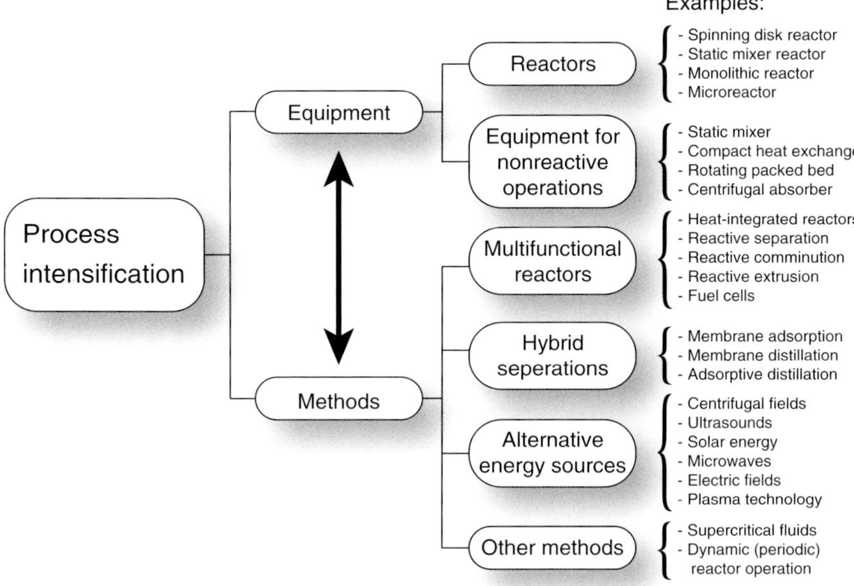

Figure 1.4 Process intensification as a "toolbox" with equipment and methods. Source: Adapted from Stankiewicz and Moulijn 2000 [14].

- *Process-intensifying methods*: such as new or hybrid separations, integration of reaction and separation and/or heat exchange and/or phase transition (in so-called multifunctional reactors), techniques using alternative energy sources (light, ultrasound, etc.), and new process control methods (e.g. intentional unsteady-state operation).

Obviously, in some cases, overlaps between these two domains can be observed as new methods often require novel types of equipment to be developed and vice versa, novel apparatuses already developed make sometimes use of new, unconventional processing methods.

The above-described classical "toolbox" view had functioned for almost a decade and was adopted or referred to in numerous publications. However, in the course of time, it became apparent that the "toolbox/examples" approach to process intensification was insufficient for a comprehensive analysis of existing and systematic development of new, intensified processes. It could be used for an incidental improvement of one or another process step or equipment, but it lacked the generic dimension, which is needed for a multiscale, holistic process design and development. To allow the latter, process intensification needs to be approached and understood in a more fundamental way. This is also the primary aim of the present book.

We will start the book by analyzing *the generic principles of PI*. Earlier approaches (e.g. [31]) defined the ultimate goal of process intensification as achieving a purely kinetics-limited process without any limitations resulting from the momentum, heat, and mass transfer. This is surely a valid approach, but it addresses only a part of the problem as, as you will read further in this book, reaction kinetics can also be intensified. By saying "intensified," we do not mean operating at higher temperatures or using catalysts. Fundamental research in the field of chemical physics shows that one can boost the kinetics of a chemical reaction by orders of magnitude providing the right form of energy to chemical bonds, without using a catalyst and without increasing the macroscopic temperature even by one degree! This proves that kinetic limitation is indeed a relative concept. Furthermore, the earlier approaches to process intensification neglected two other elements that are of great importance for an intensified process, namely the uniformity of the processing history for all molecules and the synergistic effects resulting from interactions and interrelations between various process steps at different scales.

In view of the above considerations, an ideal intensified processing system (and the *ultimate goal of PI*) is a system in which reactions proceed at a maximum achievable efficiency, all molecules undergo the same processing history, the hydrodynamic, heat and mass transfer limitations are removed, and the synergies resulting from interrelations between various operations and steps are fully utilized. Such a definition of the ultimate goal of PI leads us directly to the following four guiding principles, which we call *generic principles of process intensification*:

- Maximize the effectiveness of intra- and intermolecular events;
- Give each molecule the same processing experience;
- Optimize the driving forces and resistances at every scale and maximize the specific surface areas to which these forces or resistances apply;
- Maximize the synergistic effects from partial processes.

The above principles, in one form or another, are obviously not entirely new to chemical engineering. In process intensification, however, they result from the explicit definition of the goal that an intensified process aims to reach. Besides, the PI interpretation of these principles often goes beyond the boundaries of the classical chemical engineering. This can be seen, for instance, in the first principle, where process intensification looks at the molecular-scale methods for improving the intrinsic kinetics of chemical reactions, such as molecular alignment, orientation, and selective bond excitation.

In the realization of the above guiding principles, process intensification operates in *four elementary domains*: spatial, thermodynamic, functional, and temporal (Figure 1.5). A detailed analysis and discussion of those four domains and the corresponding PI approaches constitute the main part of this book.

Finally, one needs to stress that the above-described principles and approaches should address all *relevant time and length scales* (Figure 1.5). As you will see further in the book, this is indeed the case. Looking more into the future, we believe that, whereas process intensification until recent years has mostly been focused on the larger scales (from films and particles up to processing units), the major breakthroughs to come will arise from process intensification on the molecular scale.

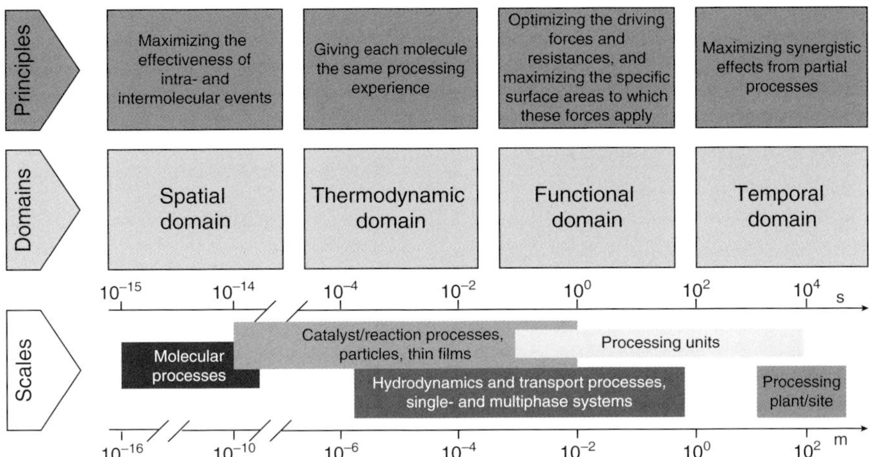

Figure 1.5 Principles, approaches, and scales of process intensification. Source: Adapted from Van Gerven and Stankiewicz 2009 [32].

References

1 Ramshaw, C. (1983). "Higee" distillation – an example of process intensification. *Chem. Engr. (London)* (389): 13–14.

2 Ramshaw, C. ed. (1995). *1st International Conference on Process Intensification for the Chemical Industry*, Antwerp, Belgium (6–8 December 1995). BHR Group Conference Series. Publication No. 18. London: Mechanical Engineering Publications Limited.

3 Stankiewicz, A. and Moulijn, J.A. (eds.) (2003). *Re-Engineering the Chemical Processing Plant*. New York: Marcel Dekker.

4 Keil, F.J. (ed.) (2007). *Modeling of Process Intensification*. Weinheim: Wiley-VCH.

5 Reay, D., Ramshaw, C., and Harvey, M. (2008). *Process Intensification: Engineering for Efficiency, Sustainability and Flexibility*. Oxford: Butterworth-Heinemann.

6 European Roadmap for Process Intensification (2007). Creative energy – energy transition. http://efce.info/EUROPIN.html (accessed 23 February 2019).

7 Górak, A. and Stankiewicz, A. (eds.) (2012). Special Issue: Delft Skyline Debates. *Chem. Eng. Proc. Proc. Intens.* 51: 1–149.

8 Górak, A. and Stankiewicz, A. (eds.) (2011). *Research agenda for process intensification – towards sustainable world of 2050*. Amersfoort: Institute for Sustainable Process Technology.

9 Dagle, R.A. and Wegeng, R.S. (2008). Microchannel CO methanation reactors for martian and lunar ISRU. *Proceedings of the 6th International Energy Conversion Engineering Conference (IECEC)*, Cleveland, Ohio (28–30 July 2008). AIAA 2008-5748, American Institute of Aeronautics and Astronautics, Inc.

10 Heggs, P. (1983). Process intensification. *Chem. Engr. (London)* (394): 13.

11 Ramshaw, C. (1985). Process intensification: a game for n players. *Chem. Engr. (London)* 416: 30–33.

12 Cross, W.T. and Ramshaw, C. (1986). Process intensification – laminar-flow heat-transfer. *Chem. Eng. Res. Des.* 64: 293–301.

13 Green, A. (1998). Process intensification: the key to survival in global markets? *Chemistry & Industry* 5: 168–172.

14 Stankiewicz, A. and Moulijn, J.A. (2000). Process intensification: transforming chemical engineering. *Chem. Eng. Progr.* 96 (1): 22–34.

15 Dautzenberg, F.M. and Mukherjee, M. (2001). Process intensification using multifunctional reactors. *Chem. Eng. Sci.* 56: 251–267.

16 Swamy, K.M. and Narayana, K.L. (2001). Intensification of leaching process by dual-frequency ultrasound. *Ultrason. Sonochem.* 8 (4): 341–346.

17 BHR Group (2003). http//:www.bhrgroup.co.uk/pi/index.htm (accessed 8 August 2010).

18 Tsouris, C. and Porcelli, J.V. (2003). Process intensification – has its time finally come? *Chem. Eng. Progr.* 99 (10): 50–55.

19 Costello, R.C. (2004). Process intensification: think small. *Chem. Eng.* 111 (4): 27–31.

20 Franke, R. (2007). Process intensification – an industrial point of view. In: *Modeling of Process Intensification* (ed. F.J. Keil), 9–23. Weinheim: Wiley-VCH.

21 Becht, S., Franke, R., Geisselman, A., and Hahn, H. (2009). An industrial view on process intensification. *Chem. Eng. Proc. Proc. Intens.* 48 (1): 329–332.

22 Stitt, E.H. (2002). Alternative multiphase reactors for fine chemicals. A world beyond stirred tanks? *Chem. Eng. J.* 90: 47–60.

23 Mae, K. (2007). Advanced chemical processing using microspace. *Chem. Eng. Sci.* 62: 4842–4851.

24 Huang, K., Wang, S.J., Shan, L. et al. (2007). Seeking synergistic effect – a key principle in process intensification. *Sep. Pur. Technol.* 57: 111–120.

25 Arizmendi-Sánchez, J.A. and Sharatt, P.N. (2008). Phenomena-based modularisation of chemical process models to approach intensive options. *Chem. Eng. J.* 15: 83–94.

26 Freund, H. and Sundmacher, K. (2008). Towards a methodology for the systematic analysis and design of efficient chemical processes – part 1: from unit operations to elementary process functions. *Chem. Eng. & Proc. – Proc. Intens.* 47 (12): 2051–2060.

27 Moulijn, J.A., Stankiewicz, A., Grievink, J., and Górak, A. (2008). Process intensification and process systems engineering: a friendly symbiosis. *Comput. Chem. Eng.* 32: 3–11.

28 Grossmann, I.E. and Westerberg, A.W. (2000). Research challenges in process systems engineering. *AIChE J.* 46: 1700–1703.

29 Agricola, G. (1556). *De Re Metallica.* Whitefish: Kessinger Publishing.

30 Stankiewicz, A. (2006). Digging for Gold in Jurassic Park: Sustainable Technologies for Non-Sustainable Mankind, Inaugural Address, Delft University of Technology.

31 Bakker, R. (2003). Process intensification in industrial practice. In: *Re-Engineering the Chemical Processing Plant* (ed. A. Stankiewicz and J.A. Moulijn), 447–470. New York: Marcel Dekker.

32 Van Gerven, T. and Stankiewicz, A. (2009). Structure, energy, synergy, time – the fundamentals of process intensification. *Ind. Eng. Chem. Res.* 48: 2465–2474.

2

The Four Principles

2.1 Principle 1 – Toward Perfect Reactions

The first of the four generic process intensification (PI) principles, *maximize the effectiveness of intra- and intermolecular events*, addresses all elementary processes in which such events play an important role, chemical reactions in particular. The control of chemical reactions at molecular level presents undoubtedly the most important scientific challenge on the way to fully sustainable, thermodynamically efficient chemical processes. The most obvious advantages of enhanced molecular reaction control are (i) higher reaction rates leading to low-temperature processes and smaller equipment, (ii) better selectivities leading to minimization or elimination of waste and the reduction of separation operations, which are responsible for c. 40% of energy consumption in chemical and related industries, and (iii) the possibility for tailored manufacturing of new, advanced products.

Let us begin with the basics. The Arrhenius equation describes simply yet remarkably accurately the dependence of the chemical reaction rate on temperature. Three parameters play a role here: the temperature itself, the activation energy E_a, and the so-called "pre-exponential" or "frequency" factor k_0. Classical chemical reaction engineering considers two ways of increasing the reaction rate: either by an increase in temperature or by a decrease of the activation energy, for which the catalysts are used (Figure 2.1). Much less attention is given to the third parameter, the earlier mentioned pre-exponential factor, which is related to the frequency and the effectiveness of collisions between the molecules.

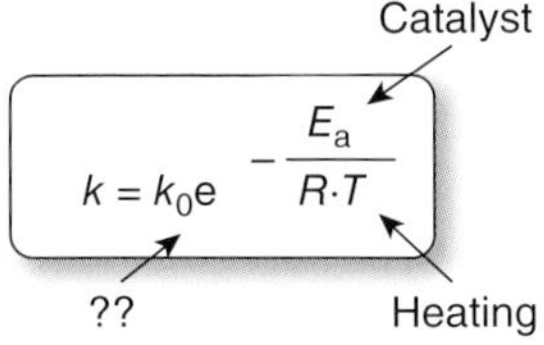

Figure 2.1 Arrhenius equation and the ways to increase the rate of a reaction.

The vast majority of the chemical reactions occur as a result of molecular collisions and the vast majority of these molecular collisions are ineffective. Even for quite simple molecules, as the ones shown in Figure 2.2, the probability of a collision leading to a reaction and product formation is much lower than the probability of an ineffective collision that results in the waste of energy.

The Fundamentals of Process Intensification, First Edition.
Andrzej Stankiewicz, Tom Van Gerven, and Georgios Stefanidis.
© 2019 Wiley-VCH Verlag GmbH & Co. KGaA. Published 2019 by Wiley-VCH Verlag GmbH & Co. KGaA.

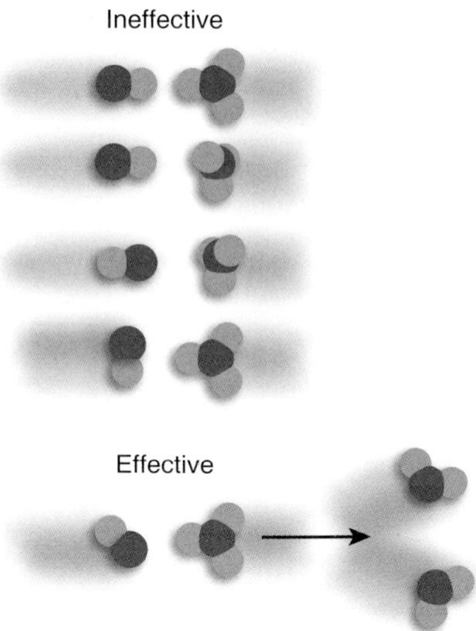

Figure 2.2 The probability of an effective collision between even simple molecules is much lower than that of an ineffective one.

Factors responsible for the effectiveness of a reaction include number/ frequency of molecular collisions, geometry of approach, mutual orientation of molecules at the moment of collision, and their energy. In this sense, the art of carrying out chemical reactions resembles the snooker game, where an experienced player remains in full control of the collisions between the balls he/she wants to get colliding, by providing them with the right amount of energy, the right geometry of approach, and the right mutual orientation at the moment of collision. Unfortunately, this is not the case in the current practice of chemical reactors as they offer a very limited degree of control of molecular-level events. Conventionally, in order to bring more molecules at the energy levels exceeding the activation energy threshold, conductive heating is applied. However, conductive heating offers only a macroscopic control upon the process and is thermodynamically inefficient. It is nonselective in nature, which means that nonreacting (bulk) molecules heat up together with the reacting ones. This way, a conductively heated reactor, instead of a well-controlled snooker game, resembles rather a pinball game where all balls move in all possible directions and randomly collide with each other. Furthermore, during conductive heating, other elements of the reactor (e.g. catalyst support) are unnecessarily heated up. Last but not least, the conductive heating generates temperature gradients, which may create a broad distribution of molecular energy levels. Full realization of the first PI principle would consist in creating a "perfect" reaction environment, in which the geometry of molecular collisions is controlled while energy is transferred selectively from the source to the required molecules in the required form, in the required amount, at the required moment, and at the required position (Figure 2.3).

Let us first look at the first of the challenges presented in Figure 2.3, i.e. control of the geometry of molecular collisions. In principle, we have two kinds of "tools" at our disposal: the nanostructures and the energy – see Figure 2.4. The nanostructure tool consists in immobilizing molecules for the time of their

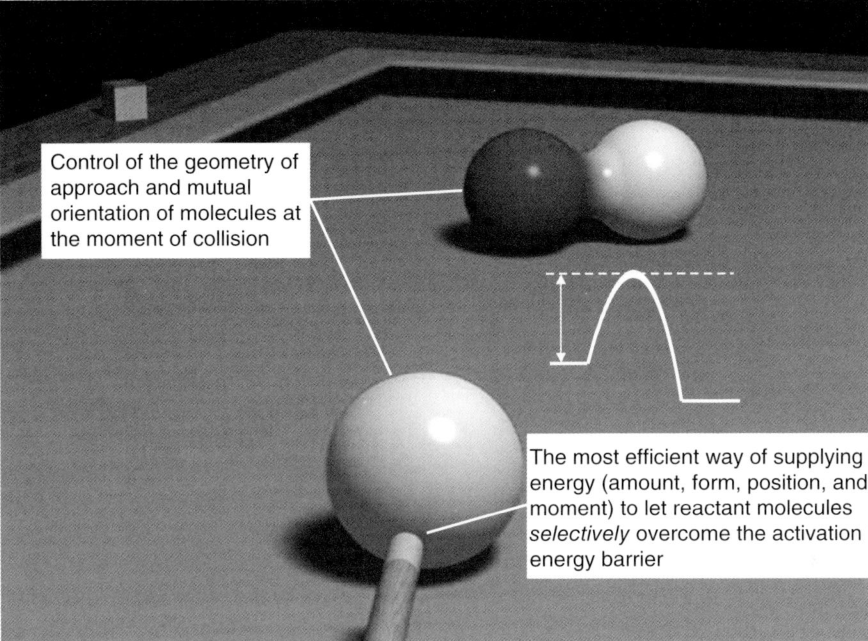

Figure 2.3 Basic challenges in creating a perfect reactor (molecular snooker table). Source: Stolte et al. 2004 [1]. Reproduced with permission of Elsevier.

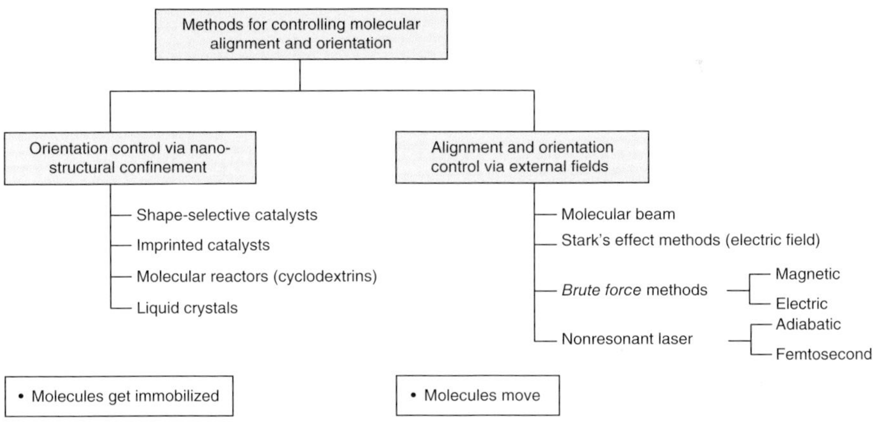

Figure 2.4 Main approaches and methods used for control of molecular orientation and alignment.

reaction in confined nanospaces by imposing on them "hard walls" of structures. This way, the reacting molecules are forced to either assume a certain position inside the offered structure or not to react ("take it or leave it"). A wide range of such systems including "molecular reactors" (cyclodextrins), shape-selective catalysts, e.g. zeolites, molecularly imprinted systems, and liquid crystals have been studied and are presented in more detail in Chapter 3. The other tool consists in acting upon molecules with spatially oriented external fields. Contrary to the previous case, here the molecules move. The basic methodology used is the Nobel Prize winning [2] molecular beam technique. Intersecting two molecular beams produce a local reaction environment with controlled collision geometry. Some simple inorganic molecules have been found to align naturally in a molecular beam. The "forced alignment" methods include the so-called brute force techniques, which utilize strong electric or magnetic fields, techniques focusing in electric fields through the Stark effect, and the optical (laser-induced) methods.

The realization of the second challenge, i.e. selective and timely energy supply to the molecules, presents an even more complex problem. Progress in this area is mostly seen in the world of fundamental chemical physics. An excellent example of such selective and timely energy supply can be, for instance, seen in the fundamental works by the group of Richard Zare at Stanford [3, 4], where the application of a laser field to excite and "stretch" the C—H bond in methane during its chlorination introduced the so-called "stripping" collisions increasing the reaction rate by a factor of more than 100, without using the "classical" means, i.e. temperature or a catalyst (Figure 2.5). This clearly proves that a targeted introduction of an alternative energy form can improve the reaction performance dramatically. It can enable getting far beyond the limits of the "conventional inherent kinetics," which are based on the macroscopic temperature, pressure, and concentrations.

In needs to be said that, next to chemical reactions, the control of intermolecular events plays an important role in crystallization processes, particularly in the nucleation stage. Also here, the transformation requires a passage over a free energy barrier and also here the rate of the process (primary nucleation) can be described by the Arrhenius-type formula [5]. Intermolecular interactions, especially hydrogen bonding, play a significant role in the formation of crystal polymorphs, i.e. different crystal forms of chemically identical material [6]. As polymorphs often exhibit different properties, the control of their formation presents a very important issue in the production of pharmaceuticals, food, pigments, or energetic materials. Figure 2.6 summarizes the elements that play the role in controlling the polymorphic crystallization [7].

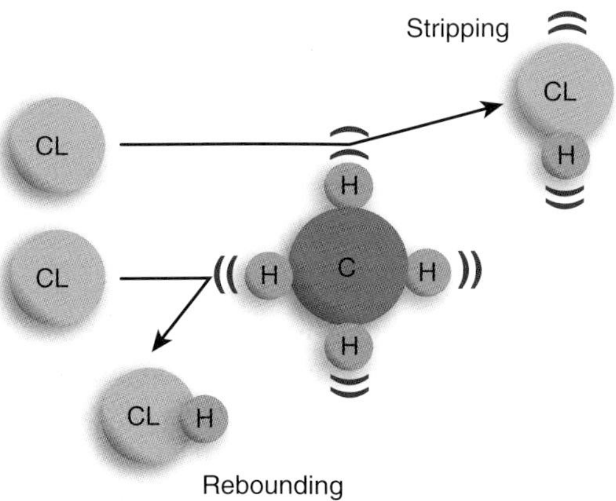

Figure 2.5 Introduction of stripping collisions by selective exciting and stretching of the C—H bond in methane molecule with nonresonant laser pulses. Source: Adapted from Hoffmann 2000 [4].

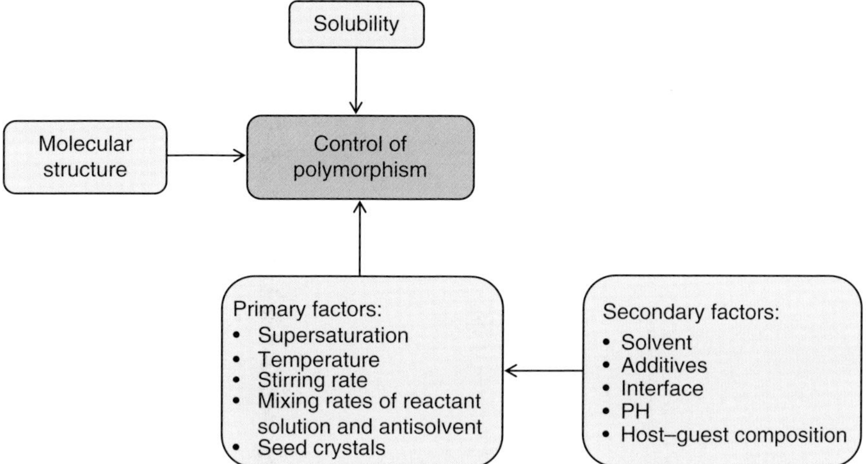

Figure 2.6 Elements playing the role in controlling polymorphic crystallization. Source: Adapted from Kitamura 2009 [7].

2.2 Principle 2 – What Experience Molecules?

The reason to *give each molecule the same processing experience* is simple: processes, in which all molecules undergo the same history, deliver ideally uniform products with minimum waste. Here, not only residence time distribution, dead zones, or bypassing but also meso- and micromixing as well as temperature gradients play an important role.

Obviously, a plug flow device is much closer to the realization of the above principle than a stirred tank unit, in which the distribution of residence times is broad. However, one must not forget that even a plug flow device does not guarantee real uniformity at the molecular level. Plug flow in chemical engineering describes a certain flow characteristic at the level of fluid elements, measured by the macroscopic concentration distribution. Molecules of a gas moving in a plug flow are not in a "plug flow" themselves as they move freely with high velocities in all directions. In order to have a molecular "plug flow," one needs to apply special techniques, such as the earlier mentioned molecular beam, which is in fact a collection of noncharged molecules moving in the same direction at very high translational velocity and under collision-free (supersonic expansion into vacuum), extremely low-temperature conditions (Figure 2.7).

The second principle also targets the uniformity of the thermal conditions in the equipment. Temperature gradients or local hotspots should be avoided. Unfortunately, the conductive heating is inherently connected with temperature differences (driving force for heat flow). We illustrate that problem with a simplified example of a jacket-heated stirred tank reactor inspired by Schwalbe et al. [8] and shown in Figure 2.8. Here, the molecules entering the vessel take random trajectories through its volume encountering colder and hotter zones. The energy of molecules is therefore distributed. In the case of a parallel reaction scheme of the type shown in the figure, where P is the required product, a

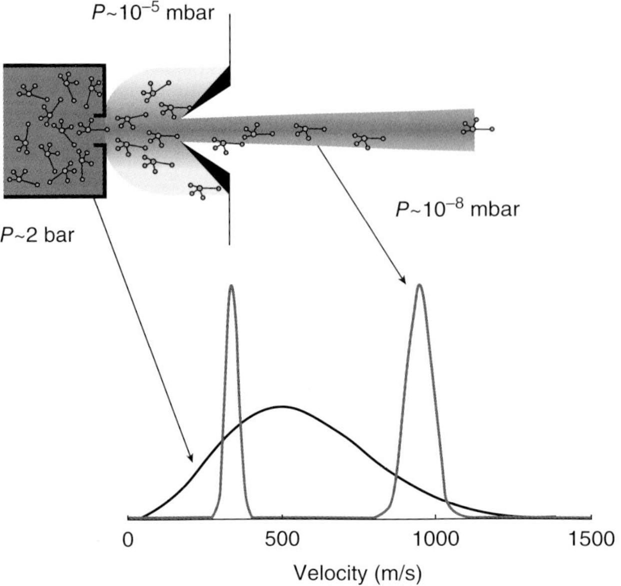

Figure 2.7 Principle of molecular beam: supersonic expansion of diluted reactant + inert gas mixture ensures high translational energy with very small spread leading to narrowly defined velocities of molecules.

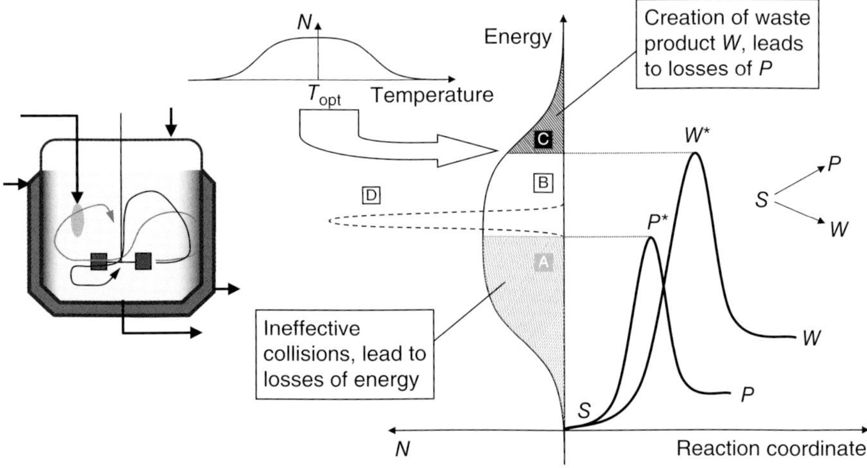

Figure 2.8 Illustration of the energy distribution problem in molecules, in relation to the yield of simple parallel reactions. A stirred tank reactor with conductive heating generates a wide energy distribution due to temperature gradients, which translate to both material and energy losses.

portion of molecules (A) has insufficient energy to pass the transition state P^*. Another portion (B) has sufficient energy to get over the threshold and form product P. A large number of these molecules has in fact more energy than it is needed to form P. Finally, there are molecules in the pool (C) that possess enough energy to also generate the transition state W^*, which eventually leads to the formation of the unwanted waste product W. Ideally, one should provide all molecules with a narrowly distributed amount of energy, just exceeding the energy level of P^*, as it is illustrated by the dashed curve D.

Obviously, the above picture is simplified. However, many chemical accidents are caused by a local temperature increase, which gives rise to uncontrollable highly exothermic side reactions leading to the so-called "thermal runaway" in the process. Clearly, the thermal uniformity in the equipment has not only an energy- or material-related but also a safety-related aspect.

2.3 Principle 3 – Driving Forces, Resistances, and Interfaces

The third principle, *optimize the driving forces and resistances at every scale and maximize the specific surface areas to which these forces or resistances apply*, is about the transport rates across interfaces. The word "optimize" is used here on purpose, as not always the maximization of the driving force (e.g. concentration difference) or minimization of the resistance (e.g. in mass transfer film) is required. On the other hand, the resulting effect always needs to be maximized, and this is done by the maximization of the specific surface area, to which that driving force or resistance applies.

Generally speaking, enhanced specific surface areas (or surface-to-volume ratios) can be realized via the structuring of the processing environment or by application of various forms of energy. Such enhanced surface areas can be seen, for example, in advanced packing structures used in mass transfer operations, such as extraction, absorption, or distillation. One can increase the specific surface areas by decreasing the characteristic dimensions of the system. In liquid–liquid or gas–liquid systems, it can, for instance, be realized in the form of microdroplets of microbubbles. As an example, the generation of oxygen microbubbles in an industrial fermenter increased its production capacity by a factor of c. 2 because of an enormous enhancement of the specific area for mass transfer [9] (see Section 4.6.3).

In channel-based systems (e.g. hollow fiber membranes and microreactors), the specific surface areas for mass or heat transfer is enhanced by moving from the millimeter to the micrometer scales of channel diameters. A microchannel of 100 μm in diameter delivers a very high specific area of c. 40 000 m^2/m^3. The nature, however, still leads the race: the capillaries in our bodies (Figure 2.9) are about 10 μm in diameter, have specific areas of c. 400 000 m^2/m^3, and (most of the time) do not clog!

2.4 Principle 4 – About the Synergies

It is evident that in a chemical process, synergistic effects should be sought and utilized, whenever possible and at all possible scales. Looking for and *maximizing the synergistic effects from partial processes* presents the fourth

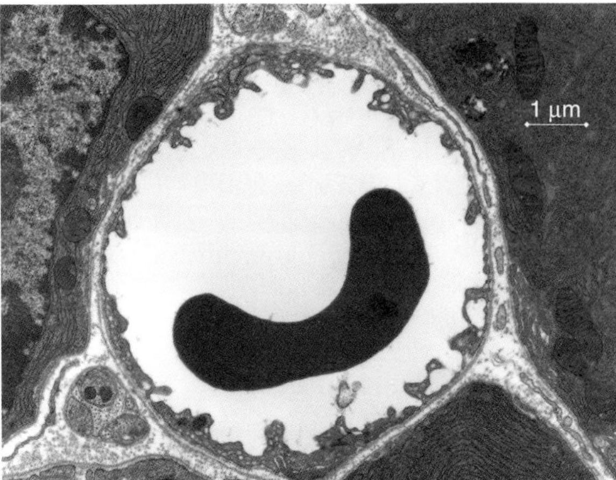

Figure 2.9 Still out of reach for microengineering: TEM image of a human pancreas capillary with a red blood cell in it. Source: By Louisa Howard – http://remf .dartmouth.edu/ imagesindex.html, Public Domain, https://commons .wikimedia.org/w/index.php? curid=1246750.

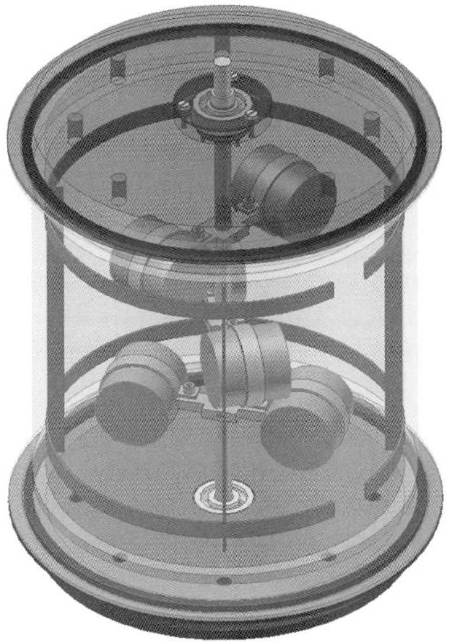

Figure 2.10 Monolithic stirrer reactor. Source: Hoek 2004 [10].

generic principle of PI and multifunctionality is the key term here. Bringing multiple functions together in one component (a molecule, a phase, and a reactor) often leads to significantly better performance than the separate functions executed sequentially. For example, the monolithic stirrer reactor (Figure 2.10) developed at Delft University of Technology not only combines the function of a heterogeneous catalyst with that of a stirrer but also eliminates the troublesome slurry catalyst separation step and improves process safety [10, 11].

Synergies can sometimes be realized on the nanoscale and multifunctional catalysts are a good example of such small-scale realization of this principle. Most commonly, however, the realization of synergies occurs on the macroscale, for instance, in reactive separation units, where the reaction equilibrium is shifted by removing the products in situ from the reaction environment, or in hybrid separations, where two or more separation techniques are combined with each other to deliver better separation efficiency or energy savings. Chapter 5 discusses in detail the various ways of realizing synergies in the functional domain.

References

1 Stolte, S., Suits, A., and Linnartz, H. (2004). Preface. *Chem. Phys.* 301: 159–160.
2 Lee, Y.T. (1986). *Molecular Beam Studies of Elementary Chemical Processes.* Nobel Prize Lecture.
3 Kandel, S.A. and Zare, R.N. (1998). Reaction dynamics of atomic chlorine with methane: importance of methane bending and torsional excitation in controlling reactivity. *J. Chem. Phys.* 109: 9719–9727.
4 Hoffmann, R. (2000). Exquisite control. *Am. Sci.* 88: 14–25.
5 De Yoreo, J.J. and Vekilov, P. (2003). Principles of crystal nucleation and growth. In: *Biomineralization*, vol. 54 (ed. P.M. Dove, J.J. De Yoreo and S. Weiner), 57–93. Washington: Mineralogical Society of America.
6 Caira, M.R. (1998). Crystalline polymorphism of organic compounds. *Topics Curr. Chem.* 198: 164–208.
7 Kitamura, M. (2009). Strategy for control of crystallization of polymorphs. *CrystEngComm* 11: 949–964.
8 Schwalbe, T., Autze, V., Hohmann, M., and Stirner, W. (2004). Novel innovation systems for a cellular approach to continuous process chemistry from discovery to market. *Org. Proc. Res. Dev.* 8: 440–454.
9 Groen, D.J., Noorman, H.J., and Stankiewicz, A. (2005). Improved method for aerobic fermentation intensification. In: *Papers Presented at Sustainable (Bio)Chemical Process Technology Incorporating 6th International Conference Process Intensification*, Delft, The Netherlands: 27–29 September 2005 (ed. P. Jansens, A. Stankiewicz and A. Green), 105–112. Cranfield: BHR Group Ltd.
10 Hoek, I. (2004). Towards the catalytic application of a monolithic stirrer reactor. PhD Dissertation. Delft University of Technology.
11 Edvinsson-Albers, R.K., Houterman, M.J.J., Vergunst, T. et al. (1998). Novel monolithic stirrer reactor. *AIChE J.* 44: 2459–2464.

Part II

Domains

3

STRUCTURE – PI Approaches in Spatial Domain

3.1 Randomness and Order: Why Structure?

Randomness accompanies the chemical processes at all scales. The Brownian motion is random in nature. Collisions between the molecules in a reactor have also a random character (unless special measures are undertaken). On the mesoscale, pore distribution in common porous materials (e.g. catalyst particles) exhibits a high degree of randomness because of the manufacturing techniques used. Typical examples of the random systems on macroscale are beds of particulate catalysts in industrial catalytic reactors.

Let us take, for instance, a multitubular fixed bed reactor. Such a reactor (Figure 3.1) can contain more than 30 000 tubes, each tube filled with a bed of catalyst particles. The loading of fresh catalyst into the tubes delivers in principle a random configuration of the bed, as we are not able to exactly control the positioning of particles with respect to each other during that operation. As a result, different tubes may exhibit different properties, such as voidage distribution or hydraulic resistance. If these differences were large enough, the entire reactor would malfunction during its operation, as the reactants would preferentially flow through the tubes with lower hydraulic resistance. This is why all the tubes in such reactors are individually checked for pressure drop after being filled with fresh catalysts. Tubes deviating from the required average by more than the allowed limit need to be discharged and reloaded again. Such procedure in large reactors can take weeks, which means that multitubular reactors should only be used if the expected catalyst lifetime is long.

Trickle-bed reactors on the other hand are often used in solid-catalyzed gas–liquid reactions. In these reactors, gas and liquid flow co-currently downward through a single bed of particulate catalyst (Figure 3.2a). The bed can be of large dimensions (often several meters in diameter). Obviously, loading such a large catalyst bed into the reactor has, as in the previous case, a highly random character. The distribution of local bed porosity (voidage) will each time be different and nonuniform. The nonuniformity may result in a secondary flow maldistribution inside the bed, where liquid will take preferential flow paths through the high-voidage (low resistance) areas, leaving other areas dry. Such a situation is presented in Figure 3.2b, showing liquid retention (%) in a 60-cm trickle-bed, measured by γ-ray tomography [1].

The Fundamentals of Process Intensification, First Edition.
Andrzej Stankiewicz, Tom Van Gerven, and Georgios Stefanidis.
© 2019 Wiley-VCH Verlag GmbH & Co. KGaA. Published 2019 by Wiley-VCH Verlag GmbH & Co. KGaA.

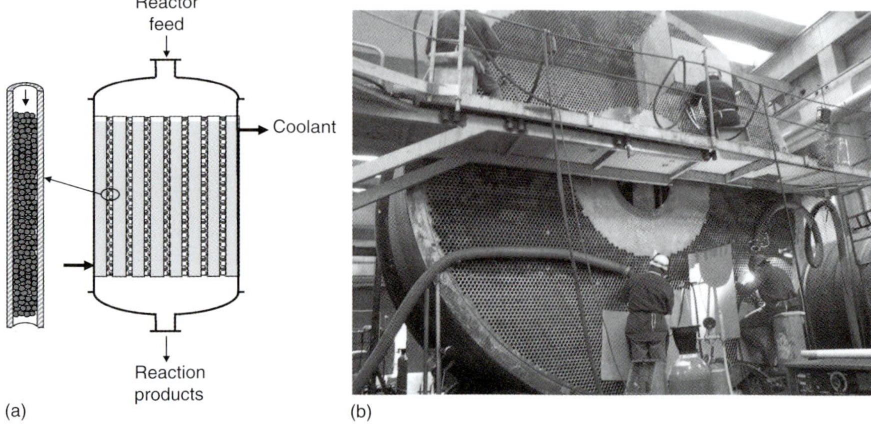

Figure 3.1 Multitubular fixed bed reactor: (a) scheme and (b) manufacturing. Source: Photograph courtesy of RolleChim S.r.l., www.rollechim.com.

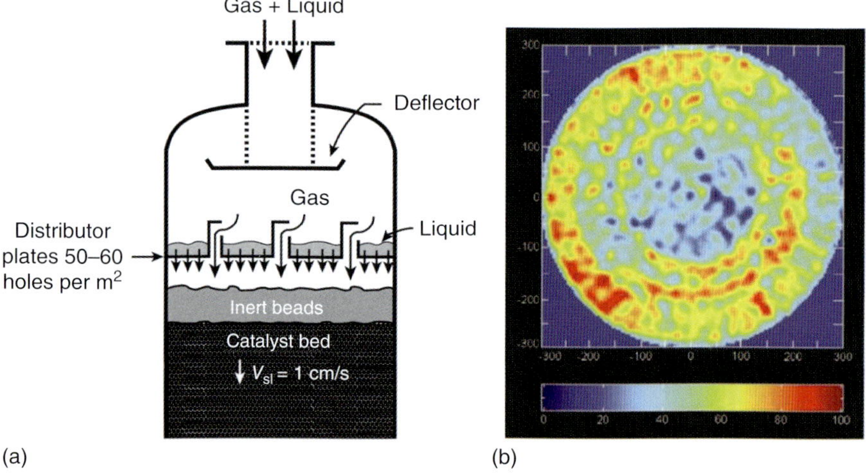

Figure 3.2 Trickle-bed reactor: (a) operational scheme and (b) liquid retention γ-ray tomography picture. Source: Boyer and Fanget 2002 [1]. Reproduced with permission of Elsevier.

"Dry" zones in a trickle-bed reactor not only decrease the efficiency of the device in terms of product yield but also present a serious safety problem in the case of exothermic processes. The absence of the liquid locally changes the reaction system from gas–liquid–solid to gas–solid, dramatically reducing the heat removal capacity there (no liquid). Numerous severe accidents in industrial trickle-bed units have been reported, where the above-described situation led to a local thermal runaway resulting in fire and destruction of the reactor.

The two just mentioned examples let us draw the first and most important conclusion: randomness in the organization of spatial domain at any scale in chemical processing leads to a reduced *predictability and control of the system behavior*. On the other hand, a purposeful introduction of a *reproducible structure* in that domain should result in the improvement of both. A structure is much *easier to understand* than a randomly organized space. Also, *mathematical descriptions (models) of structured systems are simpler* and can be done at *less human/computer time and effort*. For example, modeling of transfer phenomena in fixed bed of particulate catalysts is complex and the reliability of the models and correlations is limited by the randomness of the bed configuration. This obviously has consequences for the reliability of the reactor scale-up. Structured monolithic catalysts (see 3.4.2), on the other hand, consist of a multitude of straight, parallel channels through which the reactants flow. Modeling of hydrodynamics or heat/mass transfer in a straight channel is quite straightforward and the behavior of a single channel is fully representative for the entire reactor. Therefore, the scale-up of such reactors is also straightforward. Other possible advantages of structured spatial domain in chemical processing systems include more specific surface area (hence smaller equipment and/or higher productivity), less energy consumed, lower cost, and ease of operation.

Spatial structures can have very different forms. The most common classes of structures seen in chemical processing include

- cage-type structures
- channels
- corrugated structures
- twisted/folded structures
- network-type structures
- fractal structures.

Examples of these basic types of structures are shown in Figure 3.3.

Spatial structures can be divided into four classes depending on the elements of chemical processes they primarily target. These elements include molecular events, heat transfer, mass transfer, fluid flow, and mixing. Obviously, in some cases, two or more of the above elements are simultaneously addressed. The corresponding ranges of characteristic dimensions of those classes of spatial structures are shown in Table 3.1.

3.2 Structures Targeting Molecular Events

3.2.1 Molecular Imprints

Molecularly imprinted systems [2] are nowadays investigated mainly in relation to racemic or biomolecular separations as selective adsorbents in solid-phase extraction, as stationary phases for chromatographic separations, as antibody mimics for immunoassays, and as selective sensor layers in chemosensors. They can also be potentially applied for improving the stereoselectivity of the reaction events as selective enzyme mimics and catalysts.

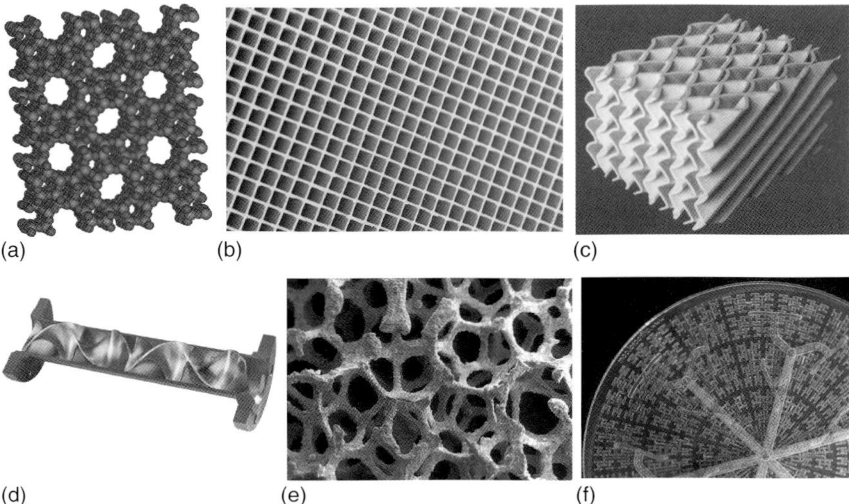

Figure 3.3 Different forms of structures: (a) cage-type (zeolite). Source: Reproduced with permission of Johnson Matthey. (b) Channel-type (monolithic catalyst). Source: Reproduced with permission of Johnson Matthey. (c) Corrugated (Sulzer packing). Source: Reproduced with permission of Sulzer Chemtech Ltd., Switzerland, www.sulzer.com. (d) Twisted/folded (Kenics static mixer). Source: Reproduced with permission of National Oilwell Varco, L. P., USA, www.nov.com. (e) Network-type (catalytic foam). Source: https://upload.wikimedia.org/wikipedia/commons/d/df/Image-Metal_Foam_in_Scanning_Electron_Microscope%2C_magnification_10x_b.GIF. and (f) Fractal (fractal fluid distributor. Source: Reproduced with permission of Arifractal, Amalgamated Research LLC, USA, www.arifractal.com.

Table 3.1 Classification of spatial structures according to the targeted process elements.

Target	Characteristic dimensions	Common forms of structures
Molecular events	Å–nm	Cage
Heat transfer	μm–cm	Channels, corrugated
Mass transfer	μm–cm	Channels, corrugated, network
Fluid flow and mixing	μm–m	Channels, corrugated, twisted/folded, fractal

Simply said, these systems involve surfaces/polymer matrices, the "structure" of which has the shape of the target molecules, which need to be separated or reacted. Formation of such imprinted matrices begins with a covalent or non-covalent binding of template molecules with appropriate monomers containing functional groups, after which the monomers are copolymerized in the presence of a cross-linker and template molecules are removed (Figure 3.4). In such a way, molecularly imprinted polymers (MIPs) become essentially plastic casts of specific molecules. The cavities left by the template molecules have shape and functional groups complementary to the structure of the template, which can be used for binding or for catalysis.

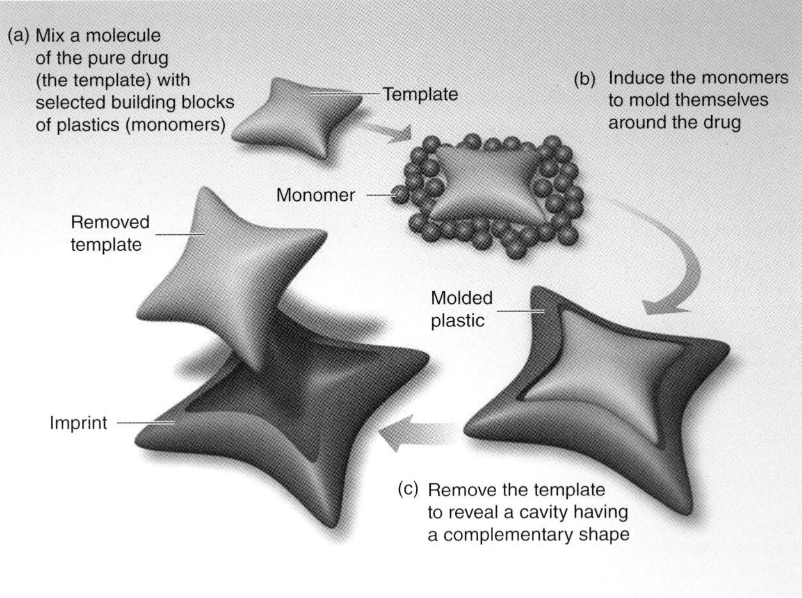

Figure 3.4 Preparation of a molecularly imprinted polymer (MIP). Source: Górak and Stankiewicz 2011 [3]. Reproduced with permission of Annual Reviews.

A simplified scheme of a selective separation of a pharmaceutical molecule on the molecularly imprinted beads is shown in Figure 3.5. When a batch of a drug with impurities is brought on a bed of such beads, the imprinted molecule will be retained on the beads, as it "fits" into the imprints. The imprinted cavities in the beads will therefore capture the targeted drug but will ignore all other substances. The retained drug can be washed away afterward as a pure substance.

To illustrate the separation intensification effect of molecularly imprinted systems, we bring here as an example the preferential separation of tetracycline hydrochloride over a similar compound by a molecularly imprinted membrane, as reported by Trotta et al. [4]. Figure 3.6 shows the degree of retention of tetracycline hydrochloride and another antibiotic – chloramphenicol, on a nonimprinted membrane and a membrane imprinted with tetracycline as template. It can be clearly seen that the type of the membrane has no effect on separation of chloramphenicol, which does not "fit" into the templates. On the other hand, the retention of tetracycline hydrochloride on the imprinted membrane is more than three times higher than on the nonimprinted one, effectively reversing the selectivity toward the two compounds.

Molecularly imprinted systems are highly selective because of three structural characteristics of the cavities they incorporate: size, shape, and functional group orientation. In theory, use of such systems should lead to 100% yield in separation and synthesis processes. In practice, however, there are still diverse problems associated with synthesis and use of molecularly imprinted systems such as the difficult removal of the template during synthesis, imperfect/heterogeneous

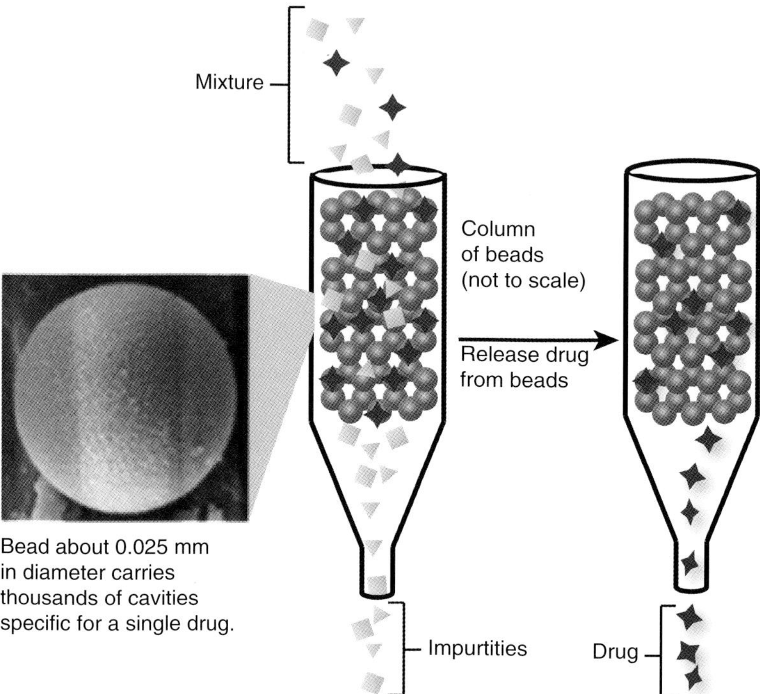

Figure 3.5 Drug purification using molecularly imprinted beads. Source: Adapted from Mosbach 2006 [2].

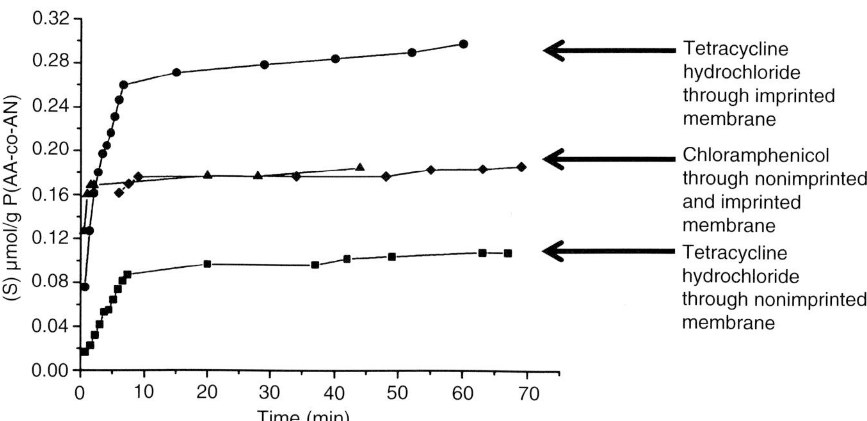

Figure 3.6 Retention of tetracycline hydrochloride and chloramphenicol on the membrane imprinted for tetracycline hydrochloride and the nonimprinted membrane. Source: Trotta et al. 2005 [4]. Reproduced with permission of Elsevier.

binding sites, the overall lack of understanding in the design and synthesis, etc. [5, 6].

Separation sorbents present the most important type of application of molecularly imprinted systems. Chen et al. [6] give an extended review of such applications reported in the literature in the field of environmental protection, bioanalysis, and food and pharmaceutical processing.

Suriyanarayana et al. [7] review MIP-based chemosensors, including optical chemosensors, piezoelectric chemosensors, and electrochemical sensors. Possible applications of the molecular imprinting technology in catalysis have been reviewed by Davis et al. [8]. MIPs are also investigated in relation to crystallization. In the work by Saridakis et al. [9], introduction of MIPs yielded protein crystals in conditions that do not give crystals otherwise and also increased the rate of crystal formation in other conditions.

3.2.2 Molecular Reactors

As already mentioned in Chapter 2, molecular reactors present one of the ways for controlling the orientation of molecules during reactions by immobilizing them in confined nanostructures. In the case of molecular reactors, such structures are provided by cyclodextrins. Cyclodextrins are cyclic oligosaccharides that have been known for more than a century, and they are now also used extensively in the pharmaceutical industry, in household and personal care products, and as food additives [10]. All these applications are related to the shape of the cyclodextrins that resembles a truncated cone or a toroid (Figure 3.7a, b). Depending on the number of glucose monomers in the ring, one distinguishes

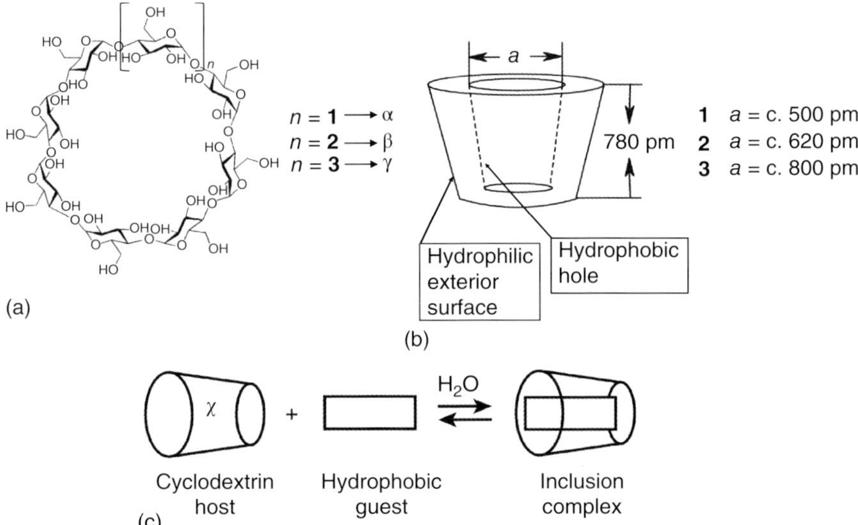

Figure 3.7 (a) Chemical and (b) spatial structure of cyclodextrin. Source: Dodziuk 2006 [11]. Reproduced with permission of Wiley-VCH Verlag GmbH & Co. KGaA; (c) inclusion of a hydrophobic guest.

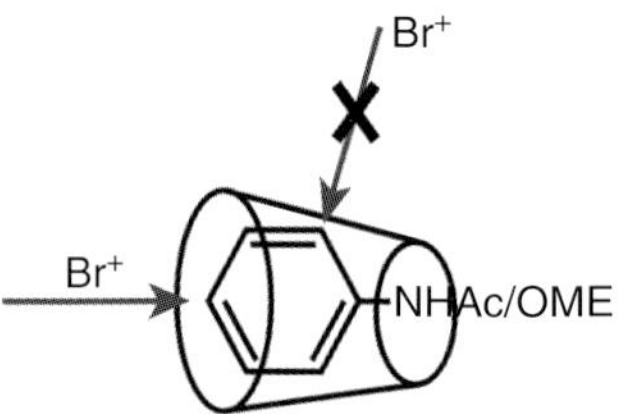

Figure 3.8 Selective bromination of a cyclodextrin-confined compound.

Table 3.2 Cyclodextrin-assisted bromination.

System	Product distribution (%)		Ratio ortho/para
	Ortho-	Para-	
No cyclodextrin	56	44	1.3
α-Cyclodextrin	79	21	3.8
β-Cyclodextrin	>98	<2	>49

Source: Easton 2005 [10]. Reproduced with permission of De Gruyter.

cyclodextrins into α-cyclodextrins (six monomers), β-cyclodextrins (seven monomers), or γ-cyclodextrins (eight monomers). The exterior surface of the cone is hydrophilic because of the presence of the hydroxyl groups, while the hole in the middle is hydrophobic, being surrounded by carbon—hydrogen bonds and ether linkages. As a result, in aqueous solution, cyclodextrins form inclusion complexes with hydrophobic guests (Figure 3.7c).

The toroid structure of cyclodextrins with hydrophilic exterior and hydrophobic interior accepts molecules only in specific orientations. This particular property can be utilized for improving the selectivity of certain reactions. Figure 3.8 presents an example of such a reaction where the cyclodextrin "shell" blocks the access of a brominating agent to the ortho- positions of the reagent, whereas the para-brominated product is selectively synthesized.

The results of the above cyclodextrin-assisted bromination are given in Table 3.2 [10]. One can clearly see that by using the right type of cyclodextrin, the ratio between the ortho- and para-products can be increased by a factor of almost 40!

The geometrical preference of cyclodextrins can change depending on the state they are in. Figure 3.9 shows as an example the entrance of nitrophenol in a permethylated α-cyclodextrin in solution and in the solid state [11]. These properties can obviously be used to affect selectivities of reactions. Another striking example is the reaction of indoxyl and isatinsulfonate which gives, in the absence of cyclodextrins, a mixture of indigo and indirubin sulfonate in yields of 25% and 1.4%, respectively. In the presence of a cyclodextrin dimer, the yields change to <0.1% and 22%, respectively [11]. A ratio change by a factor of more than 3500!

3.2.3 Shape-Selective Catalysts

The term "shape-selective catalysis" was introduced in 1960 by Weisz and Frilette [12]. In shape-selective catalysis, the course of a heterogeneous catalytic reaction

Figure 3.9 Nitrophenol inclusion in α-cyclodextrin in solution and in solid state.

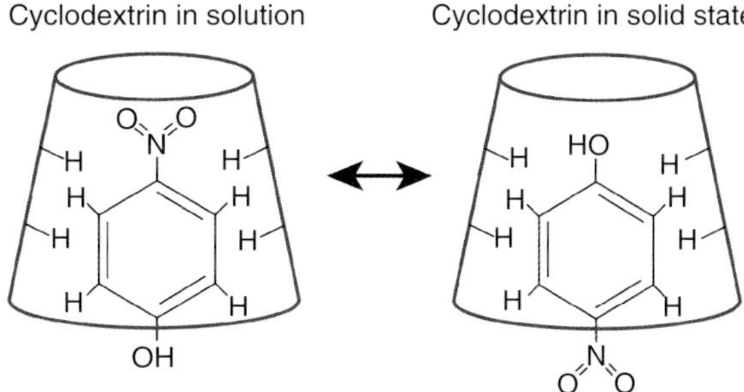

is influenced by the rigid geometrical structure of the catalyst, which imposes a hard-wall confinement on the reacting molecules. There are three basic types of shape-selectivity, as shown in Figure 3.10 [13]. In *reactant selectivity*, the catalyst acts as a molecular sieve preventing the "nonfitting," too bulky, or too branched molecules from entering the internal structure. Figure 3.10a shows an example in which a synthetic zeolite (see further) permits the ingress of a straight-chain *n*-heptane but sieves out its branched isomer. As a result, only *n*-heptane is cracked in the catalyst. In *product selectivity*, the sterically less hindered forms of product molecules, such as the *para* form in Figure 3.10b, can leave the catalyst cage, whereas the bulkier forms remain there. Selective removal of one product form from the catalyst structure may shift the equilibrium toward that form and make the entire reaction process more selective. Finally, in the *restricted transition state-type selectivity*, certain shapes/sizes of the transition state geometries cannot be well accommodated within the structure and therefore products from those geometries cannot be formed.

Zeolites present the most important category of shape-selective catalysts. Originally, they were defined as crystalline alumina silicates with pore sizes of c. 3–7 Å and a highly ordered crystalline structure. That definition has been extended to other oxidic materials, e.g. aluminophosphates, metal silicates, etc. Nowadays, the term is widely applied to materials exhibiting "zeolitic" properties, i.e. selective sorption (molecular sieves), ion exchange, large surface areas, or perfectly defined pore sizes. Zeolites exist in nature or can be synthesized. Examples of the more important synthetic zeolites are shown in Figure 3.11.

The history of zeolites in chemical industries is long and dates back to 1930s. The first synthetic zeolites were developed at Union Carbide and soon found applications in oil refining. According to Yilmaz and Müller [14], it resulted in c. 30% increase in gasoline yield, thus more efficient utilization of petroleum feedstocks. By far, the largest application area of zeolites as catalysts is fluid catalytic cracking (FCC). Other zeolite-catalyzed processes include hydroxylation (e.g. phenol), alkylation (e.g. benzene, cumene), oximation (e.g. cyclohexanone oxime), and epoxidation (e.g. propylene oxide) [14], as well as methanol-to-gasoline (MTG), Mobil olefin-to-gasoline-and-distillate (MOGD) and methanol-to-olefins (MTO) processes [15]. More recent investigations concern possible uses of zeolites as shape-selective catalysts in biomass conversion [16].

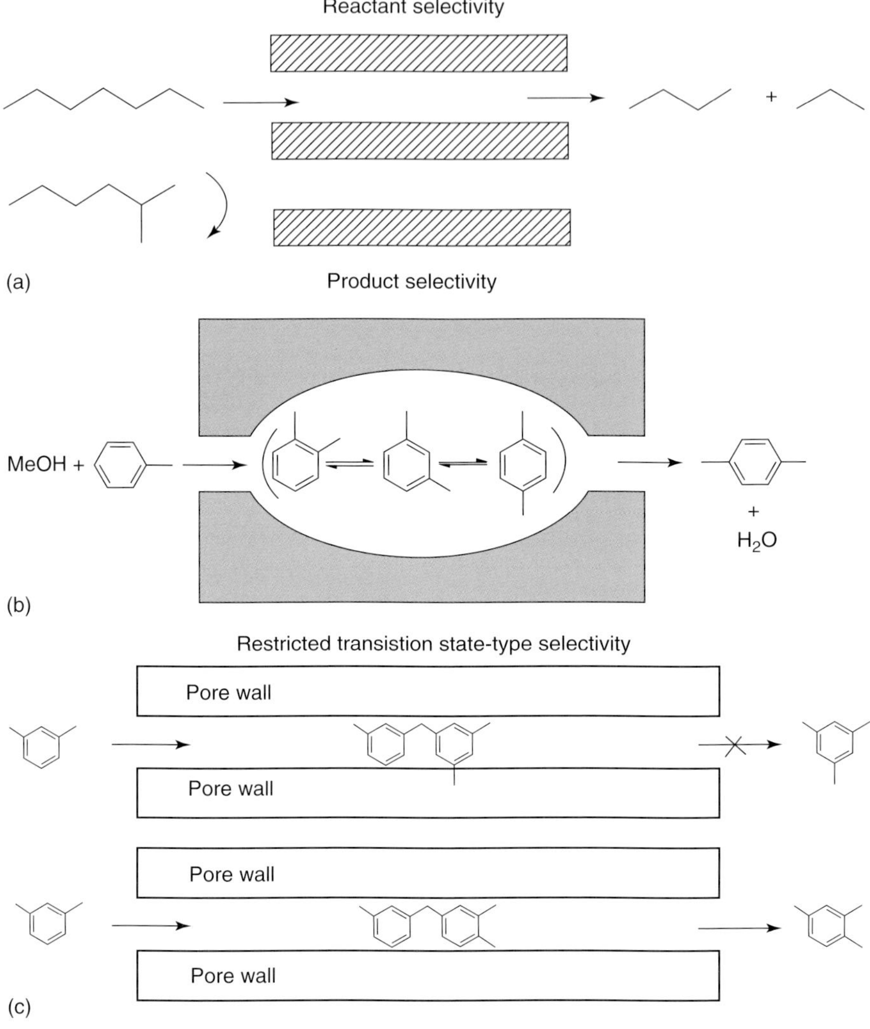

Figure 3.10 Various types of shape selectivity in catalysis: (a) reactant selectivity, (b) product selectivity, and (c) restricted transition state-type selectivity. Source: Lercher and Jentys 2002 [13]. Reproduced with permission of Wiley-VCH Verlag GmbH.

Another, more recent family of structures that can be used for shape-selective operations are *Metal–Organic Frameworks* (MOFs, Figure 3.12). These are crystalline hybrid materials whose crystal structure is built from a three-dimensional network of metal ions or small discrete clusters connected by multidentate organic molecules. The pore size and geometry of MOFs can be tuned to suit various applications.

Hydrogen storage is one of the most important applications of MOFs investigated in the literature [18, 19]. Another large group of (potential) applications resulting from MOFs' unique structural properties include gas separations via selective adsorption, e.g. for CO_2 capture or isomer separation [20–22]. The

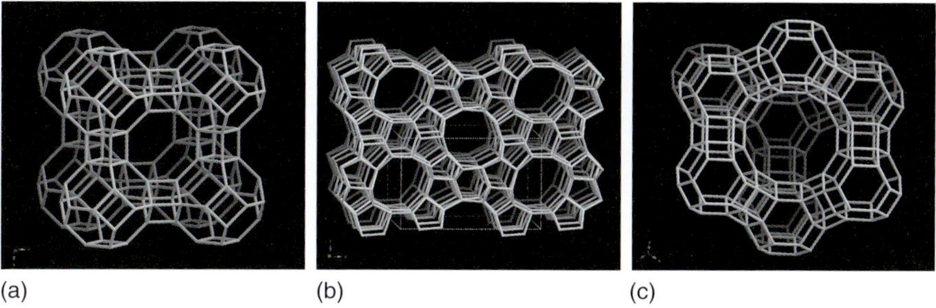

(a)   (b)   (c)

Figure 3.11 Structures of three common types of zeolites: small-pore A-zeolite (a), medium-pore zeolite ZSM-5 (b), and large-pore Y-zeolite (c). Source: Courtesy of www.iza-structure.org/databases/.

Figure 3.12 Crystal structure of MOF-5 with metal clusters (blue) at the corners of a cube and organic linkers at the edges of a cube. Source: Kaye et al. 2007 [17]. Reproduced with permission of American Chemical Society.

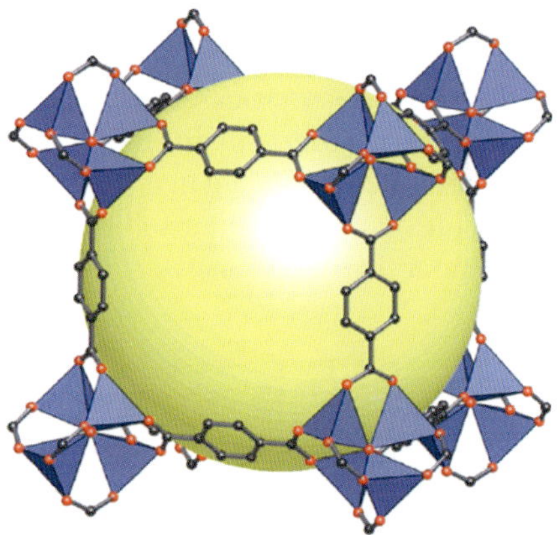

application of MOFs in catalysis still has to be proven. Despite their high metal content, the use of MOFs in catalysis is hampered by problems such as low thermal stability and low resistance to chemicals because of the presence of the organic component. Furthermore, the accessibility of the metal sites to reactants in MOFs is quite limited. Nevertheless, the research in that area progresses, particularly focusing on catalyst design for acidic shape-selective catalysis, sequential asymmetric catalysis, and photocatalysis [23–28]. The potential is definitely there as can be seen in Table 3.3, showing superior selectivity

Table 3.3 Selectivity of *tert*-butylation reaction on different catalysts.

Catalyst	Toluene			Biphenyl		
	para	*ortho*	*dialkylated*	*para*	*ortho*	*dialkylated*
IRMOF-1	82	18	0	96	3	1
IRMOF-2	84	16	0	95	3	2
H-BEA	72	28	0	55	22	23
AlCl$_3$	46	54	0	51	38	11

Source: Ravon et al. 2008 [24]. Reproduced with permission of Royal Society of Chemistry.

performance of MOFs as compared to beta-zeolite and $AlCl_3$ catalysts in toluene and biphenyl alkylation with *tert*-butylchloride [24].

3.2.4 Semirigid Structures: Liquid Crystals

Liquid crystals have been known and applied in various areas of technology for many years. They have properties intermediate to those of conventional liquids and those of solid crystals. A liquid crystal may therefore flow like a liquid, but its molecules may be oriented in a crystal-like way. Liquid crystals are therefore both fluid and anisotropic. Generally, liquid crystals can be divided into four basic categories: nematic, smectic, chiral (or cholesteric), and discotic (Figure 3.13). Nematic, smectic, and chiral molecules are rodlike, whereas discotic molecules are plate-like. Nematic phases have generally one degree of orientational order and therefore the molecules are parallel to each other. Chiral phases have one degree of orientational order locally, due to which they are sometimes referred to as "twisted nematics." Smectic phases on the other hand have one degree of orientational order and one degree of translational order, which results in parallel molecules being stacked in groups on top of each other. Finally, in discotic phases, molecules can orient themselves in a layer-like fashion, sometimes forming stacks (columnar discotic phases).

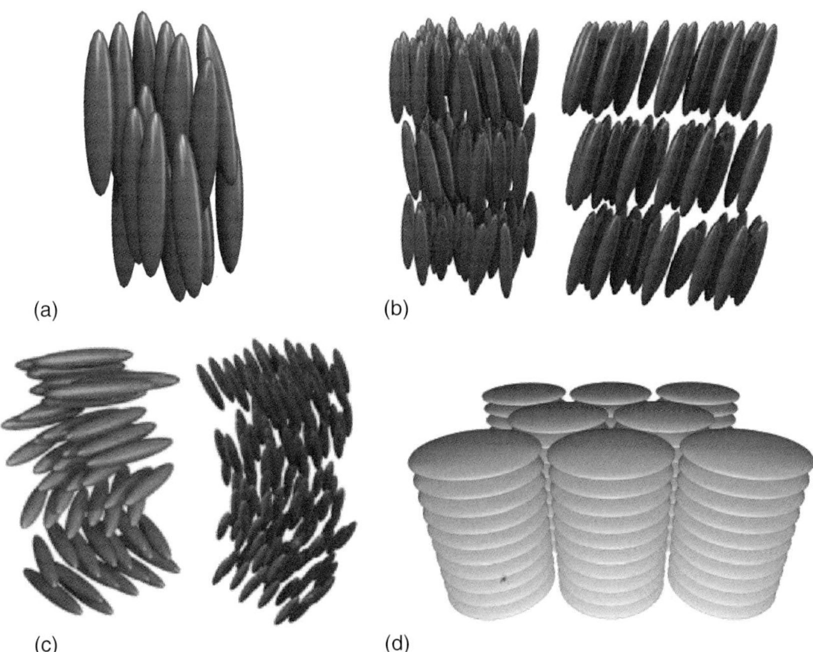

Figure 3.13 Basic categories of liquid crystal phases: (a) nematic, (b) smectic, (c) chiral. Source: (a), (b), and (c) – Kebes~commonswiki, Creative Commons Attribution-Share Alike 3.0 Unported (https://en.wikipedia.org/wiki/Liquid_crystal; license: https://creativecommons.org/licenses/by-sa/3.0/deed.en, and (d) discotic columnar Source: Dr. Kristiaan Neyts, http://lcp.elis.ugent.be/tutorials/lc/lc1.

Liquid crystals miss the rigidity of a solid matrix. At the same time, however, the order of the crystalline phase restricts the randomness in motion of the dissolved molecules. Because of that feature, liquid crystals can be expected to influence the behavior of the chemical reactions carried out in those media. Most important classes of reactions investigated in liquid crystals as reaction media include (photo)dimerizations, isomerizations, and cycloadditions [29–38]. In the vast majority of studies, a clear effect of the liquid crystal and its type on the selectivity of the reaction has been observed. For example, in photochemical dimerization of acenaphthylene in a cholesteric liquid crystal, the reported ratio between trans- and cis-isomers reached 4.73 at >98% conversion, whereas in a nematic liquid crystal that ratio was 0.33 at 70% conversion and in benzene (no liquid crystal present), the same ratio was only 0.14 at 40% conversion [29].

3.3 Structures Targeting Heat Transfer

3.3.1 Microstructured Reactors

Microchemical processing systems are often seen as the most extreme form and the hallmark of process intensification. These are the systems that usually have a stack structure, consisting of a number of slices (platelets) with small channels (in the range of c. 50–400 μm in diameter, see example in Figure 3.14). The channels are fabricated using various micromachining techniques, lithography, etching, or laser radiation. They can be applied as reactors, heat exchangers, mixers, or separators.

Excellent heat exchange properties make microchannel reactors particularly attractive for processes involving fast, highly exothermic reactions. This is due to two reasons. Firstly, the specific surface for heat exchange (channel walls) is inversely proportional to the characteristic channel dimension, d:

$$a = \frac{4}{d}\ (\mathrm{m^2/m^3}).\tag{3.1}$$

Table 3.4 presents values of the specific surface area of square channels of different dimensions. As mentioned in the previous chapter, the Nature was able to

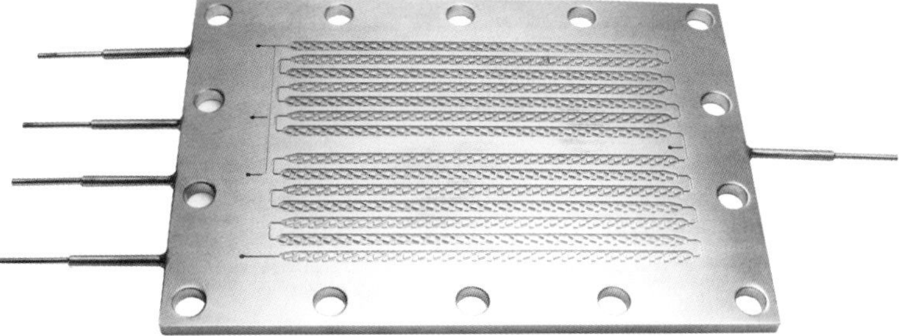

Figure 3.14 Microchannel reactor platelet. Source: Courtesy of Amar Equipments Pvt. Ltd., India, www.amarequip.com.

Table 3.4 Specific heat exchange surface area of square channels of different dimensions.

L	$a\ (\mathrm{m^2/m^3})$
1 dm	40
1 cm	400
1 mm	4 000
400 µm	10 000
100 µm	40 000
50 µm	80 000

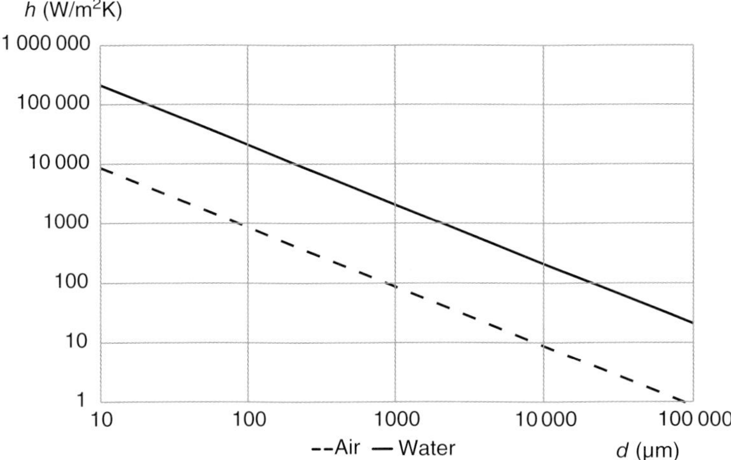

Figure 3.15 Heat transfer coefficients in fully developed laminar flow of air and water in a square channel, as a function of channel dimension.

generate channels (capillary veins) having specific wall surface areas of up to c. $400\,000\ \mathrm{m^2/m^3}$.

On the other hand, the heat transfer coefficient in fully developed laminar flow is inversely proportional to the characteristic dimension of the channel. In such a fully developed laminar flow without temperature gradients in square channels, the Nusselt number is constant and equal to 3.61. Figure 3.15 shows values of heat transfer coefficients in fully developed laminar flow of air and water, as a function of the channel dimension.

The rates of transfer processes can also be approached via characteristic times, which for both heat and mass transfer are proportional to the square of characteristic dimension. Small channel dimensions mean high transfer rates. This makes microchannel reactors superior in comparison with other types of reactors (Figure 3.16).

The research on microstructured reactors started in 1990s [39–41] and was initially oriented toward the applications in high-throughput screening (HTS)

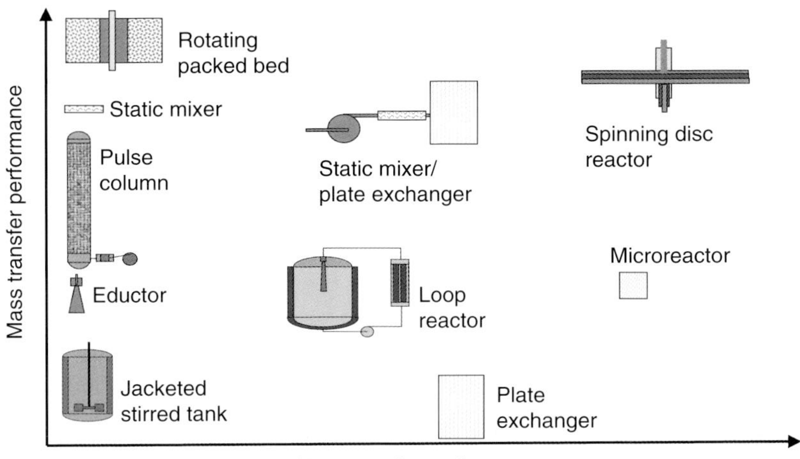

Figure 3.16 Mass and heat transfer in various reactor types. Source: Courtesy of Britest Ltd, UK, www.britest.co.uk.

techniques and in catalyst testing. Meanwhile, hundreds of papers on microreactors are published each year, and there are some comprehensive reviews and books on that subject [42–46]. Currently, various designs of microchannel reactors are commercially available, both for the laboratory and for production applications. They are made from different types of materials that include polymers, glass, silicon carbide, and metals. Table 3.5 presents the relevant properties of those materials.

Obviously, the application of microchannel reactors for production purposes requires very large numbers of microchannels (from 10^5–10^6 for production capacity of $1\,kT/a$, up to 10^7–10^8 for $100\,kT/a$). The strategy to get there is called "numbering-up" or "scale-out." Contrary to scale-up, in numbering-up the critical dimensions (in this case channel diameter) remain unchanged (Figure 3.17).

Table 3.5 Properties of materials used for making microchannel reactors.

	Polymer	Glass	SiC	Metal
Corrosion resistance	High	Alkaline	High	Acid, halogen
Solvent resistance	Low	High	High	High
Thermal conductivity (W/m K)	0.3	1.4	140	14
High-temperature resistance	Low	High	High	High
High-pressure resistance	Low	Medium	Medium	High
Price of the material	Low	Low	Medium	Medium
Ease of manufacturing	High	Low	Low	Medium

Source: After [47].

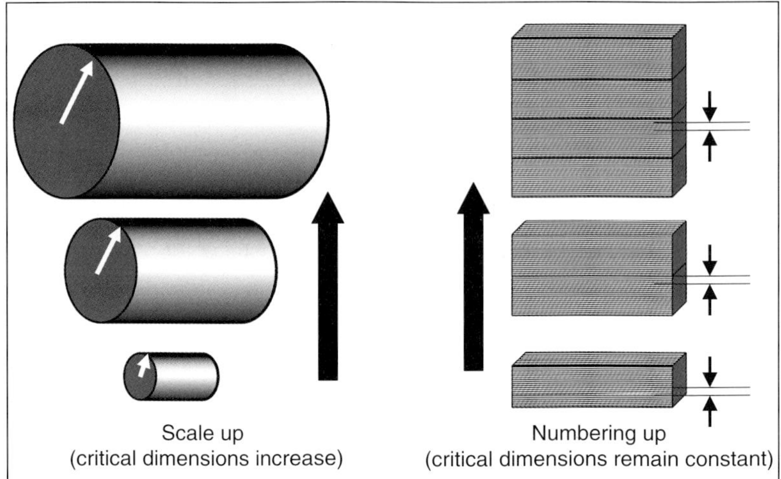

Figure 3.17 Scale-up versus numbering-up for microchannels. Source: Tonkovich and Lerou 2010 [48]. Reproduced with permission of John Wiley and Sons.

In numbering-up, two strategies can be distinguished. The first one, "external numbering-up" consists in multiplying the complete microdevices, connecting them in a parallel fashion via external connectors and tubing. In the other strategy, "internal numbering-up," the microchannel elements (platelets) are grouped and connected internally as a large stack in a common housing, using one flow manifold and one collection zone. In external numbering up, the cost and complexity of instrumentation to connect and control individual units may be considerable. Therefore, this strategy is less suitable for large production capacities requiring hundreds, if not thousands of such units. On the other hand, the internal numbering up can be applied in such cases, provided that the manifolds for uniform fluid distribution and the internal heat transfer management are properly designed. Figure 3.18a shows an example of external numbering-up, used in the microprocessing plant developed by the company Ehrfeld Mikrotechnik (Ehrfeld [49]). In Figure 3.18b, a production-scale microchannel reactor developed by the company Velocys [48] according to the internal numbering-up strategy is shown. Finally, Figure 3.18c shows a mixed concept of a microprocessing plant, developed by Corning, Inc., in which both device parallelization and plates stacking are simultaneously applied.

Microstructured reactors can be used for a very wide range of chemical reactions, from homogeneous, liquid-phase organic synthesis to heterogeneous catalytic systems [50–54]. Most industrial-scale applications can presently be seen in the fine chemical and pharmaceutical sectors. A very good analysis of microreactor applicability in these sectors has been published by Roberge et al. from Lonza Ltd. [55]. They classified the relevant reactions into three categories, depending on their kinetics:

- *Type A reactions* : very fast (<1 s) and mainly controlled by mixing;
- *Type B reactions* : rapid (10 s–20 min), but predominantly controlled by kinetics;

Figure 3.18 Numbering-up strategies for microprocessing plants: (a) external, Source: Courtesy of Ehrfeld Mikrotechnik BTS GmbH, www.ehrfeld.com; (b) internal, Source: Velocys UK, www.velocys.com; and (c) mixed, Source: Corning, Inc., www.corning.com.

- *Type C reactions* : slow (>20 min), but containing hazardous or high-value compounds and often operated batchwise.

Analyzing 86 different reactions carried out at Lonza, the authors concluded that c. 50% of these reactions could potentially benefit from a continuous process, but 63% of them could not be carried out in microreactors because of the presence of solids. In the end, only 19% of all investigated reactions could benefit, of which 8% were type A reactions, 9% type B reactions, and 2% type C reactions.

Clearly, possible solid formation presents the most important limitation of the commercial applications of microstructured reactors in fine chemical or pharmaceutical industries. Possible solutions to this problem can arise from the use of alternative energy forms such as ultrasound (see Chapter 4).

Nevertheless, in cases where such applications are possible, spectacular process improvements result. An example here is the fine chemical process developed at DSM in Linz (currently ESIM Chemicals), where a microchannel reactor of c. 3 l volume replaced a conventional 10 m^3 large stirred-tank reactor (Figure 3.19), maintaining the same production capacity of 1700 kg/h [56]. The reduction of the reactor inventory by a factor of more than 3000 is obviously a spectacular effect having direct influence on process safety. At the same time, the excellent heat removal in the microstructured unit resulted in 20% increase in the material yield and product selectivity.

In large-scale applications, the American company Velocys has developed catalytic microchannel reactor technology for Fischer–Tropsch (FT) process and for

Figure 3.19 Microstructured reactor replacing the 10 m^3 stirred tank unit at ESIM Chemicals GmbH, Linz, Austria Source: Courtesy of ESIM Chemicals, Austria, www.esim-chemicals.com.

Microchannel Fischer–Tropsch reactor core

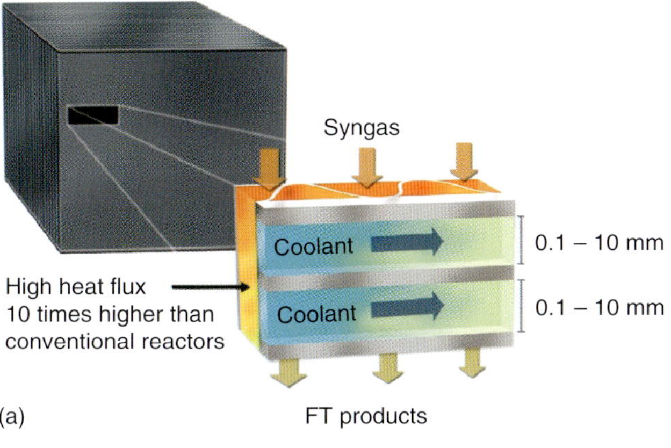

Figure 3.20 Velocys' microchannel Fischer–Tropsch reactor: (a) scheme of operation and (b) commercial unit with production capacity of 175 barrels per day. Source: Courtesy of Velocys, UK, www.velocys.com.

steam methane reforming. The Velocys FT reactor is shown in Figure 3.20. The company claims superior yield and stability compared to conventional systems, resulting in a high productivity per unit volume. Velocys FT technology exhibits conversion efficiencies of over 90% and liquid selectivities of over 87% in recycle configuration, giving over 80% yield of valuable products. The microchannel FT reactors are c. 10 times smaller than the conventional units [57]. Contrary to the conventional FT units, the microchannel reactors maintain economic feasibility even at production capacities as small as 1000 barrels per day, therefore making small-scale, distributed, or even mobile gas-to-liquid production possible.

Another emerging application area of microstructured reactors is the synthesis and manufacturing of nanomaterials [58, 59]. Microreactors offer a much better control of mixing, nucleation, and growth processes as well as reduced use of expensive or toxic chemicals.

3.3.2 Structured Heat Exchangers

Structuring in the heat exchange equipment covers a very broad range of scales, from micrometers up to centimeters. Mehendale et al. [60] classifies heat exchangers based on the hydraulic diameter as follows:

- *Micro heat exchanger*: $1\,\mu m \leq d_h \leq 100\,\mu m$
- *Meso heat exchanger*: $100\,\mu m \leq d_h \leq 1\,mm$
- *Compact heat exchanger*: $1\,mm \leq d_h \leq 6\,mm$
- *Conventional heat exchanger*: $d_h > 6\,mm$.

Thonon and Tochon [61] describe micro heat exchangers as those with channel size lower than 1 mm.

Microchannel heat exchangers are usually fabricated in similar ways as the earlier described microchannel reactors and have similar properties. An example here are *printed circuit heat exchangers* (PCHE, Figure 3.21a). Here, the channels

Figure 3.21 (a) Printed circuits heat exchanger manufactured by company Heatric. Source: Courtesy of Heatric, UK, www.heatric.com; (b) different shapes of channel structures in PCHE; (A), (B), and (C): conventional zigzag, S-shaped, and airfoil shape. Source: Li et al. 2011 [62]. Reproduced with permission of Elsevier; and (c) footprint comparison of printed circuits heat exchanger (foreground) and shell-and-tube heat exchanger for the same cooling capacity. Source: Courtesy of Heatric, UK, www.heatric.com.

are manufactured by chemical etching into a flat plate and diffusion bonding of the plates into stacks. The structures used in the channels can have various shapes, as shown in Figure 3.21b [62]. A comparison of the PCHE footprint and that of a shell-and-tube heat exchanger is presented in Figure 3.21c.

Definitely, the most widespread type of compact heat exchangers are *plate heat exchangers* (PHEs). The structure used in those devices are corrugated plates fitted with gaskets, which seal the plates and direct the hot and the cold streams into alternate channels (Figure 3.22a). The plate pack is assembled between a pressure frame and a frame plate and compressed by the tightening bolts (Figure 3.22b). Easy assembly and disassembly of PHEs makes them particularly attractive for processes where significant fouling can be expected – plates can be regularly cleaned at low cost. This is why PHEs found their first applications in the dairy and food industries. Gasketed unit is the most common type of PHEs, and gasket material is selected depending on the application (temperature, fluid nature, etc.). In such heat exchangers, temperatures up to 200 °C and pressure up to c. 25 bar can be achieved. For applications where gaskets are undesirable (high-pressure and high-temperature or very corrosive fluids), semiwelded or totally welded heat exchangers are available. Obviously, a welded heat exchanger cannot be opened so that the chance of fouling will limit the range of applications.

Heat exchange plates can also be used in combination with conventional shells, for instance, for revamping purposes [61]. Such designs are called *plate-and-shell heat exchangers* and consist of a pack of plates inserted into a shell (Figure 3.23). On the plate side, the fluid flows inside the corrugated channels, whereas on the shell side, the flow is similar to that in classical shell-and-tube heat exchangers. Plate-and-Shell units combine the advantages of both basic types of heat exchangers. On the one hand, high heat transfer coefficients can be reached on the plate side; on the other hand, higher pressures than in standard, frame-based plate heat exchangers can be accommodated inside the shell. If necessary, the shell side can be fitted with baffles, to direct the fluid and to intensify the heat transfer there.

Plate fin heat exchangers (PFHEs) are widely used in applications such as air separation, hydrocarbon separation, or gas liquefaction [61]. These exchangers

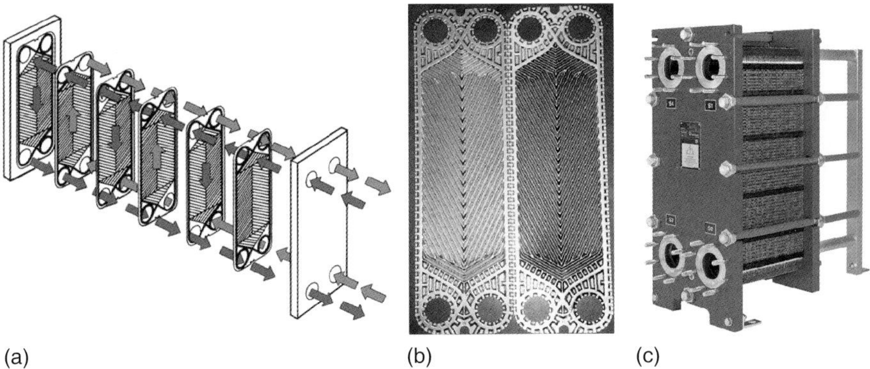

(a)   (b)   (c)

Figure 3.22 Plate heat exchanger: (a) scheme of operation; (b) corrugated plates; and (c) assembled unit. Source: Courtesy of Alfa Laval, Sweden, www.alfalaval.com.

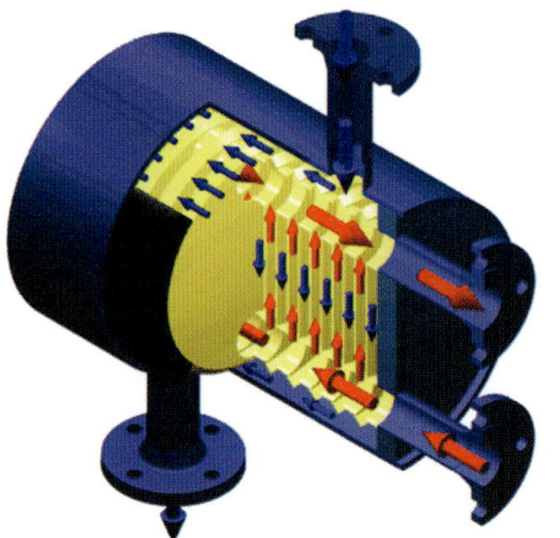

Figure 3.23 Vahterus plate-and-shell heat exchanger. Source: Courtesy of Vahterus Oy, Finland, www.vahterus.com.

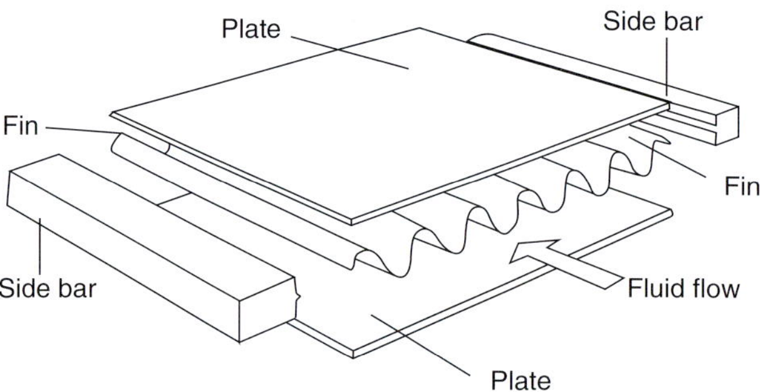

Figure 3.24 Plate-fin heat exchanger. Source: Picon-Nunez et al. 1999 [63]. Reproduced with permission of Elsevier.

consist of plates separated by the finned spaces between them (Figure 3.24). The fins may present various geometrical structures, starting from straight plain rectangular fins, trapezoidal (corrugated), wavy, up to serrated (offset), louvered, or perforated – see Figure 3.25.

In *spiral heat exchangers* (Figure 3.26), two metal sheets are welded together and then rolled to obtain spiral passages. Passages can be either smooth or corrugated, while general flow configuration can be cross-flow (single or multipass) or counter-current flow, depending on the configuration of the inlet and outlet distribution boxes. Compactness is the most important advantage of spiral heat exchangers: $100\,\text{m}^2$ of effective surface is contained in a spiral element of $1\,\text{m}$ in diameter and $1.5\,\text{m}$ long. High turbulence, low fouling, and easy access are further advantages of the spiral units.

The most important criteria in selecting the type (and the material) of a structured heat exchanger for an industrial application include pressure and temperature, number of streams, and the severity of fouling. In Table 3.6, the operational constrains of more important types of structured heat exchangers are shown [61].

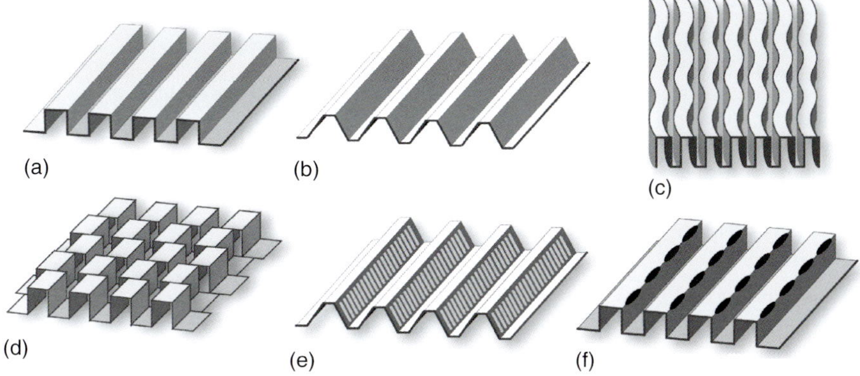

Figure 3.25 Geometrical types of fins in plate-fin heat exchangers: (a) plain rectangular, (b) plain trapezoidal, (c) wavy, (d) serrated or offset, (e) louvered, and (f) perforated.

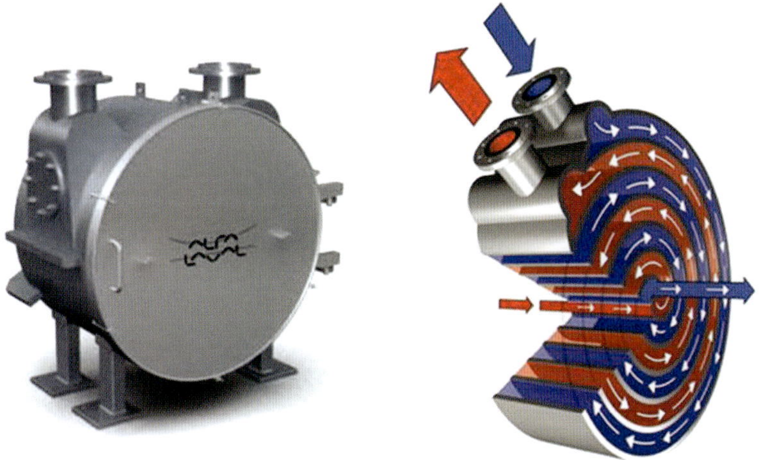

Figure 3.26 Spiral heat exchanger. Source: Courtesy of Alfa Laval, Sweden, www.alfalaval.com.

3.4 Structures Targeting Mass Transfer

3.4.1 Microstructured Separation Systems

Microstructured systems possess not only excellent heat transfer properties but they also offer interesting possibilities for carrying out intensified, small-scale mass transfer operations. In a microchannel, the predominance of the interfacial forces will cause different gas–liquid flow characteristics to occur and consequently a different mass transfer behavior than in the large channels. Indeed, Yue et al. [64] reported the liquid-side volumetric mass transfer coefficients in a square microchannel to be as high as 21/s. In another study on toluene stripping in a microfabricated stripping column [65], the overall capacity coefficients, $K_x a$, nearly an order of magnitude higher than in the conventional packed towers,

Table 3.6 Operational constraints of various types of structured heat exchangers.

Exchanger type/material	Maximal pressure (bar)	Maximal temperature (°C)	Number of streams	Applicable in presence of fouling
Aluminum plate fin heat exchanger	80–120	70–200	>10	No
Stainless steel plate fin heat exchanger	80	650	>2	No
Ceramic plate fin heat exchanger	4	1300	2	No
Diffusion bonded heat exchanger	500–1000	800–1000	>2	No
Spiral heat exchanger	30	400	2	Yes
Flat tube and Fin heat exchanger	200	200	2	No
Plate and shell heat exchanger	30–40	300–400	2	Yes/no
Welded plate heat exchanger	30–40	300–400	>2	No
Gasketed plate heat exchanger	20–25	160–200	>2	Yes
Graphite plate heat exchanger	7	180	2	Yes

Source: Thonon and Tochon 2004 [61]. Reproduced with permission of Taylor and Francis Group.

were reported (Figure 3.27). This increase was attributed to the smaller resistance to mass transfer in the liquid phase because of the reduced thickness of the liquid film.

The most important chemical system studied in *microstructured absorbers* is CO_2 absorption in sodium hydroxide or in amine solutions (monethanolamine [MEA], diethanolamine [DEA], and methyl diethanolamine [MDEA]). Various absorber designs have been investigated including simple straight microchannels [66, 67], microporous tube-in-tube contactor [68], a falling film microstructured reactor [69], and a silicon nitride mesh contactor [70].

In *distillation*, microstructured devices offer extremely high separation efficiencies. Published values of HETP (height equivalent to the theoretical plate) in such systems range from 0.5 to 10 cm and are substantially lower than in conventional distillation towers (30–60 cm). For instance, Tonkovich et al. [71] report on a microchannel distillation unit for methanol distillation, where the experimentally determined HETP for the liquid and gas side was in the order of 0.63 and 0.8 cm, respectively. Sundberg et al. [72] have developed and studied a flat microdistillation column based on a metal foam (Figure 3.28), in which *n*-hexane and cyclohexane were separated at HETP of 1.3 cm.

The group of Jensen at MIT [73] presented a microstructured device with an on-chip condenser (Figure 3.29), in which microfluidic distillation was realized

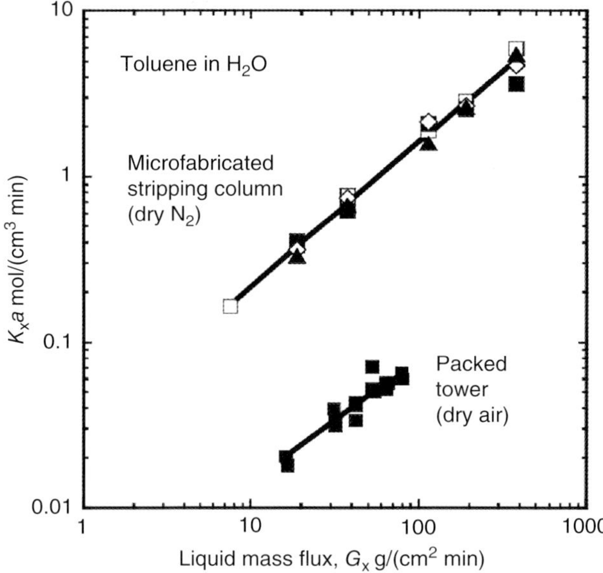

Figure 3.27 Overall mass transfer capacity coefficient, $K_x a$, as a function of the liquid mass flux, G_x, for both the microfabricated stripping column and the packed tower. The different symbols represent different values for the gas mass flux (for the microfabricated stripping column). Source: Cypes and Engstrom 2004 [65]. Reproduced with permission of Elsevier.

Figure 3.28 Metal foam-based flat microdistillation column. Source: Adapted from Sundberg et al. 2009 [72].

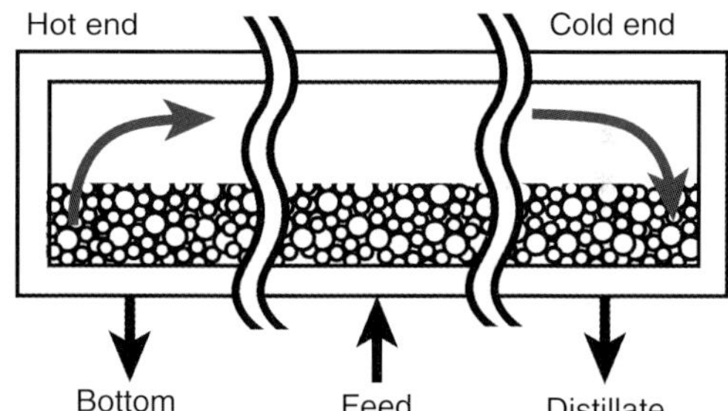

by establishing vapor–liquid equilibrium during segmented flow. Enriched vapor was separated using capillary forces enabling a single-stage distillation operation.

MacInnes et al. [74] proposed a rotating spiral microchannel distillation column, in which centrifugal force is used to create a continuous interface between the gas and the liquid phase (Figure 3.30). They studied the 2,2-dimethylbutane/2-methyl-2-butene system and found out that each equilibrium stage required 5.3 mm of channel length. By increasing the total channel length to, say, 1 m, an excess of 100 distillation stages could be realized. Another type of device, a membrane distillation-based microseparator, has been presented by Adiche and Sundmacher [75].

Finally, Ziogas et al. [76] investigated a novel microrectification apparatus for analytical and preparative applications (Figure 3.31). Several different mixtures

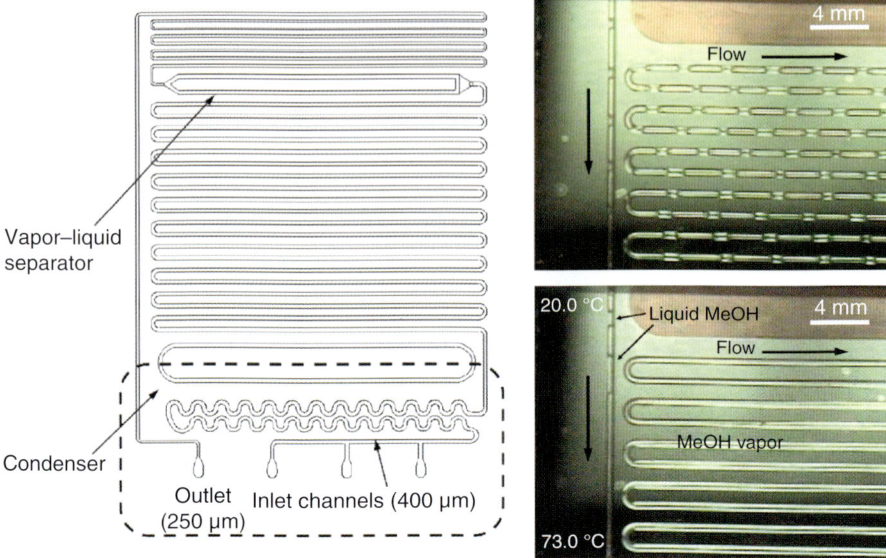

Figure 3.29 Microfluidic distillation device: (a) scheme and (b) flow of methanol–nitrogen mixture below and above the boiling point of methanol. Source: Hartman et al. 2009 [73]. Reproduced with permission of The Royal Society of Chemistry.

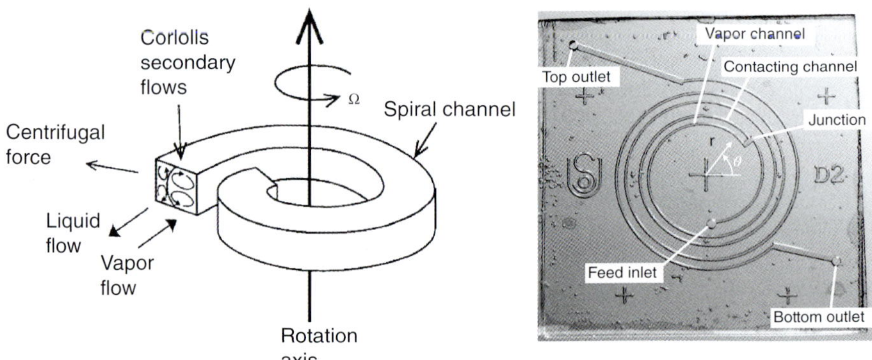

Figure 3.30 Rotating spiral microchannel distillation column. Source: MacInnes et al. 2010 [74]. Reproduced with permission of Elsevier.

were studied and a theoretical separation stage number of 12 was obtained with a HETP of 1.08 cm.

Microstructured systems can also be applied in solvent extraction [77, 78]. They are used as liquid–liquid contactors [78–82] or as settlers/coalescers [83, 84]. When the microchannels are used in liquid–liquid extraction, the high interfacial area and intensity of internal circulation (vortex flow) significantly affect the mass transfer rate. Application of two nozzles injecting two different aqueous phases into microchannels can give liquid–liquid slug (drop) flows [85, 86]. The increased interfacial mass transfer area and internal circulation flow in each slug

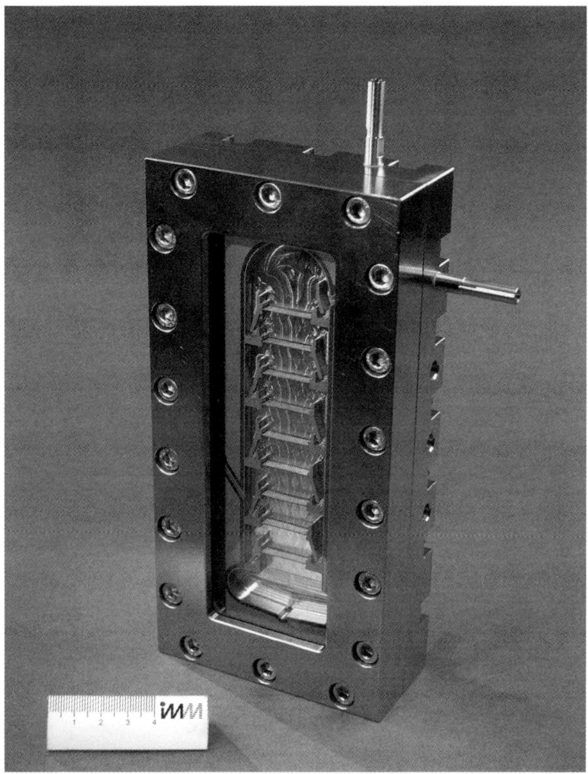

Figure 3.31 Microrectification apparatus developed by IMM (www.imm.fraunhofer.de/).

(drop or bubble) cause an increase in mass transfer coefficient and extraction efficiency, which are measurements for the performance of microchannels as extractors. In an experiment conducted by Kashid et al. [86], iodine in aqueous solution (initial concentration of $0.94\,kg/m^3$) is extracted into organic phase (kerosene). Figure 3.32 shows the volumetric mass transfer coefficient and extraction efficiency as a function of flow velocity at different sizes of the microchannels (0.5–1 mm) with equal flow rate of both phases. Extraction efficiency >90% is achieved within less than five seconds.

A microchannel plate-type coalescer has been developed by Kolehmainen and Turunen [84] to minimize the manufacturing cost and to keep the structure simple. The device can be connected to an extraction unit to form an integrated extractor–separator system. It was observed that under suitable conditions, the plate-type coalescer offered efficient phase separation in stable, continuous operation. The flat channel, 100 or 200 µm, enabled the fluid–surface interaction. By using PTFE and stainless steel materials as channel surfaces and at a total flow rate of less than 180 ml/h, stable operation and complete separation of aqueous and organic phases were possible. The phase separation in the coalescer took place considerably faster than conventionally in a decantation container.

Figure 3.33 presents a microfluidic extraction system developed by Kralj et al. [79], in which the mixing, contacting, and phase separating stages have

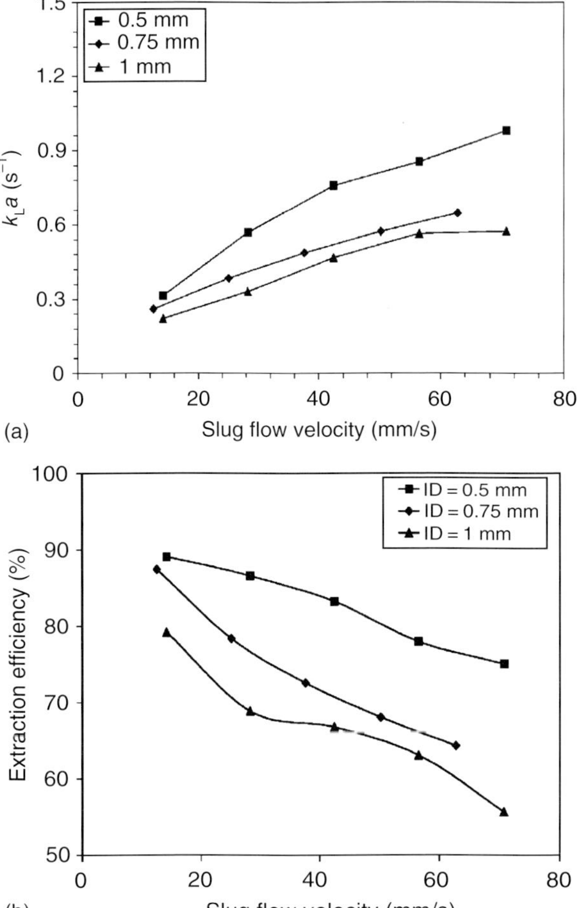

(a)

(b)

Figure 3.32 The effect of total flow velocity on the mass transfer coefficient and extraction efficiency of iodine extraction into kerosene from its aqueous solution, with equal flow rate of each phase and channel length of 100 mm. Source: Kashid et al. 2007 [86]. Reproduced with permission of American Chemical Society.

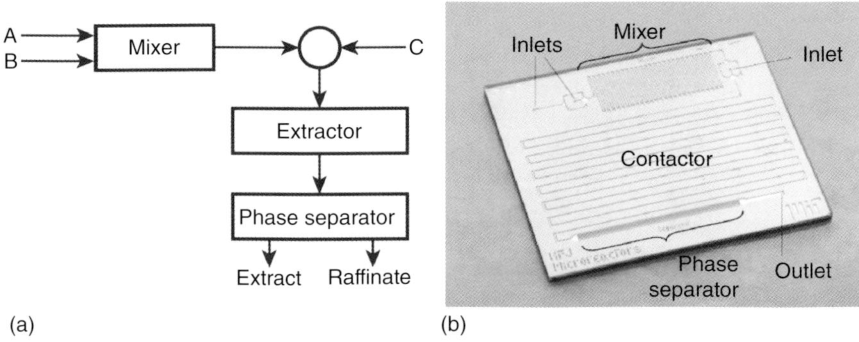

(a)

(b)

Figure 3.33 Integrated microstructured extraction system. Source: Kralj et al. 2007 [79]. Reproduced with permission of The Royal Society of Chemistry.

been integrated in a single device. The device was manufactured using silicon micromachining and packaged with the membrane using compression sealing. In this device, solvent extraction of *N,N*-dimethylformamide (DMF) from dichloromethane (DCM) and diethyl ether to an aqueous phase was performed and found to be equivalent to one equilibrium stage. According to the authors, this technique can be used for a wider range of chemistries and operating conditions because of device robustness, chemical compatibility, and ease of use.

3.4.2 Structured Internals for Reactions and Separations

Mesoscale structures are broadly used as internals, both in reaction and in separation systems. They serve either as catalysts or catalyst supports, or just as inert packings, in each case addressing the interphase or intraphase mass transfer resistances.

Probably, the simplest form of structures used in chemical reactors is *gauzes*. The concept of catalytic reactions over metallic gauzes at millisecond contact time has been known for decades. The most widespread process using metallic Pt/Rh gauzes as catalysts is ammonia oxidation in nitric acid manufacturing (see Figure 3.34). Another well-known gauze-based process of industrial importance is hydrogen cyanide (HCN) production via methane ammoximation (Andrussow process). Processes investigated in the literature include partial oxidation [87–89] and catalytic cracking [90] of methane as well as cyclohexane oxidation [91].

Monolithic catalysts present a very important type of structured internals for chemical reactor applications. In fact, monolithic reactors can be considered the most widespread catalytic reactors nowadays, as the exhausts of almost

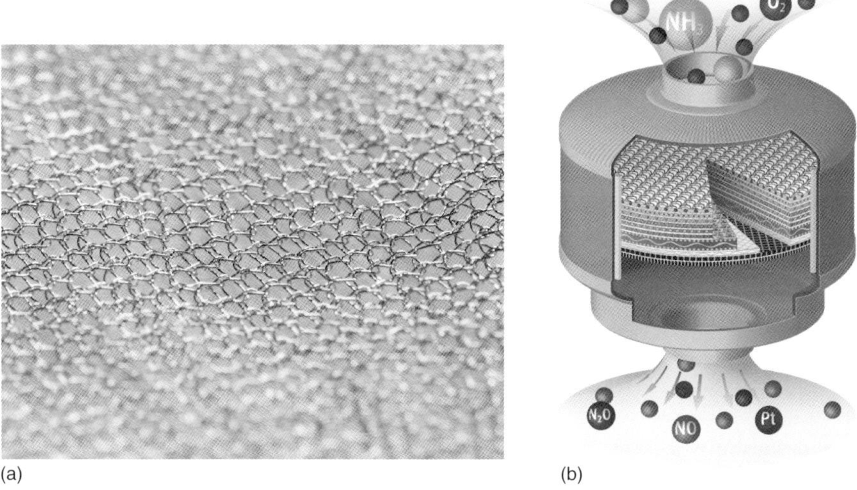

(a)   (b)

Figure 3.34 (a) Catalytic gauze for ammonia oxidation in nitric acid plant. Source: Courtesy of Johnson Matthey, https://matthey.com/ and (b) gauze-based reactor for ammonia oxidation. Source: Courtesy of Umicore AG & Co. KG – Platinum Engineered Materials, http://www .umicore.com/.

all modern cars are equipped with monolithic catalysts for the exhaust gas treatment [92].

As structures, monoliths are ceramic or metallic bodies consisting of a multitude of straight, parallel channels (Figure 3.35). The inner walls of the channels are usually covered with a thin layer of a washcoat (e.g. $\gamma\text{-Al}_2\text{O}_3$), in which the nanoclusters of catalytic material are placed (Figure 3.36). Cell density in monoliths is usually expressed in "cpsi" (cells-per-square-inch) that may vary in a very broad range, from c. 20 to more than 1600.

Monolithic catalysts have been widely applied in catalytic gas-phase reactions, especially in environmental technologies, such as the already mentioned automotive exhaust gas treatment, deNO$_x$ing, SCR, or VOC incineration. In

Figure 3.35 Ceramic (left) and metallic (right) monolithic catalysts. Source: Courtesy of Continental Emitec GmbH, www.continental-corporation.com.

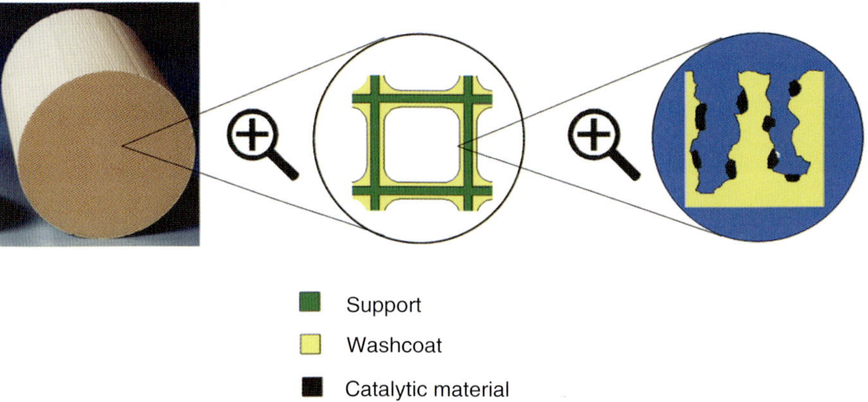

Figure 3.36 Zooming into a single channel of a monolithic catalyst.

these applications, metal monoliths with twisty passages and/or interconnected passages (MTIP) are often used. A typical MTIP consists of alternate layers of crimped/corrugated and flat strips, possibly perforated. When arranged together, they form twisty passages that are interconnected. Exemplary methods to arrange the twisty flow through the passages or to form the possibility of exchange between neighboring passages are illustrated by metal structures that were designed by the company EMITEC for car catalysts, see Figure 3.37. The interconnecting holes improve mixing and lateral heat transfer in the structure and provide the possibility of the whole of the catalytic surface being used, even if flow distribution at the catalyst front is not uniform.

Monoliths consisting of straight capillary channels are characterized by large specific geometrical surface areas and very low pressure drop values compared to the conventional, tortuous beds of particulate catalysts. This is presented in Figure 3.38 [93], which compares monoliths with beds made of rings or spherical particles. One can clearly see that monoliths offer up to c. 1 order of magnitude

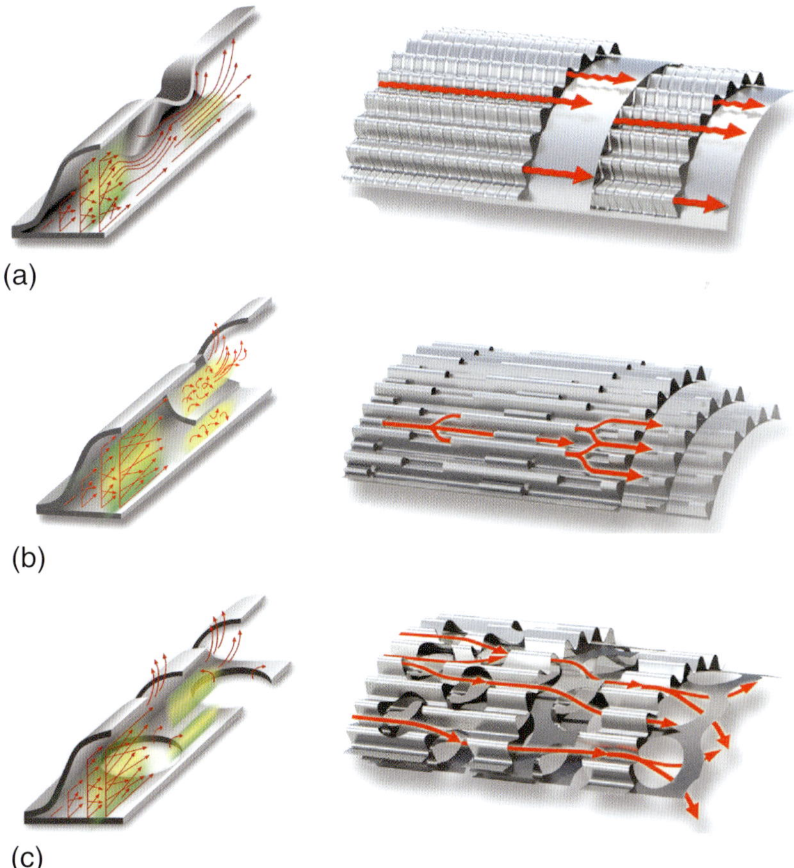

(a)

(b)

(c)

Figure 3.37 EMITEC monolithic structures for car catalysts: (a) type TS, (b) type LS, and (c) type PE. Source: Courtesy of Continental Emitec GmbH, www.continental-corporation.com.

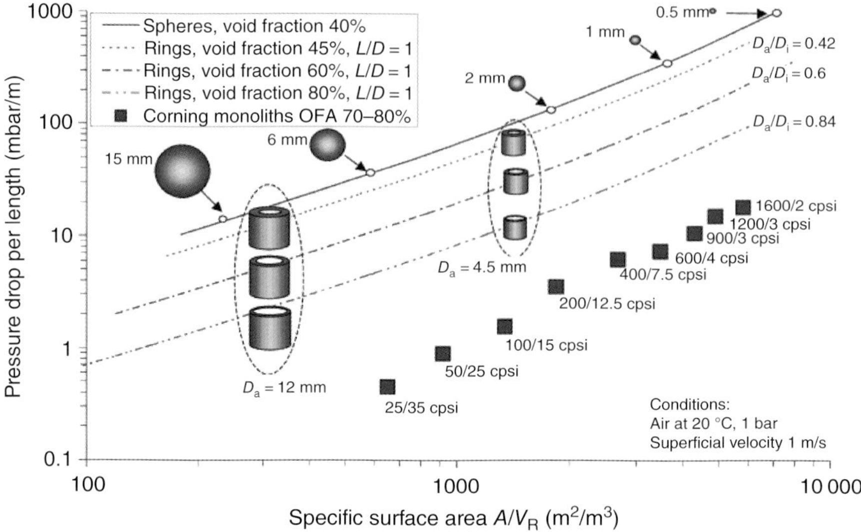

Figure 3.38 Pressure drop versus specific geometrical surface area of monolithic structures, spheres, and rings. Data are for air at 20 °C and 1 bar and a superficial gas velocity of 1 m/s. For monoliths, cell densities in cpsi and wall thickness in 1/1000 in. are given. Source: Boger et al. 2004 [93]. Reproduced with permission of American Chemical Society.

higher specific geometrical surface areas for a given pressure drop or up to 2 orders of magnitude lower pressure drop for a given surface area.

Next to the gas-phase environmental applications, a lot of attention in the recent years has been paid to monolithic catalysts and reactors for gas–liquid reactions [93–97]. The flow regime usually applied in those reactors is the Taylor flow, in which individual gas bubbles are separated by liquid slugs (Figure 3.39).

Taylor flow delivers high gas-to-liquid mass transfer rates. This is due to two reasons. First, the liquid layer between the bubble and the catalyst coating is thin, increasing mass transfer. Secondly, the liquid slugs show a vigorous internal mixing during their travel through a channel. Because of this, radial transfer of mass is increased. Moreover, the gas bubbles push the liquid slugs forward as a piston and a type of plug flow is created. The gas bubbles in Taylor flow do not get a chance to coalesce and therefore high interfacial areas are maintained without any extra energy expenses. This is not the case in conventional systems (stirred tanks, bubble columns, etc.) where extra power input is needed to keep the solid catalyst suspended and the gas bubbles dispersed. That difference can be seen in Figure 3.40, which compares the volumetric mass transfer coefficients as a function of specific power input, for monoliths and conventional reactors.

Another important advantage of monolithic catalysts lies in the possibility of minimization of the diffusional resistance inside the catalyst without paying the price of increased pressure drop. In the conventional particulate catalysts, the intraparticle and extraparticle characteristic lengths are coupled to each other. In order to decrease the diffusional resistance and increase the effectiveness factor in such a catalyst, the intraparticle length should be decreased, i.e. particles

Figure 3.39 Taylor flow through a single tube. (a) Picture of air–water flow, (b) schematic representation of the gas and liquid slugs, (c) CFD velocity pattern in a liquid slug showing the liquid recirculation. Source: Moulijn et al. 2004 [92]. Reproduced with permission of Taylor and Francis Group LLC Books.

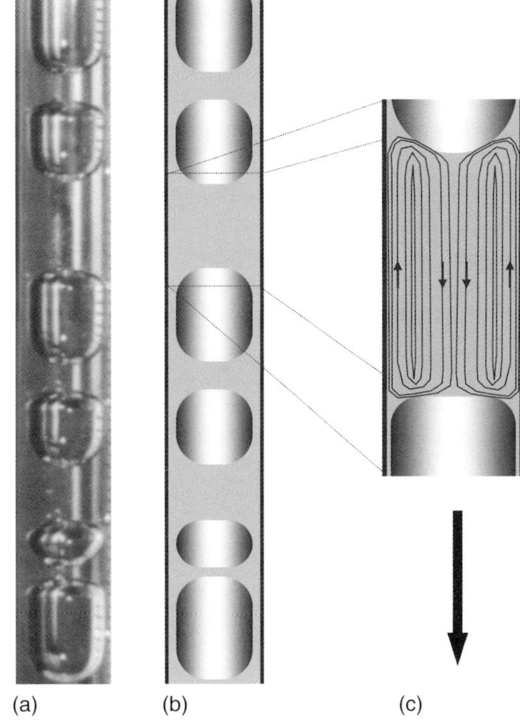

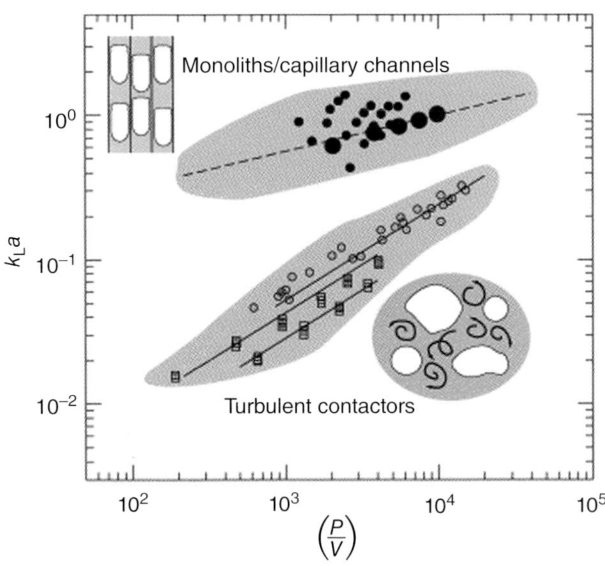

Figure 3.40 Volumetric mass transfer coefficients as a function of specific power input in monoliths and in conventional systems. Source: Kreutzer et al. 2006 [98]. Reproduced with permission of Elsevier. All data for oxygen/water is mentioned in [98] to be adapted from [97]).

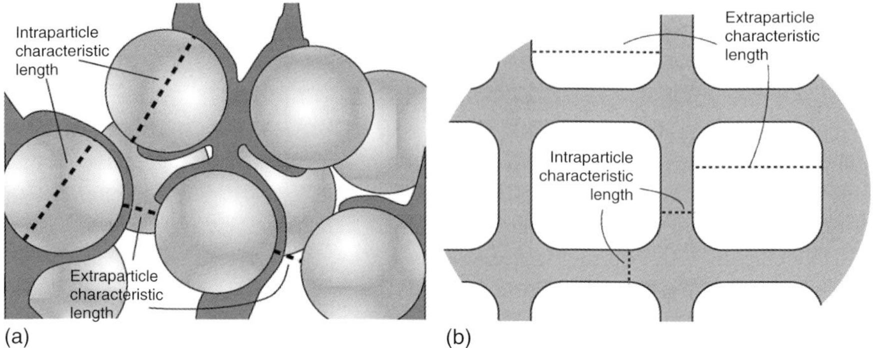

Figure 3.41 Coupling and decoupling in catalytic reactors. In a random packing (a), the extraparticle and intraparticle lengths scale are coupled. The ratio of these lengths is fixed. In a structured packing (b), the intraparticle and the extraparticle length can be varied independently. The ratio of these lengths is a new degree of freedom that can be optimized in a design. Source: Kreutzer et al. 2006 [98]. Reproduced with permission of Elsevier.

should be made smaller. This, however, simultaneously decreases the extraparticle length, the bed porosity decreases, and pressure drop increases (Figure 3.41a). In monolithic catalysts, on the other hand, the intraparticle characteristic length (channel wall or washcoat thickness) and the extraparticle characteristic length (channel diameter) are not coupled to each other and therefore high effectiveness factors at low pressure drops can be achieved. Stankiewicz [99] presented a case study of a hydrogenation process, in which the so-called "in-line monolithic reactor" delivered effectiveness factors up to 40 times higher than in the conventional packed-bed unit.

A major drawback of monolithic catalysts is their poor heat transfer characteristics. With the exception of a few special designs (e.g. Figure 3.37b, c), there is no communication between the channels in the radial direction and therefore the heat transfer in that direction occurs via conduction only. This limits the application of monoliths to highly exothermic processes. For such processes, the use of *catalytic foams* can be considered. Catalytic foams, shown in Figure 3.42, are available in metals (Al, Ni, Cu, etc.), ceramics (mullite, TiO_2, SiC, etc.), carbon, or even polymers [100–103]. They are characterized usually by the "density" of the pores network, referred to as "pores per linear inch" or PPI number.

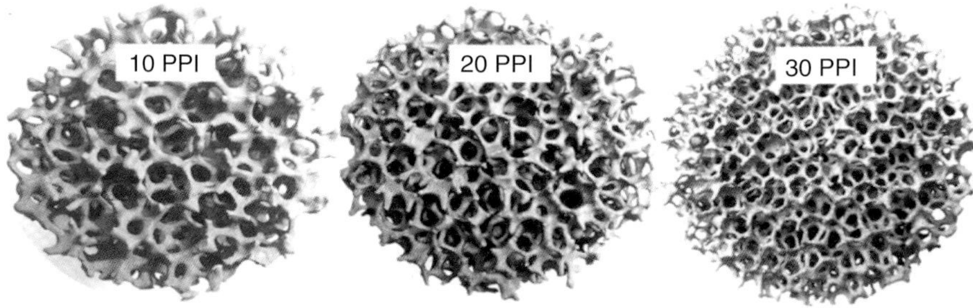

Figure 3.42 SiC foams. Source: Inayat et al. 2011 [100]. Reproduced with permission of Elsevier.

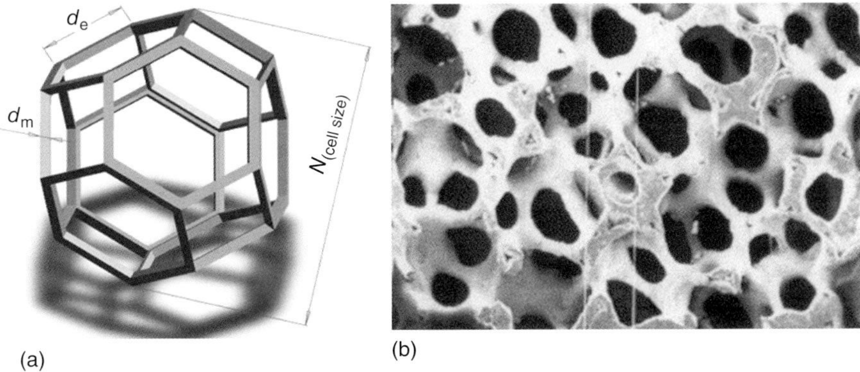

Figure 3.43 Catalytic foam as a "random structure" – the regularity of the idealized model and the reality. Source: (a) Stemmet 2008 [104]. (b) Richardson et al. 2000 [105]. Reproduced with permission of Elsevier.

Strictly speaking, foams can be called "random structures" as the spatial distribution of pore sizes is random (Figure 3.43). Their open structure in all directions allows, in contrast to the monoliths, the heat transfer via convection. Despite this fact, in metal foams, conductive heat transfer remains the dominant mechanism [103, 106].

Initially, foam catalysts were investigated in gas-phase reactions, such as CO_2 reduction, methane reforming, CO oxidation, or methane combustion [101, 107–110]. More recently, also the applications of foams in gas–liquid and liquid–liquid systems have been investigated [104, 111] and authors report very high energetic efficiency of foam structures in those systems. For example, according to Yu et al. [111], a 50 PPI metallic foam-based reactor for biodiesel production exhibited energy consumption per gram biodiesel merely 1.69% and 0.77% of energy consumption in a zigzag microchannel and stirred reactor, respectively.

The so-called *open cross-flow structures (OCFS)* or corrugated structures are widely applied both in reactors and separators. They contain superimposed individual corrugated sheets with the corrugations in opposed orientation, such that the resulting unit is characterized by an open cross-flow structure pattern (Figure 3.44). The angle of channels to the axis of flow is in principle variable between 0° and 90°, whereby the optimum value lies between 30° and 45°. The geometrical surface area of the OCFS used in catalytic reactors ranges from 300 to 1800 m^2/m^3, whereas the void fraction is c. 90%. Pressure drop in OCFSs is 10–100 lower than in fixed beds of conventional catalysts.

OCFS are also widely applied in separations, distillation, and extraction in particular. The specific geometrical areas in those packings are slightly lower than in the ones used for catalytic reactions and usually vary between 200 and 500 m^2/m^3 [112].

Another interesting category of structured packings are so-called sandwich structures (SS). These structures are formed via embedding catalyst particles between two (permeable) screens (perforated plates, web of microfibers, and

Figure 3.44 OCFS packing for distillation column. Source: Courtesy of Luigi Chiesa.

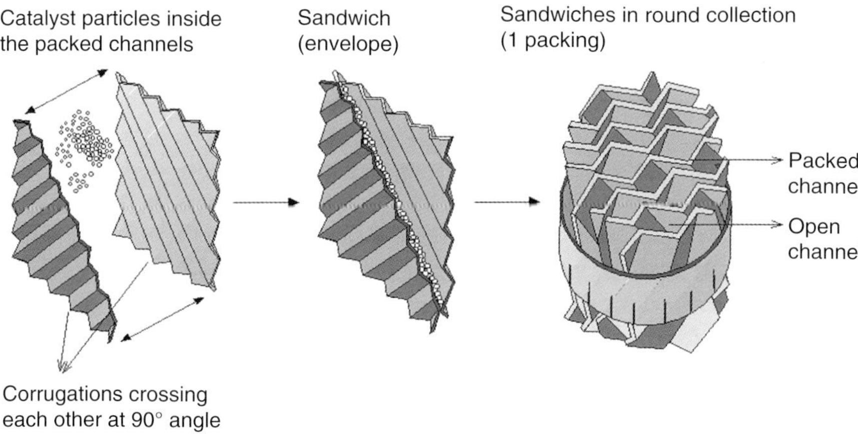

Figure 3.45 Scheme of the Sulzer Katapak-S structure. Source: Courtesy of Sulzer Chemtech Ltd., Switzerland, www.sulzer.com.

gauze). Sandwiches are then stacked. Sandwich structures are particularly suitable for three-phase heterogeneous reactions that are carried out in trickle-beds or in reactive distillation columns. A well-known example of a SS is the Katapak packing developed by Sulzer (Figure 3.45). It consists of two pieces of rectangular crimped wire gauzes sealed around the edge, thereby forming a pocket of the order of 10–50 mm wide between the two screens. These catalyst "sandwiches" or "wafers" are then bound together. The modules are arranged into a cubical or a round collection.

Summarizing, the structured internals for reactions or separations, such as monoliths, foams, or open cross-flow structures, clearly surpass the random packings, either on pressure drop or on mass transfer intensity or both. This can be seen in Figure 3.46, which shows a comparison of voidage and specific surface areas in various random and structured systems [104]. The data for foam

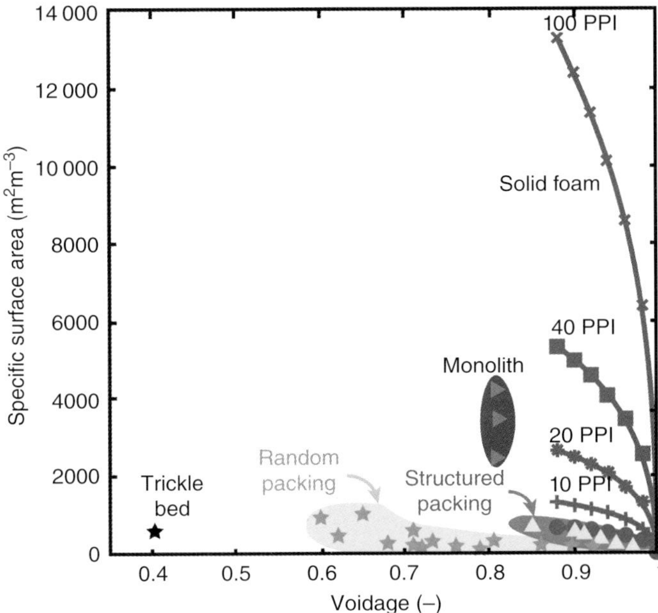

Figure 3.46 Specific surface area versus voidage in various random and structured internals. Source: Adapted from Stemmet 2008 [104].

structures have been calculated based on an idealized tetrakaidecahedra model to represent the interconnected cell structure; all other data represent experimental values. As can be seen, structures can offer up to c. 1 order of magnitude larger specific surface area at higher voidage, thus lower pressure drop.

3.5 Structures Targeting Mixing and Fluid Flow

3.5.1 Micromixers

Microstructured mixing devices can be used as elements of an integrated microsystem, such as microreactor for fast exothermic reaction or a microanalyzer, but they can also be applied as independent, "stand-alone" processing units, e.g. in the production of microemulsions. Micromixers can be divided into two broad categories, depending on the basic principles exploited to induce mixing. In *passive mixers*, the pumping energy is the sole source of the mixing, whereas in *active mixers*, the mixing can be accomplished by introducing various kinds of disturbances, such as pulsing flow, electrical fields, acoustic fluid shaking, ultrasound, electrowetting-based droplet shaking, microstirrers, and others [113]. One can therefore say that in passive mixers, the STRUCTURE approach plays the primary role (see Figure 3.47), whereas in active mixers, other PI approaches, such as ENERGY (acoustic, electrical, magnetic, etc.) or TIME (pulsing and oscillations), become dominant. Passive mixers can be further divided into six subcategories [114, 115]:

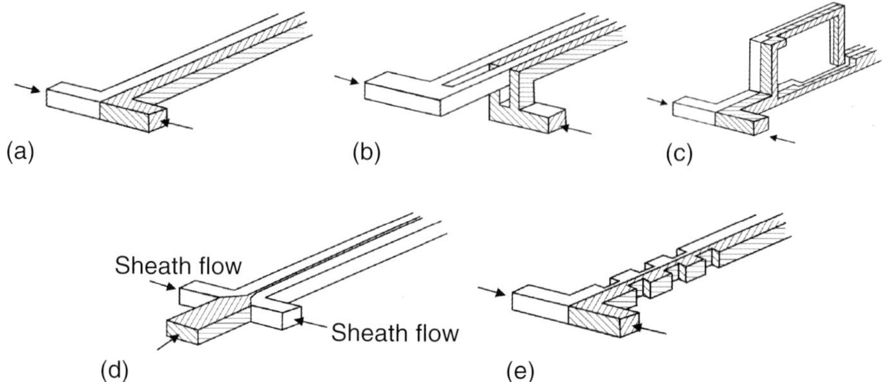

Figure 3.47 Schematic representation of mixing principles in various types of passive micromixers: (a) T-shape; (b) parallel lamination; (c) sequential lamination join-split-join; (d) flow focusing; and (e) chaotic advection with obstacles on wall. Source: Nguyen and Wu 2005 [114]. Reproduced with permission of Institute of Physics Publishing.

1. *T- and Y-shaped micromixers*: The simplest design consisting of a T- or Y-shaped microchannel.
2. *Parallel lamination micromixers*: in which the inlet main stream is split into parallel substreams that rejoin to form a laminate stream. Here, the mixing is enhanced by decreasing the diffusion length (the mixing time decreases with the square of the diffusion length) and increasing the contact surface area between the mixed fluids.
3. *Sequential lamination micromixers*: where the mixing is achieved by sequential splitting and recombining the fluids, similarly to static mixers (see Section 3.5.2.).
4. *Flow focusing micromixers*: in which the width of one fluid stream is decreased ("focused") by the sheath of the outer fluid. The focusing decreases the diffusion path length and what follows, the mixing time.
5. *Chaotic advection micromixers*: in which a transverse flow is induced, e.g. by placing obstacles on the channel walls. The generated transverse flow components result in exponential increase of the interfacial area and simultaneous decrease in the striation thickness.
6. *Droplet micromixers*: in which droplets of the mixed liquids are formed.

Similarly, active micromixers can be categorized, depending on the type of the external energy producing a disturbance in the mixing channel or chamber [115]:

1. *Pressure field*: where a pulsing alteration of the flow rate creates a segmented flow or stretches the interface between the fluids.
2. *Electrokinetic*: where a fluctuating electric field causes rapid stretching and folding of the fluid interfaces.
3. *Dielectrophoretic*: where a polarization of particle is induced by a nonuniform electric field, which causes that particle to move to and from the electrode.
4. *Electrowetting*: where droplets are electrically forced to coalesce, shake, and split, acting as virtual mixing chambers.

5. *Magnetohydrodynamic*: where a Lorentz body force is created, generating deformations and stretching the interfaces, hence enhancing the mixing.
6. *Ultrasound*: where the acoustic stirring is created by ultrasonic waves.

The above list is not exhaustive as various other types of active micromixers are being developed. Some of them, for instance, include microsized moving elements acting as stirrers, others are based on thermal disturbances. An interesting type of a micromixer has been developed at Freiburg University [116]. This centrifugal micromixer utilizes Coriolis force to fold the interface between the mixed liquids, thereby increasing the contact surface between them. The liquid is pumped through radial microchannels of a device rotating with frequencies between 20 and 120 Hz and reaching high-gravity conditions of up to $10\,000\,g$. This allows fast mixing at high throughputs up to 1 ml/s per channel (Figure 3.48).

With such a variety of available designs and working principles, the evaluation and comparison of the performance of various micromixers is a complex task. Obviously, the relation between the mixing times achieved and the specific power dissipation gives important clues with regard to the energetic efficiency of the mixing. In fact, Falk and Commenge [117] found out that whatever the internal geometry of the mixer is, the energy dissipation was the only relevant parameter to design an efficient mixer. Figure 3.49, reproduced from their work, presents the mixing time as a function of specific power dissipated for several types of micromixers. On the other hand, it is extremely complex to predict the influence of mixing and what follows, mixer design, upon product selectivity in the reaction applications, where often complex kinetics and heat transfer play an important role. As stated by Falk and Commenge [117], the general idea that the better mixing, the higher selectivity, although often true in practice, is not always the case and there are some statements of no improvement using micromixers; however, cases with unfavorable effects are unfortunately not reported.

Mixing times and achievable Reynolds numbers can also provide the first selection criteria regarding the applicability of a given micromixer design to one process or another. Figure 3.50 illustrates the micromixing application field according to Reynolds number versus the mixing time [118].

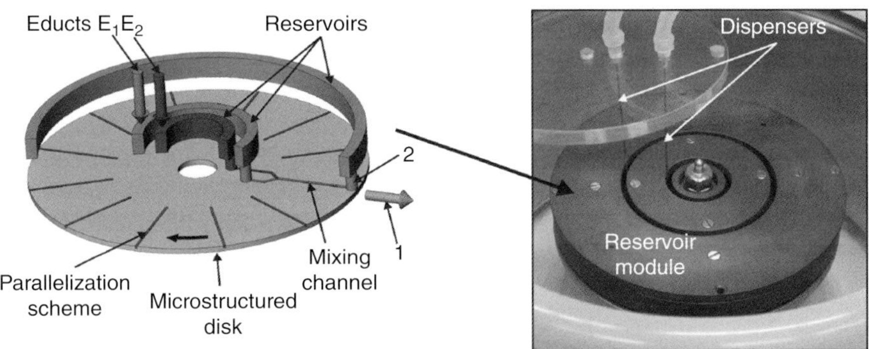

Figure 3.48 Centrifugal Coriolis micromixer. Source: Haeberle et al. 2005 [116].

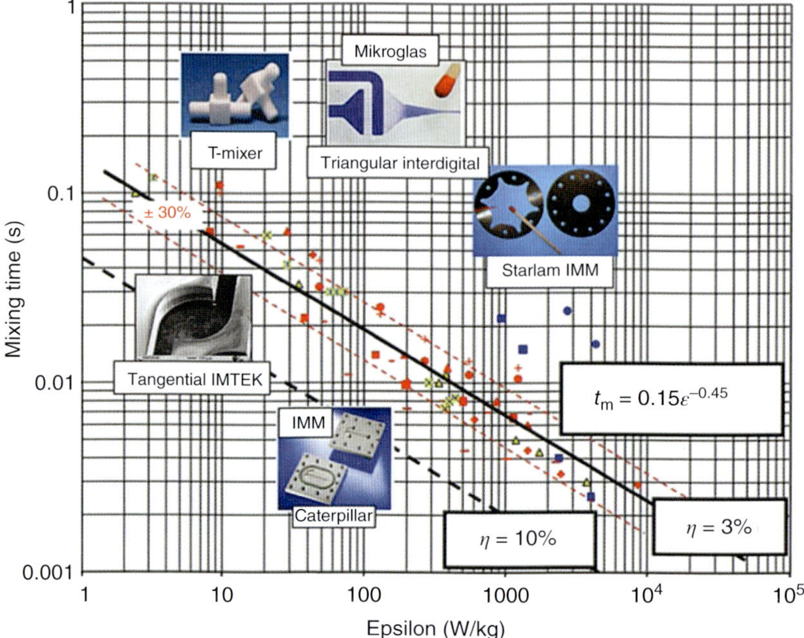

Figure 3.49 Mixing time as a function of specific power dissipation for various types of micromixers. Source: Falk and Commenge 2010 [117]. Reproduced with permission of Elsevier.

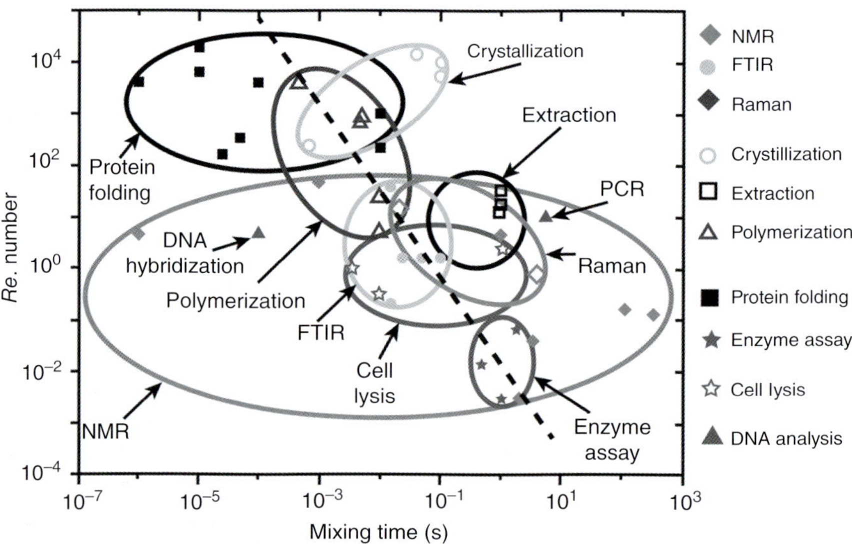

Figure 3.50 Micromixing application field, according to Reynolds number and mixing time. Source: Jeong et al. 2010 [118]. Reproduced with permission of The Royal Society of Chemistry.

3.5.2 Static Mixers

Static or motionless mixers present perhaps the most classical example of process intensification, in which the technology of mixing or dispersing the fluids has been fundamentally changed. Instead of a stirrer, a structure is used while the energy necessary for the mixing is not coming from the motor driving the impeller but from the pumping of the mixed fluids. Static mixers consist of a series of precisely configured motionless elements (inserts) that impose splitting the fluid stream, reorientation, and recombination of the substreams (Figure 3.51). In this way, the elements generate mixing of fluid(s) flowing through the mixer, impose intense movement across the pipe, produce good conditions for heat and mass transfer, and increase interfacial surface area for immiscible fluids. At turbulent flow mixing, heat and mass transfer are enhanced by the formation of eddies. Configurations of the elements are designed to suit different applications.

Energy dissipation rates in motionless mixers are high, with typical values between 10 and 1000 W/kg compared to an upper value of around 5 W/kg in conventional stirred tanks. Mass transfer coefficients ($k_L a$) can be 10–100 times higher than in conventional systems, which can be clearly seen in Figure 3.52. An intensive radial mixing results in leveling of the fluid velocity profile over the mixer cross-section. Consequently, the residence time distribution (RTD) is narrow, i.e. a nearly plug flow is reached.

The structures used for static mixing can be divided into several categories [120]:

(a)

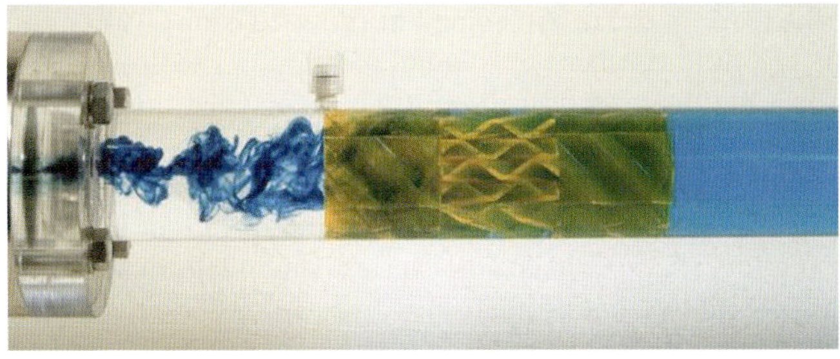

(b)

Figure 3.51 (a) Mixing principle in a static mixer. Source: Courtesy of Primix BV, Netherlands, www.primix.eu and (b) blending of miscible components in a static mixer. Source: Courtesy of Sulzer Chemtech Ltd., Switzerland, www.sulzer.com.

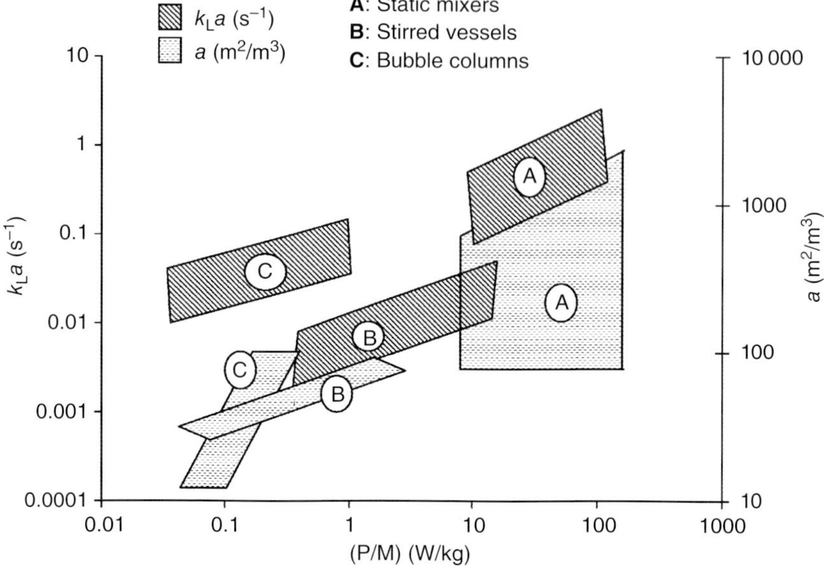

Figure 3.52 Comparison of interfacial area and mass transfer coefficient between static mixers and other classical gas–liquid contactors. Source: Heyouni et al. 2002 [119]. Reproduced with permission of Elsevier.

1. Open designs with twisted strips or ribbons
2. Open designs with blades and baffles of high aspect ratio
3. Open designs with blades and baffles of low aspect ratio
4. Channels formed by corrugated or folded plates
5. Multilayer designs such as those formed by crossing bars
6. Closed designs with channels or holes.

Examples of static mixers representing each of those categories are presented in Figure 3.53. The KM Kenics mixer (Figure 3.53a) is the most used and studied static mixer. It consists of rectangular plates that are twisted 180° to either the right direction or the left direction. The right- and left-hand helical elements are connected alternately with a connecting angle of 90°. A fluid introduced to the mixer is divided at the front of the mixing element (flow division). When the flow passes the element, the fluid is mixed by a radial rotation of the flow within the mixing element (radial mixing). The flow direction is reversed at the connection point between the elements (flow reversal). The KM mixer is used mostly for blending or dispersion involving liquids and gases, as well as for enhancement of heat transfer. It is suitable for operations in both laminar and turbulent flow.

The Ross LPD mixer (Figure 3.53b) consists of a series of semielliptical plates, which are discriminately positioned in a tubular housing. A single element consists of two plates perpendicular to each other. The mixing operation is based on splitting and then diverting input streams. The LPD mixer is used in laminar and turbulent flow regimes and for fluids of different viscosities.

Mixing in the Chemineer HEV mixer (Figure 3.53c) is accomplished by controlled vortex structures generated by the low-profile geometry. The HEV mixer

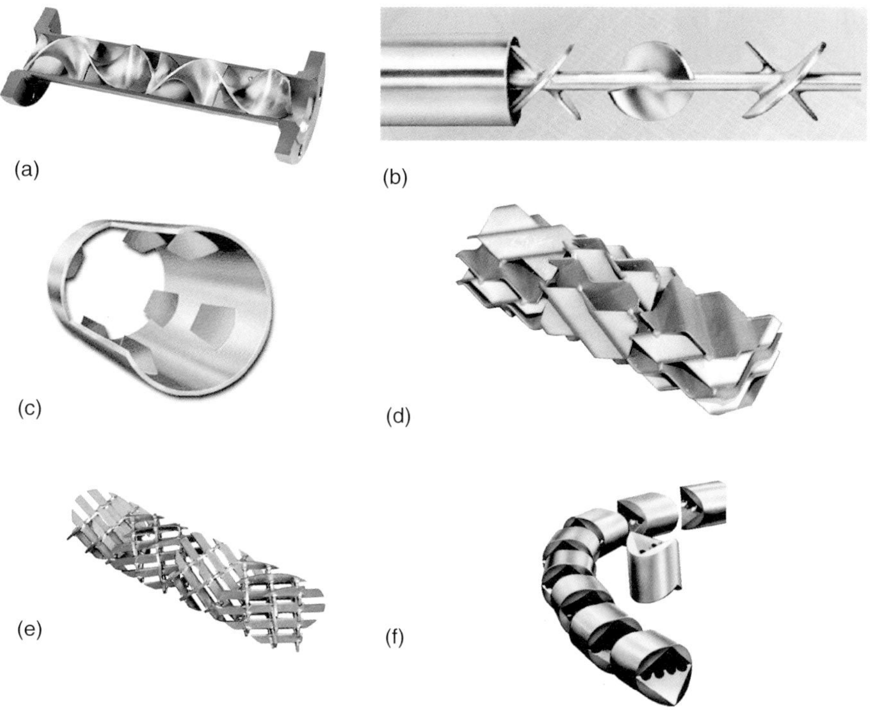

Figure 3.53 Examples of various categories of static mixers: (a) helical mixer; (b) open design with blades and baffles of high aspect ratio; (c) open design with blades and baffles of low aspect ratio (Chemineer HEV mixer); (d) channels formed by corrugated or folded plates; (e) multilayer design; and (f) closed design with holes. Source: (a, c) National Oilwell Varco, L.P., USA, www.nov.com; (b, f) Charles Ross and Son Company, USA, www.mixers.com; (d, e) StaMixCo-AG Switzerland, www.stamixco.com.

is typically used for low-viscosity liquid–liquid blending processes, as well as gas–gas mixing.

Mixing elements of the Sulzer SMV mixer (Figure 3.53d) consist of intersecting inclined corrugated plates and channels, which encourage a rapid mixing action in combination with plug flow progression through the mixer. Elements are rotated with regard to each other. Mass transfer in the SMV mixer is very effective. The SMV mixers are recommended for the use in turbulent regime to handle with liquids of low viscosity, dispersing immiscible liquids, and contacting liquids with gases.

The mixing elements of Sulzer SMX mixer (Figure 3.53e) are complex networks of angled guide blades, positioned at an angle between 30° and 45° to the pipe axis. Mixing occurs through the continuous redirecting, splitting, stretching, and diffusion of the fluids as they pass through the available openings. The SMX static mixers are mainly used for difficult homogenization and mixing of highly viscous fluids (polymerizations and polymer processing), admixing low viscous media to the highly viscous liquids, and dispersing in laminar flow.

The ISG (interfacial surface generator) mixer (Figure 3.53f) consists of mixing elements enclosed in a pipe housing. The ends of the elements are shaped so that the adjacent elements form a tetrahedral chamber. Four holes bored through each element provide the flow paths. If two input streams enter an ISG mixer, the number of layers emerging from the first, second, and third elements are 8, 32, and 128, respectively. This exponential progression generates over two million layers in just 10 elements. This high theoretical efficiency comes at the price of higher pressure drop. The ISG mixer is used mostly for viscous fluids and in the laminar flow regime.

Obviously, the performance of static mixers in formation and recombination of substreams, and what follows, their mixing efficiency depends roughly on the number of passages that are formed at the edge of (modules of) mixing elements. A measure of (macro)mixing performance is the coefficient of variation, CoV, defined as

$$
\mathrm{CoV} = \frac{\sqrt{\dfrac{1}{k-1}\displaystyle\sum_{i=1}^{k}(c_i - c_{\mathrm{mean}})^2}}{c_{\mathrm{mean}}}, \tag{3.2}
$$

where k is the number of measurements over the cross section of the pipe (usually $k \geq 9$) and c_i is the time-averaged concentration of the ith probe. The lower the CoV, the better is mixing of the streams. A CoV of 0.01 or 0.05 is considered standard.

In summary, static mixers offer a number of important advantages over the conventional systems, which include

- higher mass transfer coefficients and higher specific interfacial areas;
- compactness, requiring smaller footprint and a lower capital expenditure (CAPEX); and
- improved safety because of a smaller inventory and to the lack of moving parts that reduce the sealing problems.

These advantages have brought thousands of commercial applications of static mixers so far, ranging from mixing and blending of chemicals or food, through phase dispersing and emulsification, up to (exothermic) reactions and extraction processes [120–123].

On the other hand, static mixers have also several limitations. They are not suited for slow reactions (with reaction time longer than, say, a few minutes), as a very long mixer would mean extensive pumping cost. Static mixers may cause problems in systems where solids are present or formed, although the resistance to clogging by solids is different in different types of mixers and special "nonclog" designs have been commercialized in recent years that can handle even fibrous slurries, such as sludge. Finally, static mixers are less flexible than stirred tanks; each time they are designed specifically for given flow rates, residence times, volumetric proportions between the components to be mixed, and their properties.

3.5.3 Fractal Systems

Fractals are a type of structure that needs a separate mentioning. Fractal branching is often seen in nature, see for example our vascular network, lungs, kidneys, or tree leaves. That is why fractal structures in chemical engineering applications have been called "nature-inspired chemical engineering" (NICE, [124]).

However, what actually are fractal structures? Very simply speaking, fractal structures are formed by progressively adding copies of the same structure at smaller and smaller scales, where each iteration results in smaller and exact reproduction of the geometry. A very simple fractal structure is shown in Figure 3.54b, where the basic "H" shape has been branched down into smaller and smaller scales. After only five iterations, a network shown in Figure 3.54c is obtained.

As the surfaces of most materials are fractals on the molecular scale [125], the earliest chemical engineering interest in fractal structures concerned surfaces and porous catalysts [126, 127]. Fractals were brought for the first time in

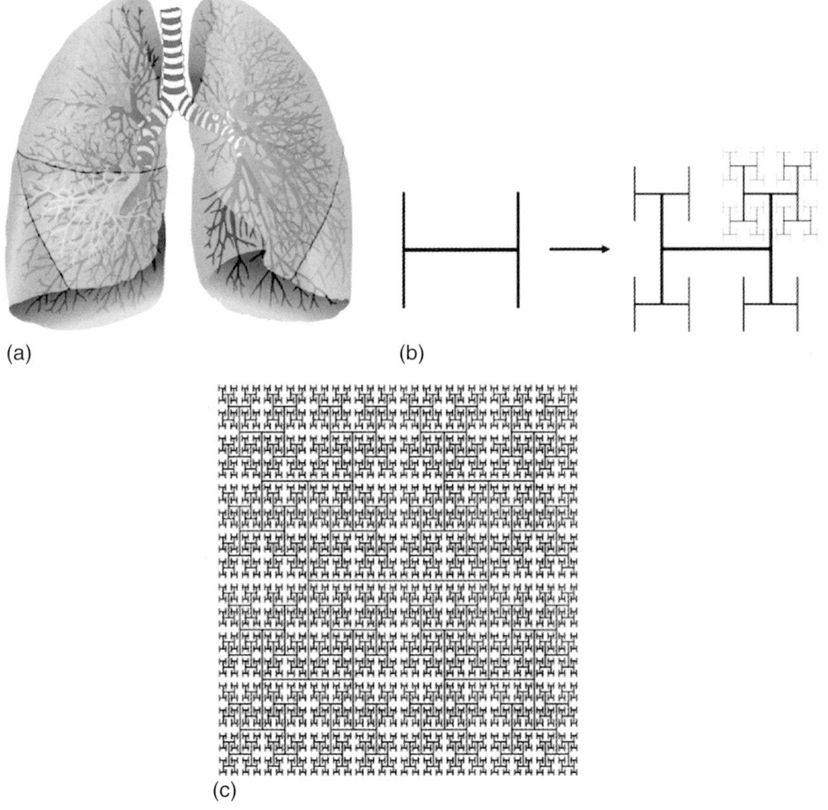

(a)

(b)

(c)

Figure 3.54 (a) Fractal structure of human lungs; (b) formation of a simple two-dimensional fractal structure; (c) the same structure after only five iterations. Source: (a) Mikael Häggström (https://commons.wikimedia.org/wiki/File%3ALungs_(animated).gif). (b, c) Arifractal, Amalgamated Research LLC, USA, www.arifractal.com.

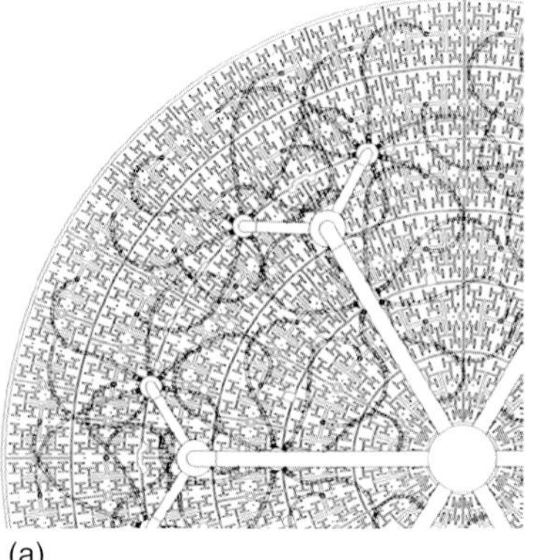
(a)

(b)

Figure 3.55 Concept of two-dimensional fractal fluid distributor (a) and fractal distributor-based juices softening unit (b). Source: Kochergin and Kearney 2006 [129]. Reproduced with permission of Springer.

connection with process intensification by Kearney [128], who advocated their use as almost perfect distributors of fluids, thus in the role similar to the one fractal structures play in the nature. A concept of a fractal distributor is shown in Figure 3.55a. Advantages of such systems include uniform fluid distribution, easy scale-up, very low sensitivity to changes in the feed flow rate, and very low pressure drop. First commercial application of fractal fluid distributor was reported by Kochergin and Kearney [129], in the weak cation juice softening in the sugar beet industry. The use of the fractal fluid distributor (see Figure 3.55b) allowed utilization of shallow resin beds operating at very high flow rates of 500 bed volumes per hour. Energy requirements for pumping were very significantly decreased because of a small bed height, whereas the overall capital cost was reduced by 35–40%. A comparison between the conventional and the fractal-intensified process for juice softening is presented in Table 3.7.

Other possible applications of two-dimensional fractal distributors investigated in the literature include microchannel nets for cooling of electronic chips [130], catalytic monoliths, and multitubular heat exchangers [131], as well as proton-exchange membrane (PEM) fuel cells [132]. Also, the concepts of *three-dimensional* fractal distributors have been studied [124, 133, 134]. Such

Table 3.7 Comparison of conventional and fractal distributor-based weak cation juice softening.

	Conventional ion exchange	Fractal ion exchange	Intensification factor
Resin bed depth (inches)	40	6	6.5
Exhaustion flow rate (bed volumes/h)	50	500	10
Maximum resin bed pressure drop (bar)	3.5–5.6	0.1	>35
Regeneration flow rate (bed volumes/h)	30	150	5
Relative process size	10	1	10

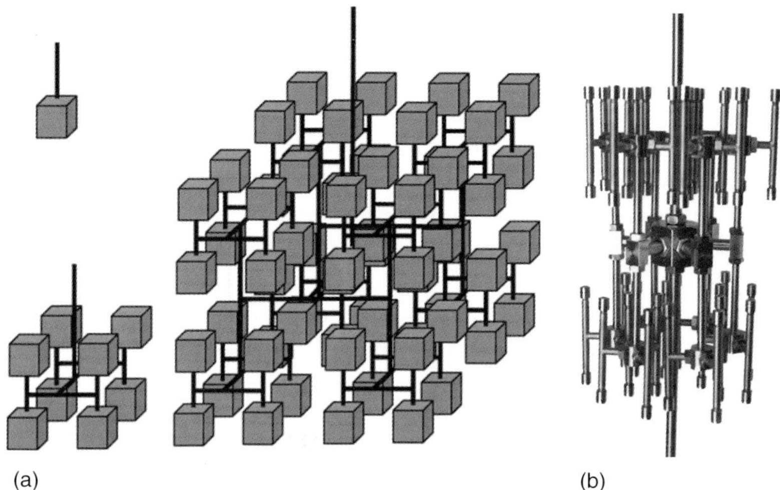

Figure 3.56 (a) Concept of a three-dimensional fractal distributor. Source: Coppens 2005 [124]. Reproduced with permission of American Chemical Society; (b) fractal gas injector for fluidized bed systems Van Ommen et al. 2009 [134]. Republished with permission of American Institute of Chemical Engineers; permission conveyed through Copyright Clearance Center, Inc.

distributors (see an example in Figure 3.56) may be applied in the fluidized-bed units, in order to distribute the gas evenly "in volume," avoiding radial nonuniformity and compensating for axial gradients in the gas flow and reactant concentrations.

Finally, an interesting example of a three-dimensional, quasi-fractal structure is the distillation packing Super X-Pack developed in late 1990s by Nagaoka International Corporation. This packing consists of branched bundles of wires. The liquid flows downward in the form of a film along the wires, in counter-current to the continuous vapor phase. Super X-Pack was reported to be able to reduce the height of a distillation column by a factor of 5, compared to the conventional tray design. Also, up to 80% energy saving, due to a substantially lower pressure drop, was claimed by the manufacturer. However, no industrial applications have been reported for the Super X-Pack structures so far, which may be due to narrow operating regions and problems with even distribution of the liquid on each packing element.

References

1 Boyer, C. and Fanget, B. (2002). Measurement of liquid flow distribution in trickle bed reactor of large diameter with a new gamma-ray tomographic system. *Chem. Eng. Sci.* 57: 1079–1089.

2 Mosbach, K. (2006). The promise of molecular imprinting. *Sci. Am.* 295 (4): 86–91.

3 Górak, A. and Stankiewicz, A. (2011). Intensified reaction and separation systems. *Ann. Rev. Chem. Biomol. Eng.* 2: 431–451.

4 Trotta, F., Baggiani, C., Luda, M.P. et al. (2005). A molecular imprinted membrane for molecular discrimination of tetracycline hydrochloride. *J. Membr. Sci.* 254: 13–19.

5 Kandimalla, V.B. and Ju, H. (2004). Molecular imprinting: a dynamic technique for diverse application in analytical chemistry. *Anal. Bioanal. Chem.* 380: 587–605.

6 Chen, L., Xuab, S., and Lia, J. (2011). Recent advances in molecular imprinting technology: current status, challenges and highlighted applications. *Chem. Soc. Rev.* 40: 2922–2942.

7 Suriyanarayana, S., Cywinski, P.J., Moro, A.J. et al. (2010). Chemosensors based on molecularly imprinted polymers. In: *Molecular Imprinting. Topics in Current Chemistry*, vol. 325 (ed. K. Haupt). Berlin–Heidelberg: Springer-Verlag.

8 Davis, M.E., Katz, A., and Ahmad, W.R. (1996). Rational catalyst design via imprinted nanostructured materials. *Chem. Mater.* 8: 1820–1839.

9 Saridakis, E., Khurshid, S., Govada, L. et al. (2011). Protein crystallization facilitated by molecularly imprinted polymers. *Proc. Nat. Acad. Sci. USA* 108: 18566–18566.

10 Easton, C.J. (2005). Cyclodextrin-based catalysts and molecular reactors. *Pure Appl. Chem.* 77 (11): 1865–1871.

11 Dodziuk, H. (2006). Molecules with holes – cyclodextrins. In: *Cyclodextrins and Their Complexes* (ed. H. Dodziuk), 1–30. Weinheim: Wiley-VCH.

12 Weisz, P.B. and Frilette, V.J. (1960). Intracrystalline and molecular-shape-selective catalysis by zeolite salts. *J. Phys. Chem.* 64 (3): 382–382.

13 Lercher, J.A. and Jentys, A. (2002). Application of microporous solids as catalysts. In: *Handbook of Porous Solids* (ed. F. Schüth, K.S.W. Sing and J. Weitkamp). Weinheim, Germany: Wiley-VCH Verlag GmbH.

14 Yilmaz, B. and Müller, U. (2009). Catalytic applications of zeolites in chemical industry. *Top. Catal.* 52: 888–895.

15 Olsbye, U., Svelle, S., Bjørgen, M. et al. (2012). Conversion of methanol to hydrocarbons: how zeolite cavity and pore size controls product selectivity. *Angew. Chem. Int. Ed.* 21: 5810–5831.

16 Jae, J., Tompsett, G.A., Foster, A.J. et al. (2011). Investigation into the shape selectivity of zeolite catalysts for biomass conversion. *J. Catal.* 279: 257–268.

17 Kaye, S.S., Dailly, A., Yaghi, O.M., and Long, J.R. (2007). Impact of preparation and handling on the hydrogen storage properties of $Zn_4O(1,4\text{-benzenedicarboxylate})_3$ (MOF-5). *J. Am. Chem. Soc.* 129: 14176–14177.

18 Purewal, J., Liu, D., Sudik, A. et al. (2012). Improved hydrogen storage and thermal conductivity in high-density MOF-5 composites. *J. Phys. Chem.* 116: 20199–20212.

19 Goldsmith, J., Wong-Foy, A.G., Cafarella, M.J., and Siegel, D.J. (2013). Theoretical limits of hydrogen storage in metal–organic frameworks: opportunities and trade-offs. *Chem. Mater.* 25 (16): 3373–3382.

20 Britt, D., Furukawa, H., Wang, B. et al. (2009). Highly efficient separation of carbon dioxide by a metal-organic framework replete with open metal sites. *Proc. Nat. Acad. Sci.* 106 (49): 20637–20640.

21 Lia, J.-R., Mab, Y., McCarthyb, M.C. et al. (2011). Carbon dioxide capture-related gas adsorption and separation in metal-organic frameworks. *Coord. Chem. Rev.* 225: 1791–1823.

22 Li, J.-R., Sculley, J., and Zhou, H.C. (2012). Metal organic frameworks for separations. *Chem. Rev.* 12 (2): 869–932.

23 Llabrés i Xamena, F.X., Abad, A., Corma, A., and Garcia, H. (2007). MOFs as catalysts: activity, reusability and shape-selectivity of a Pd-containing MOF. *J. Catal.* 250: 294–298.

24 Ravon, U., Domine, M.E., Gaudillère, C. et al. (2008). MOFs as acid catalysts with shape selectivity properties. *New J. Chem.* 32: 937–940.

25 Lee, J.-Y., Farha, O.K., Roberts, J. et al. (2009). Metal–organic framework materials as catalysts. *Chem. Soc. Rev.* 38: 1450–1459.

26 Song, F., Zhang, T., Wang, C., and Lin, W. (2012). Chiral porous metal-organic frameworks with dual active sites for sequential asymmetric catalysis. *Proc. Royal Soc. A.* 468: 2035–2052.

27 Gascon, J., Hernández-Alonso, M.D., Almeida, A.R. et al. (2008). Isoreticular MOFs as efficient photocatalysts with tunable band gap: an operando FTIR study of the photoinduced oxidation of propylene. *ChemSusChem* 1: 981–983.

28 Wang, J.-L., Wang, C., and Lin, W. (2012). Metal-organic frameworks for light harvesting and photocatalysis. *ACS Catal.* 2: 2630–2640.

29 Nakano, T. and Hirata, H. (1982). Liquid crystals as reaction media. I. photochemical dimerization of acenaphthylene in cholesteric liquid crystal. *Bul. Chem. Soc. Jpn.* 55: 947–948.

30 Leigh, W.J., Frendo, D.T., and Klawunn, P.J. (1985). Organic reactions in liquid crystalline solvents. 1. The thermal *cis-trans* isomerization of a bulky olefin in cholesteric liquid crystalline solvents. *Can. J. Chem.* 63: 2131–2138.

31 Leigh, W.J. (1985). Organic reactions in liquid crystalline solvents. 2. An investigation into the use of liquid crystalline solvents to effect stereochemical control in the Diels-Adler reaction. *Can. J. Chem.* 63: 2736–2741.

32 Ramamurthy, V. (1986). Organic photochemistry in organized media. *Tetrahedron* 42 (21): 5753–5839.

33 Ramesh, V. and Labes, M.M. (1987). Nematic lyotropic liquid crystals as media for chemical reactions. *Mol. Cryst. Liq. Cryst.* 152: 57–73.

34 Weiss, R.G. (1988). Thermotropic liquid crystals as reaction media for mechanistic investigations. *Tetrahedron* 44 (12): 3413–3475.

35 Hiraoka, S., Yoshida, T., Kansui, H., and Kunied, T. (1992). Strong regiochemical control of bimolecular thermochemical reactions in cholesteric liquid crystalline solvents. *Tetrahedron Lett.* 33 (30): 4341–4344.

36 Kansui, H., Hiraoka, S., and Kunieda, T. (1996). Liquid crystal control of bimolecular thermal reactions. Highly regioselective pericycloaddition of fumarates to 2,6-dialkoxyanthracenes in liquid-crystalline media. *J. Am. Chem. Soc.* 118: 5346–5352.

37 Ishida, Y., Kai, Y., Kato, S. et al. (2008). Two-component liquid crystals as chiral reaction media: highly enantioselective photodimerization of an anthracene derivative driven by the ordered microenvironment. *Angew. Chem. Int. Ed.* 47: 8241–8245.

38 Ishida, Y., Achalkumar, A.S., Kato, S. et al. (2010). Tunable chiral reaction media based on two-component liquid crystals: regio-, diastereo, and enantiocontrolled photodimerization of anthracenecarboxylic acids. *J. Am. Chem. Soc.* 132: 17435–17446.

39 Ehrfeld, W., Hessel, V., Möbius, H. et al. (1995). Potentials and realization of microreactors. In: *Microsystem Technology for Chemical and Biological Microreactors: Papers of the Workshop on Microsystem Technology, Mainz, 20–21 February*, vol. 132, 1–28. Frankfurt: DECHEMA.

40 Lerou, J.J., Harold, M.P., Ryley, J. et al. (1995). Microfabricated minichemical systems: technical feasibility. In: *Microsystem Technology for Chemical and Biological Microreactors: Papers of the Workshop on Microsystem Technology, Mainz, 20–21 February*, vol. 132, 51–69. Frankfurt: DECHEMA.

41 Quiram, D.J., Jensen, K.F., Schmidt, M.A. et al. (2000). Integrated microchemical systems: opportunities for process design. *AIChE Symp. Ser.* 96 (123): 147–162.

42 Jensen, K.F. (2001). Microreaction engineering: is small better? *Chem. Eng. Sci.* 56: 293–303.

43 Ehrfeld, W., Hessel, V., and Löwe, H. (2000). *Microreactors. New Technology for Modern Chemistry*. Weinheim: Wiley-VCH.

44 Hessel, V., Hardt, S., Löwe, H. et al. (2005). *Chemical Micro Process Engineering*. Weinheim: Wiley-VCH.

45 Kockmann, N. (2008). *Transport Phenomena in Micro Process Engineering*. Berlin: Springer-Verlag.

46 Kashid, M.N. and Kiwi-Minsker, L. (2009). Microstructured reactors for multiphase reactions: state of the art. *Ind. Eng. Chem. Res.* 48: 6465–6485.

47 Reintejns, R. (2013). Industrial micro reactors. OSPT Course "Fundamentals and Practice of Process Intensification", Delft.

48 Tonkovich, A.L. and Lerou, J.J. (2010). Microstructures on macroscale: microchannel reactors for medium- and large-scale processes. In: *Novel Concepts in Catalysis and Chemical Reactors* (ed. A. Cybulski, J.A. Moulijn and A. Stankiewicz), 239–260. Weinheim: Wiley-VCH.

49 Ehrfeld, W. (2004). Process intensification through microreaction technology. In: *Re-Engineering the Chemical Processing Plant: Process Intensification* (ed. A. Stankiewicz and J.A. Moulijn), 167–190. New York: Marcel Dekker.

50 Jähnisch, K., Hessel, V., Löwe, H., and Baerns, M. (2004). Chemistry in microstructured reactors. *Angew. Chem. Int. Ed.* 43: 406–446.

51 Protasova, L.N., Bulut, M., Ormerod, D. et al. (2013). Latest highlights in liquid-phase reactions for organic synthesis in microreactors. *Org. Proc. Res. Dev.* 17: 760–791.

52 Kolb, G. and Hessel, V. (2004). Micro-structured reactors for gas-phase reactions. *Chem. Eng. J.* 98: 1–38.

53 Frost, C.G. and Muton, L. (2010). Heterogeneous catalytic synthesis using microreactor technology. *Green Chem.* 12: 1687–1703.

54 Elvira, K.S., Casadevall i Solvas, X., Wootton, R.C.R., and DeMello, A.J. (2013). The past, present and potential for microfluidic reactor technology in chemical synthesis. *Nat. Chem.* 5: 905–915.

55 Roberge, D.M., Ducr, L., Beler, N. et al. (2005). Microreactor technology: a revolution for the fine chemical and pharmaceutical industries? *Chem. Eng. Technol.* 28: 318–323.

56 Poechlauer, P., Vorbach, M., Kotthaus, M. et al. (2009). Microreactor plant for the large-scale production of a fine chemical intermediate – a technical case study. In: *Micro Process Engineering: A Comprehensive Handbook Vol. 3* (ed. V. Hessel, A. Renken, J.C. Schouten and J. Yoshida), 249–254. Weinheim: Wiley-VCH.

57 Lerou, J.J., Tonkovich, A.L., Silva, L. et al. (2010). Microchannel reactor architecture enables greener processes. *Chem. Eng. Sci.* 65: 380–385.

58 Singh, A., Malek, C.K., and Kulkarni, S.K. (2010). Development in microreactor technology for nanoparticle synthesis. *Int. J. Nanoscience* 9: 93–112.

59 Zhao, C.-X., He, L., Qiao, S.Z., and Middelberg, A.P.J. (2011). Nanoparticle synthesis in microreactors. *Chem. Eng. Sci.* 66: 1463–1479.

60 Mehendale, S.S., Jacobi, A.M., and Ahah, R.K. (2000). Fluid flow and heat transfer at micro- and meso-scales with application to heat exchanger design. *Appl. Mech. Rev.* 53: 175–193.

61 Thonon, B. and Tochon, P. (2004). Compact multifunctional heat exchangers: a pathway to process intensification. In: *Re-Engineering the Chemical Processing Plant: Process Intensification* (ed. A. Stankiewicz and J.A. Moulijn), 121–165. New York: Marcel Dekker.

62 Li, Q., Flamant, G., Yuana, X. et al. (2011). Compact heat exchangers: a review and future applications for a new generation of high temperature solar receivers. *Renew. Sustain. Energy Rev.* 15: 4855–4875.

63 Picon-Nunez, M., Polley, G.T., Torres-Reyes, E., and Gallegos-Munoz, A. (1999). Surface selection and design of plate–fin heat exchangers. *Appl. Therm. Eng.* 19 (9): 917–931.

64 Yue, J., Chen, G., Yuan, Q. et al. (2007). Hydrodynamics and mass transfer characteristics in gas–liquid flow through a rectangular microchannel. *Chem. Eng. Sci.* 62: 2096–2108.

65 Cypes, S.H. and Engstrom, J.R. (2004). Analysis of a toluene stripping process: a comparison between a microfabricated stripping column and a conventional packed tower. *Chem. Eng. J.* 101: 49–56.

66 Niu, H., Pan, L., Su, H., and Wang, S. (2009). Effects of design and operating parameters on CO_2 absorption in microchannel contactors. *Ind. Eng. Chem. Res.* 48: 8629–8634.

67 Ye, C., Chen, G., and Yuan, Q. (2012). Process characteristics of CO_2 absorption by aqueous monoethanolamine in a microchannel reactor. *Chin. J. Chem. Eng.* 20: 111–119.

68 Gao, N.-N., Wang, J.-X., Shao, L., and Chen, J.-F. (2011). Removal of carbon dioxide by absorption in microporous tube-in-tube microchannel reactor. *Ind. Eng. Chem. Res.* 50: 6369–6374.

69 Zanfir, M., Gavriilidis, A., and Wille, Ch. And Hesel, V. (2005). Carbon dioxide absorption in a falling film microstructured reactor: experiments and modeling. *Ind. Eng. Chem. Res.* 44: 1742–1751.

70 Constantinou, A., Barrass, S., Pronk, F. et al. (2012). CO_2 absorption in a high efficiency silicon nitride mesh contactor. *Chem. Eng. J.* 207-208: 766–771.

71 Tonkovich, A.L., Jarosch, K., Arora, R. et al. (2008). Methanol production FPSO plant concept using multiple microchannel unit operations. *Chem. Eng. J.* 135S: S2–S8.

72 Sundberg, A., Uusi-Kyyny, P., and Alopaeus, V. (2009). Novel micro-distillation column for process development. *Chem. Eng. Res. Des.* 87: 705–710.

73 Hartman, R.L., Sahoo, H.R., Yen, B.C., and Jensen, K.F. (2009). Distillation in microchemical systems using capillary forces and segmented flow. *Lab Chip* 9: 1843–1849.

74 MacInnes, J.M., Ortiz-Osorio, J., Jordan, P.J. et al. (2010). Experimental demonstration of rotating spiral microchannel distillation. *Chem. Eng. J.* 159: 159–169.

75 Adiche, C. and Sundmacher, K. (2010). Experimental investigation on a membrane distillation based micro-separator. *Chem. Eng. Proc. Proc. Intens.* 49: 425–434.

76 Ziogas, A., Cominos, V., Kolb, G. et al. (2012). Development of a microrectification apparatus for analytical and preparative applications. *Chem. Eng. Technol.* 35: 58–71.

77 TeGrotenhuis, W.E., Cameron, R.J., Butcher, M.G. et al. (1999). Microchannel devices for efficient contacting of liquids in solvent extraction. *Sep. Sci. Technol.* 34: 951–974.

78 Wojik, A. and Marr, R. (2005). Mikroverfahrenstechnische Prinzipien in der Flüssig/Flüssig-Extraktion. *Chem.-Ing.-Tech.* 77: 653–668.

79 Kralj, J.G., Sahoo, H.R., and Jensen, K.F. (2007). Integrated continuous microfluidic liquid–liquid extraction. *Lab on a Chip* 7: 256–263.

80 Mary, P., Studer, V., and Tabeling, P. (2008). Microfluidic droplet-based liquid-liquid extraction. *Anal. Chem.* 80: 2680–2687.

81 Assmann, N. and Rudolf von Rohr, P. (2011). Extraction in microreactors: intensification by adding an inert gas phase. *Chem. Eng. Proc. Proc. Intens.* 50: 822–827.

82 Tang, J., Zhang, X., Cai, W., and Wang, F. (2013). Liquid–liquid extraction based on droplet flow in a vertical microchannel. *Exp. Therm. Fluid Sci.* 49: 185–192.

83 Okubo, Y., Toma, M., Ueda, H. et al. (2004). Microchannel devices for the coalescence of dispersed droplets produced for use in rapid extraction processes. *Chem. Eng. J.* 101: 39–48.

84 Kolehmainen, E. and Turunen, I. (2007). Micro-scale liquid–liquid separation in a plate-type coalescer. *Chem. Eng. Proc.* 46: 834–839.

85 Baret, J.-C., Lucas, F., Blouwolff, J., and Griffiths, A.D. (2008). Microfluidic production of droplet pairs. *Langmuir* 24: 12073–12076.

86 Kashid, M.N., Harshe, Y.M., and Agar, D.W. (2007). Liquid-liquid slug flow in a capillary: an alternative to suspended drop or film contactors. *Ind. Eng. Chem. Res.* 46: 8420–8430.

87 Hofstad, K.H., Rokstad, O.A., and Holmen, A. (1996). Partial oxidation of methane over platinum metal gauze. *Catalysis Letters* 36: 25–30.

88 Quiceno, R., Pérez-Ramírez, J., Warnatz, J., and Deutschmann, O. (2006). Modeling the high-temperature catalytic partial oxidation of methane over platinum gauze: detailed gas-phase and surface chemistries coupled with 3D flow field simulations. *Appl. Catal. A.* 303: 166–176.

89 Quiceno, R., Deutschmann, O., Warnatz, J., and Pérez-Ramírez, J. (2007). Rational modeling of the CPO of methane over platinum gauze. Elementary gas-phase and surface mechanisms coupled with flow simulations. *Catal. Today* 119: 311–316.

90 Monnerat, B., Kiwi-Minsker, L., and Renken, A. (2001). Hydrogen production by catalytic cracking of methane over nickel gauze under periodic reactor operation. *Chem. Eng. Sci.* 56: 633–639.

91 O'Connor, R.P., Schmidt, L.D., and Deutschmann, O. (2002). Simulating cyclohexane millisecond oxidation: coupled chemistry and fluid dynamics. *AIChE J.* 48: 1241–1256.

92 Moulijn, J.A., Kapteijn, F., and Stankiewicz, A. (2004). Structured catalysts and reactors: a contribution to process intensification. In: *Re-Engineering the Chemical Processing Plant: Process Intensification* (ed. A. Stankiewicz and J.A. Moulijn), 191–226. New York: Marcel Dekker.

93 Boger, T., Heibel, A.K., and Sorensen, C.M. (2004). Monolithic catalysts for the chemical industry. *Ind. Eng. Chem. Res.* 43: 4602–4611.

94 Cybulski, A., Edvinsson Albers, R., and Moulijn, J.A. (2006). Monolithic catalysts for three-phase processes. In: *Structured Catalysts and Reactors* (ed. A. Cybulski and J.A. Moulijn), 355–392. Boca Raton: Taylor & Francis.

95 Kreutzer, M.T., Kapteijn, F., Moulijn, J.A. et al. (2006). Two-phase segmented flow in capillaries and monolith reactors. In: *Structured Catalysts and Reactors* (ed. A. Cybulski and J.A. Moulijn), 393–433. Boca Raton: Taylor & Francis.

96 Edvinsson Albers, R., Cybulski, A., Kreutzer, M.T. et al. (2006). Modeling and design of monolith reactors for three-phase processes. In: *Structured Catalysts and Reactors* (ed. A. Cybulski and J.A. Moulijn), 435–477. Boca Raton: Taylor & Francis.

97 Kreutzer, M.T., Kapteijn, F., Moulijn, J.A. et al. (2005). Monoliths as biocatalytic reactors: smart gas– liquid contacting for process intensification. *Ind. Eng. Chem. Res.* 44 (25): 9646–9652.

98 Kreutzer, M.T., Kapteijn, F., and Moulijn, J.A. (2006). Shouldn't catalysts shape up? Structured reactors in general and gas–liquid monolith reactors in particular. *Catal. Today* 111: 111–118.

99 Stankiewicz, A. (2001). Process intensification in in-line monolithic reactor. *Chem. Eng. Sci.* 56: 359–364.

100 Inayat, A., Freund, H., Zeiser, T., and Schwieger, W. (2011). Determining the specific surface area of ceramic foams: the tetrakaidecahedra model revisited. *Chem. Eng. Sci.* 66: 1179–1188.

101 Twigg, M.V. and Richardson, J.T. (2007). Fundamentals and applications of structured ceramic foam catalysts. *Ind. Eng. Chem. Res.* 46: 4166–4177.

102 Zhao, C.Y. (2012). Review on thermal transport in high porosity cellular metal foams with open cells. *Int. J. Heat & Mass Transfer* 55: 3618–3632.

103 Ordomsky, V.V., Schouten, J.C., Van der Schaaf, J., and Nijhuis, T.A. (2012). Foam supported sulfonated polystyrene as a new acidic material for catalytic reactions. *Chem. Eng. J.* 207–208: 218–225.

104 Stemmet, C. (2008). Gas-liquid solid foam reactors: hydrodynamics and mass transfer. PhD Dissertation. Eindhoven University of Technology.

105 Richardson, J.T., Peng, Y., and Remue, D. (2000). Properties of ceramic foam catalyst supports: pressure drop. *Appl. Catal. A: General* 204: 19–32.

106 Bianchi, E., Heidig, T., Visconti, C.G. et al. (2012). An appraisal of the heat transfer properties of metallic open-cell foams for strongly exo−/endo-thermic catalytic processes in tubular reactors. *Chem. Eng. J.* 198–199: 512–528.

107 Patcas, F.C., Garrido, G.I., and Kraushaar-Czarnetzki, B. (2007). CO oxidation over structured carriers: a comparison of ceramic foams, honeycombs and beads. *Chem. Eng. Sci.* 62: 3984–3990.

108 Ciambelli, P., Palma, V., and Palo, E. (2010). Comparison of ceramic honeycomb monolith and foam as Ni catalyst carrier for methane autothermal reforming. *Catal. Today* 155: 92–100.

109 Gokon, N., Yamawaki, Y., Nakazawa, D., and Kodama, T. (2011). Kinetics of methane reforming over Ru/γ Al_2O_3-catalyzed metallic foam at 650-900 °C for solar receiver-absorbers. *Int. J. Hydrogen Energy* 36: 203–215.

110 Thompson, C.R., Marín, P., Díez, F.V., and Ordóñez, S. (2013). Evaluation of the use of ceramic foams as catalyst supports for reverse-flow combustors. *Chem. Eng. J.* 221: 44–54.

111 Yu, X., Wena, Z., Lin, Y. et al. (2010). Intensification of biodiesel synthesis using metal foam reactors. *Fuel* 89: 3450–3456.

112 Olujić, Ž., Seibert, A.F., and Fair, J.R. (2000). Influence of corrugation geometry on the performance of structured packings: an experimental study. *Ind. Eng. Chem. Res.* 39: 335–342.

113 Hessel, V., Löwe, H., and Schönfeld, F. (2005). Micromixers—a review on passive and active mixing principles. *Chem. Eng. Sci.* 60: 2479–2501.

114 Nguyen, N.-T. and Wu, Z. (2005). Micromixers – a review. *J. Micromech. Microeng.* 15: R1–R16.

115 Capretto, L., Cheng, W., Hill, M., and Zhang, X. (2011). Micromixing within microfluidic devices. *Top. Curr. Chem.* 304: 27–68.

116 Haeberle, S., Schlosser, H.P., Zengerle, R. and Ducrée, J. (2005). A centrifuge-based microreactor. IMRET 8, 8th International Conference on Microreaction Technology, April 10–14, Atlanta, USA, p TK-129f.

117 Falk, L. and Commenge, J.-M. (2010). Performance comparison of micromixers. *Chem. Eng. Sci.* 65: 405–411.

118 Jeong, G.S., Chung, S., Kim, C.-B., and Lee, S.-H. (2010). Applications of micromixing technology. *Analyst* 135: 460–473.

119 Heyouni, A., Roustan, M., and Do-Quang, Z. (2002). Hydrodynamics and mass transfer in gas–liquid flow through static mixers. *Chem. Eng. Sci.* 57: 3325–3333.

120 Cybulski, A. (2007). Static mixers. In: *European Roadmap for Process Intensification – Technology Report*. Utrecht, The Netherlands: SenterNovem.

121 Thakur, R.K., Vial, C., Nigam, K.D.P. et al. (2003). Static mixers in process industries – a review. *Trans. IChemE* 81: 787–825.

122 Kiss, N., Brenn, G., Pucher, H. et al. (2011). Formation of O/W emulsions by static mixers for pharmaceutical applications. *Chem. Eng. Sci.* 66: 5084–5094.

123 Green, A. (2004). Inline and high intensity mixers. In: *Re-Engineering the Chemical Processing Plant: Process Intensification* (ed. A. Stankiewicz and J.A. Moulijn), 227–260. New York: Marcel Dekker.

124 Coppens, M.-O. (2005). Scaling-up and -down in a nature-inspired way. *Ind. Eng. Chem. Res.* 44: 5011–5019.

125 Avnir, D., Farin, D., and Pfeifer, P. (1984). Molecular fractal structures. *Nature* 308: 261–263.

126 Coppens, M.-O. (1999). The effect of fractal surface roughness on diffusion and reaction in porous catalysts – from fundamentals to practical applications. *Catal. Today* 53: 225–243.

127 Sheintuch, M. (2001). Reaction engineering principles of processes catalyzed by fractal solids. *Catal. Rev. Sci. Eng.* 43: 233–289.

128 Kearney, M. (2000). Engineering fractals enhance process applications. *Chem. Eng. Prog.* 96 (12): 61–68.

129 Kochergin, V. and Kearney, M. (2006). Existing biorefinery operations that benefit from fractal-based process intensification. *Appl. Biochem. Biotechnol.* 129–132: 349–360.

130 Chen, Y. and Cheng, P. (2002). Heat transfer and pressure drop in fractal tree-like microchannel nets. *Int. J. Heat Mass Transfer* 45: 2643–2648.

131 Tondeur, D. and Luo, L. (2004). Design and scaling laws of ramified fluid distributors by the construal approach. *Chem. Eng. Sci.* 59: 1799–1813.

132 Kjelstrup, S., Coppens, M.-O., Pharoah, J., and Pfeifer, P. (2005). Nature-inspired energy and material efficient design of a polymer electrolyte membrane fuel cell. *Energy Fuels* 24: 5097–5108.

133 Christensen, D., Nijenhuis, J., Van Ommen, J.R., and Coppens, M.-O. (2008). Influence of distributed secondary gas injection on the performance of a bubbling fluidized-bed reactor. *Ind. Eng. Chem. Res.* 47: 3601–3618.

134 Van Ommen, J.R., Nijenhuis, J., and Coppens, M.-O. (2009). Reshaping the structure of fluidized beds. *Chem. Eng. Prog.* 105: 49–57.

4

ENERGY – PI Approaches in Thermodynamic Domain

4.1   Energy in Chemical Processes – A Broader Picture of the Present and the Future

Chemical industry, together with metal, pulp and paper, textile, and similar sectors, belongs to the energy-intensive industries. Unfortunately, the word "intensive" here has nothing to do with process intensification, i.e. efficiency improvement, but relates to the high energy consumption in those sectors. In Germany, the so-called energy-intensive industries consume c. 70% of the entire industrial energy consumption in that country [1]. In the chemical sector, the major part of energy is used in the downstream processing, which indicates the need to address better the first and second principles of process intensification, as discussed in Chapter 2.

Contemporary chemical industry is predominantly based on fossil fuels as energy source (Figure 4.1) and steam boiler is probably the most common element of today's chemical processes. Next to its negative environmental effects, the steam boiler-based heating is thermodynamically inefficient and nonselective in nature. To this end, smarter and more selective methods for energy supply in chemical processes are needed, and electricity may be one of those methods. Electricity can be generated from various renewable sources (solar, wind, geothermal, biomass, etc.) and will therefore become the widest available, most versatile energy form in the future. The National Renewable Energy Laboratory [2] predicts that by 2050, up to c. 80% of the US electricity demand could be met with the renewable electricity. Other international organizations provide similar figures. The price of some forms of renewable electricity (e.g. solar) decreases dramatically.

Hence, the long-term vision for chemical industries (similarly to other sectors) should include their gradual transition from fossil fuels to green electricity as the primary energy source. In order to realize such vision and introduce green electricity-based chemical plants, progress needs to be made not only in electricity-based processing methods but also in green electricity generation, integrated storage, and recovery (because of the intermittent character of the generation); in electricity source-dependent process instrumentation and control; and, last but not least, in region-dependent process plant design (Figure 4.2). The latter has to do with the fact that a renewable electricity-operated chemical

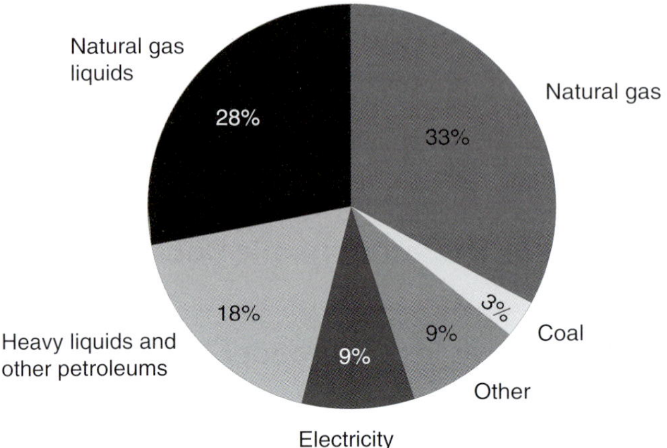

Figure 4.1 Share of total energy consumption by chemical industry, per source. Source: https://energy.gov/eere/amo/chemicals-industry-profile.

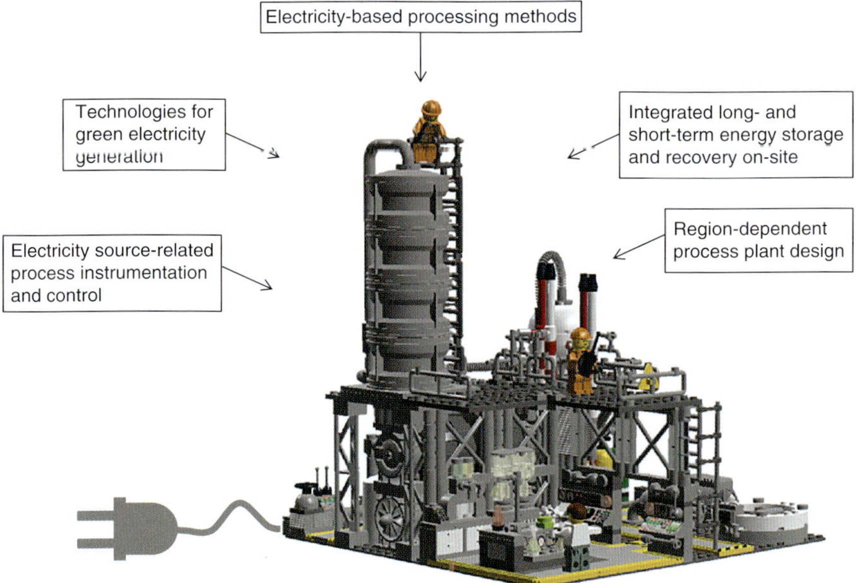

Figure 4.2 Elements needed to realize the vision of renewable electricity-based chemical processes. Source: Plant model designed by LEGO® Ideas member Ymarilego. Used by permission of ©2017 The LEGO Group.

plant in the deserts of Arabian Peninsula will be different than the one located on the north coast of Norway, simply because of the difference in renewable electricity sources there.

Electricity-based processing methods and equipment are definitely not new to chemical industries. For instance, electrothermic systems for enhancing oil

Figure 4.3 The structure of the electrothermic oil recovery process where the 480 V power is fed to a downhole contractor through an insulated production pipe. Source: Adapted from Gill 1983 [3].

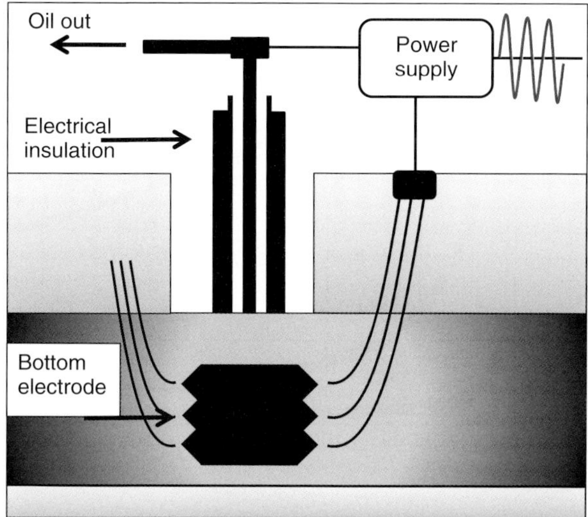

recovery have been known and applied on the commercial scale for many years (Figure 4.3). Electrical enhanced oil recovery has been found superior to the conventional one. Its advantages include efficient energy utilization in heating of a specific part of the reservoir, low cost, and no environmental impact on the geology [4].

In the present chapter, we will discuss technologies and devices that utilize alternative energy forms and transfer mechanisms, and by doing so are able to change the way the molecules are activated or separated, to increase the heat and mass transport rates or to reduce the mixing time in the system. It is worth noting that many of those technologies and devices are based on electricity. They are therefore expected to play an important role in the above discussed transition from fossil fuel-based processes to green electricity-based processes.

4.2 Electric Fields

On the fundamental level, electric fields have been investigated for many years as a means for orienting molecules, as the mutual orientation of molecules at the moment of their collision influences the effectiveness of that collision, hence the reaction outcome (see also Section 2.1). Consequently, dipolar molecules can be oriented by placing them in a uniform electric field. Neutral molecules, on the other hand, interact with electric fields through their charge distribution, via the so-called "Stark effect" [5]. However, the electric fields used for orienting molecules must be extremely strong. This is illustrated in Figure 4.4 on an example of a simple dipolar molecule – water. As can be seen from the figure, in order to get reasonable values of $\cos \theta$ (angle between the dipole and the field), field strength in excess of 10^6 V/cm and very low temperatures are needed. This obviously limits the practical applicability of the above-described method.

Closer to the industrial practice, the enhancement of selected operations by means of the electric fields has been known for many years [6]. Electric fields

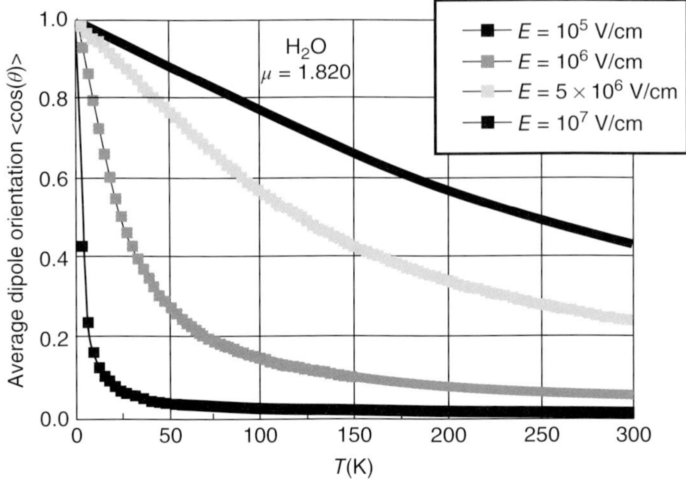

Figure 4.4 Orientation of the water dipole in the electric field, as a function of field strength and temperature. Source: Franssen et al. 2014 [5]. Reproduced with permission of John Wiley and Sons.

may lead to (additional) surface generation in multiphase systems and/or to the increase of the mass and heat transfer coefficients. According to Ref. [7], a neutral droplet elongates in the electric field because of polarization and, after reaching the instability point, disintegrates into smaller droplets creating additional interfacial surface area. Two basic methods exist for *surface area generation in liquid–liquid systems* subjected to an electric field [8]. The first one consists in the droplet formation in charged nozzles or orifices (Figure 4.5a). The second

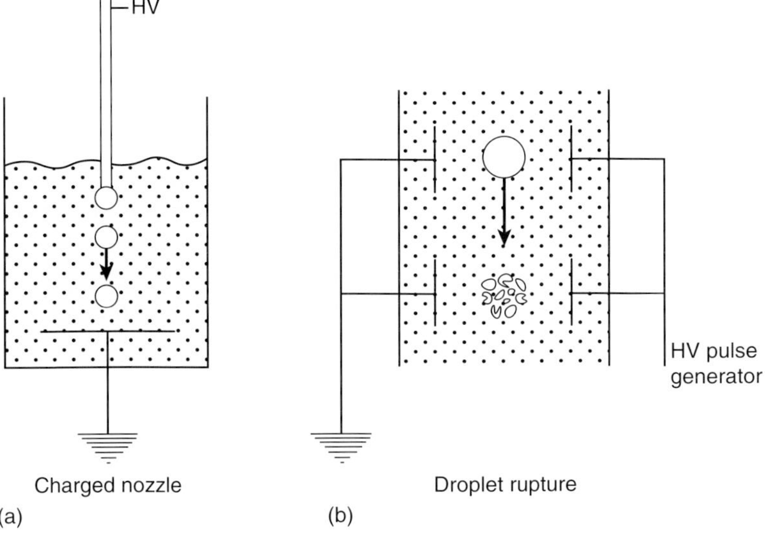

Figure 4.5 Basic methods for surface area generation in an electric field (a) via charged nozzle or orifice, (b) via droplet breakage in a strong electric field. Source: Adapted from Ptasinski and Kerkhof 1992 [8].

one uses strong electric fields (e.g. high-intensity pulsed electric fields, PEFs) to rupture droplets into extremely small droplets, sometimes smaller than 5 μm (Figure 4.5b).

Scott [9] reported enhanced surface area formation via electric field-induced emulsification. The method may lead to a 200–500-fold increase in the surface area per unit volume because of the jetting phenomenon and the formation of electrically charged micron-sized droplets.

Also, *mass transfer enhancement by electric fields* in liquid–liquid extraction by factors of up to 10 has been reported [8–12]. Ptasinski and Kerkhof [8] ascribed the observed mass transfer enhancement to a higher degree of turbulence within and around the dispersed phase, as a result of interaction between the field and the interface. Four different mechanisms are mentioned:

- Higher terminal drop velocities resulting from electrical forces of attraction exerted on the drops in the direction of motion;
- Generation of the electrically driven circulating flow in the neighborhood of the interface;
- Alteration of the velocity profiles within and around individual droplets because of the oscillations by PEFs; and
- Interfacial tension-induced surface flows (Marangoni effects) because of the presence of electric charges.

Interestingly, electric fields cannot only be used to enhance the breakage of the droplets and to create additional surface areas but they can also be used to speed up the *droplet coalescence*. The electric field strength used to coalesce the droplets is much lower than the strength used to break them up (typically c. 1 kV/cm for coalescence versus >4.5 kV/cm for breakage) [13, 14]. One of the theories to explain the coalescence under electric fields is a three-stage mechanism. In the first stage, drops that are separated by a thin film of continuous phase are approaching each other. In the next stage, the separated film is thinning with a rate that is inversely proportional to the square of drop size. In the last stage, when the film reaches a critical thickness, any disturbance will rupture the drops and coalescence occurs (Figure 4.6). The function of the applied electric field during attractive interactions is to bring two drops closer and rapidly rupture the continuous film that is separating them. Other possible mechanisms involved in the electrocoalescence and reported in literature include droplet chain formation, dipole–dipole coalescence, electrophoresis, dielectrophoresis, and random collisions. All those theories have been extensively reviewed by Eow et al. [13].

Electrocoalescers have been used in the oil and petroleum industries to remove water and contaminants from crude oil. An example of such a device is electrostatic coalescer developed by Frames Group presented in Figure 4.7.

PEFs have been shown to enhance *extraction in solid–liquid systems*. In polysaccharide extraction from linseeds, an increase of extraction rates by 50–80% has been observed [15]. Compared to the conventional technology, the extraction of polysaccharide, polyphenol, and protein from white button mushrooms with PEF [16] delivered significantly higher yields at lower temperatures and much shorter processing times (e.g. 2.5 minutes versus 60 minutes).

The research on the influence of electric fields on *heat transfer in gases and liquids* dates back to early 1930s, with the classical works by Senftleben and

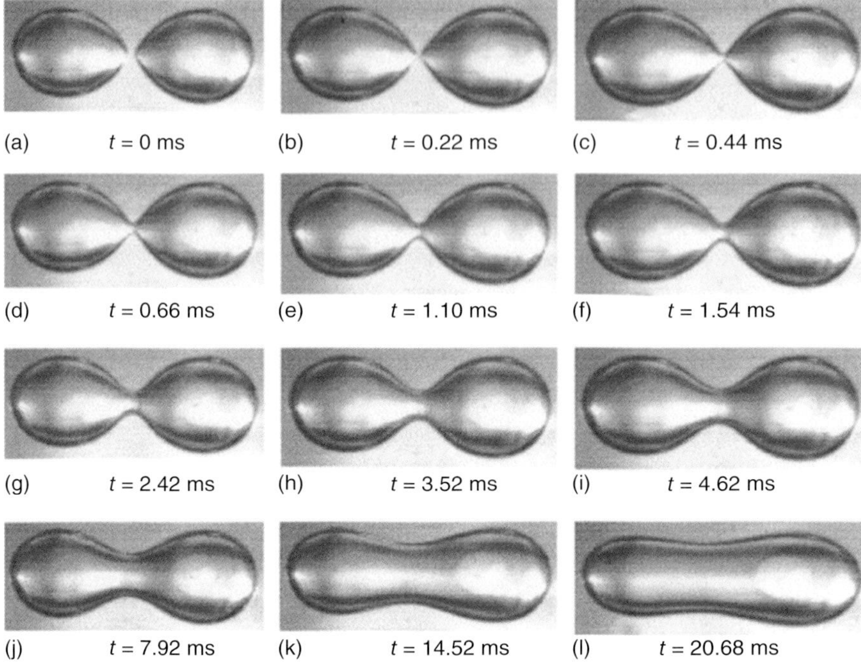

Figure 4.6 Film thinning, rupture, and water droplet coalescence under an electric field of 1 kV/cm. Source: Eow and Ghadiri 2003 [14]. Reproduced with permission of Elsevier.

coworkers [17–20], where quadratic or almost quadratic increase of the heat transfer rates with the electric field strength was reported. More recently, Ryde et al. [21] observed increased heat transfer for natural convection from a wire to hexane in both AC and DC fields, whereby the Nusselt numbers in AC fields were a factor of 4 higher than the corresponding values in the DC fields. Kaji et al. [22] observed a two- to threefold increase in the heat transfer coefficients from water droplets in silicone oil rising in the intermittent electric field.

The *heat transfer in boiling liquids* was also reported to increase by a factor of 4 to 10 under the influence of alternating electric fields with low intensity (2–10 kV/cm) [23, 24]. Markels and Durfee [25] observed that in the normal film boiling region, the surface wetting initially contributed strongly to the increased heat transfer, while at higher voltages, the local mixing and more favorable bubble shape factors prevailed.

Enhancement of heat transfer due to electric fields has also been observed in evaporation systems. Wolny and Kaniuk [26] reported an approximately eightfold increase in the average values of heat and mass transfer coefficients during evaporation from flat and cylindrical surfaces, while Darabi et al. [27] observed up to sevenfold enhancement of the heat transfer coefficient in falling film evaporators.

More recently, AC and DC electric fields were used to enhance the *mixing in microfluidic systems*. Such *electrohydrodynamic mixing* was investigated, among other authors by Tsouris et al. [28], who observed a significant shortening of the mixing lengths in a microchannel operating under strong electric field

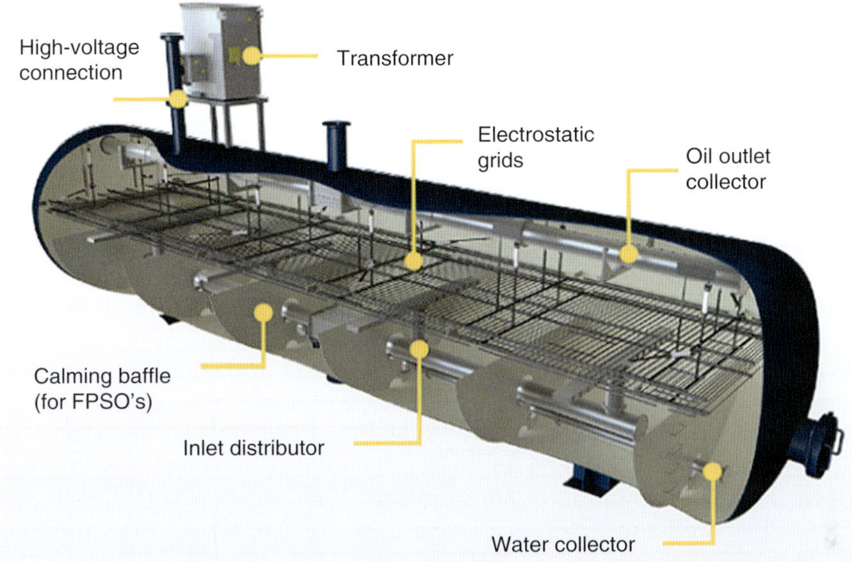

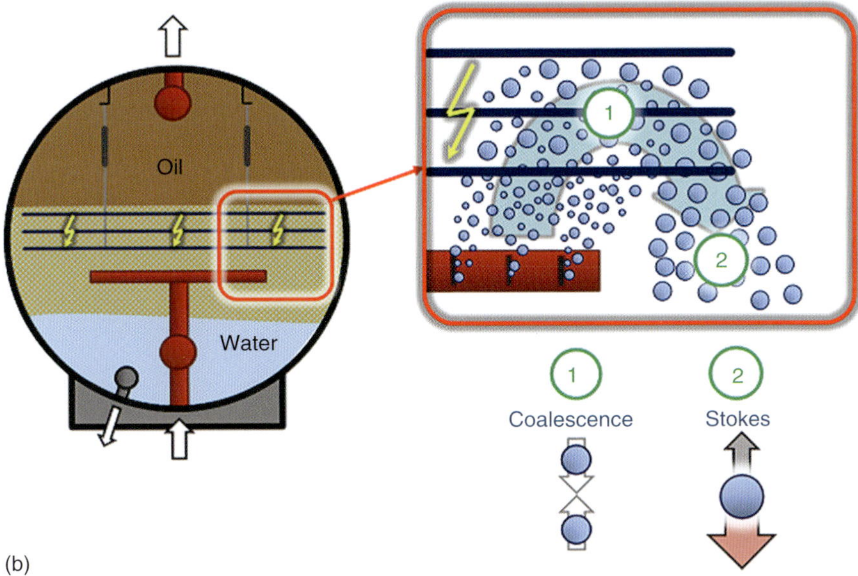

Figure 4.7 Frames electrostatic coalescer: (a) 3D model and (b) working principle. Source: Reproduced with permission of Frames Group, http://www.frames-group.com/Products/Electrostatic-Coalescers.

($>2\,kV/mm$), from $>5000\,\mu m$ (no field) down to $<150\,\mu m$. El Moctar et al. [29] studied the influence of the DC and AC electric field strength on the mixing index in a microchannel. In both cases, an increase of the field strength from c. $2E+05$ to $5E+05\,V/m$ produced a seven- to eightfold increase in the mixing index.

4.3 Magnetic Fields

Although magnetic fields are rarely mentioned as a "process intensification" technology *pur sang*, they definitely deserve a place in this book. Applications of magnetic fields in chemical processes investigated in the literature are very diverse and range from fluidization to selective bioseparations. The concept of the so-called *magnetically stabilized fluidized beds* (MSB), in which the hydrodynamics are influenced by externally imposed magnetic fields, was born in the Soviet Union [30], but attracted wide attention with the studies of Exxon Research and Engineering Company, Linden, NJ, and the resulting groundbreaking *Science* paper by Rosensweig [31].

Shortly speaking, the fluidization of magnetizable particles in the presence of a uniform magnetic field oriented parallel to the gas flow prevents the formation of the gas bubbles and enables operation in a stable emulsion up to high gas velocities (Figure 4.8).

The mechanism behind such a stabilization of the bed and elimination of gas bubbles is schematically shown in Figure 4.9. According to Ref. [32], a gas bubble represents a nonmagnetic cavity within a uniformly magnetized medium. It experiences magnetic interfacial forces that act normal to the interface and cause the bubble to collapse. A MSB is claimed to combine the advantages of both fixed bed and conventional fluidized bed, as shown in Table 4.1.

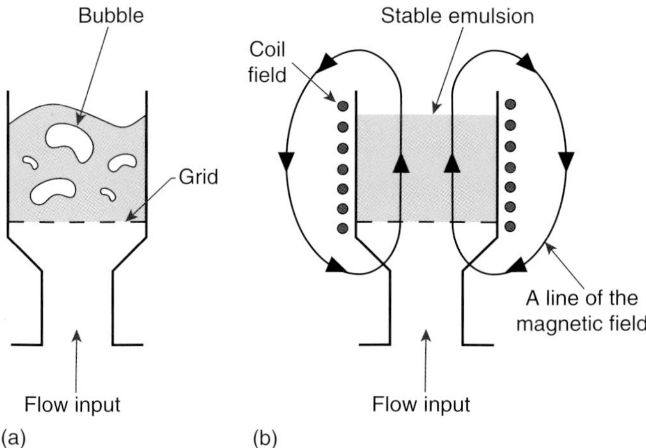

Figure 4.8 Conventional (a) and magnetically stabilized (b) fluidized beds. Source: Lucchesi et al. 1979 [32]. Reproduced with permission of World Petroleum Council.

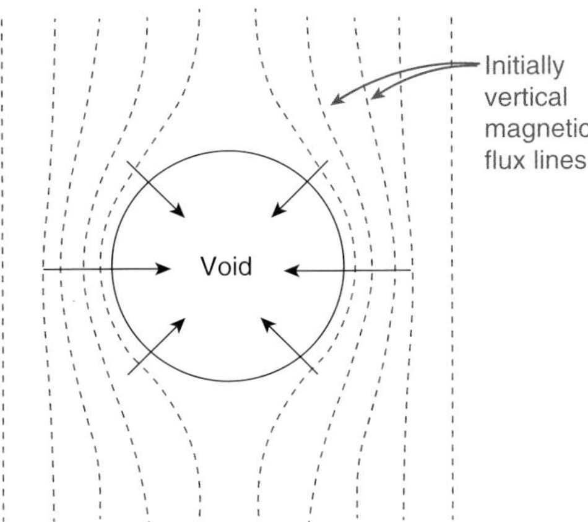

Figure 4.9 Interfacial magnetic forces opposing the formation of nonmagnetic voids (gas bubbles) in a magnetically stabilized bed. Source: Liu et al. 1991 [33]. Reproduced with permission of Elsevier.

Table 4.1 Magnetically stabilized bed combines principal advantages of fluid beds and fixed beds.

	Fluidized bed	Magnetically stabilized bed	Fixed bed
Small particle size with low ΔP	Yes	Yes	—
High reactor efficiency	—	Yes	Yes
Continuous solids throughput	Yes	Yes	—
Counter-current contacting	—	Yes	—
Avoids entrainment from bed	—	Yes	Yes

Source: Lucchesi et al. 1979 [32]. Reproduced with permission of World Petroleum Council.

The possibility of counter-current contacting presents a unique feature of the magnetically stabilized beds. Additionally, stabilization offers a greatly reduced radial dispersion with much lower axial mixing, compared to conventional bubbling systems [34, 35]. Also, magnetic stabilization allows bed operation without particles escape (entrainment) at much higher gas or liquid flow rates. Moffat et al. [36] reported a fivefold increase in the volumetric liquid throughput at the escape velocity. On the other hand, the picture with regard to the heat transfer is more complex. Although a magnetically stabilized bed appears to deliver much higher heat transfer coefficients than a packed bed, numerous authors report a strong decrease of those coefficients with the increasing field strength (e.g. [37–39]). This phenomenon can be partially explained by the formation of particle clusters at large field strengths. On the other hand, in magnetically stabilized gas–liquid–solid systems, mass transfer coefficients from gas to liquid increase with the magnetic strength, whereas the liquid–solid mass transfer coefficients are lower than in the absence of the field [40]. Nevertheless, some papers report significantly increased performance of the magnetically stabilized gas–liquid–solid reactors. Examples are ethanol fermentation process with immobilized yeast cells, where the maximum ethanol yield achieved under the magnetic field was c. 25% better than without it [41], and the purification of caprolactam via hydrogenation, where catalyst consumption decreased by 60% in the case of a magnetically stabilized system [42]. Application of magnetic fields in gas separations has also been reported to deliver significant gains in terms of the separation efficiency. For instance, Sikavitsas et al. [43] used magnetically stabilized beds in a pressure swing adsorption process for the olefin–paraffin separation. Ethylene recovery in the MSB was almost four times higher (50% versus 14%) than in the packed bed [43]. De Brito et al. [44] observed much higher efficiencies of the aromatic compounds (methylene blue, phenol) removed from water by adsorption under magnetic field.

Another large group of existing and potential applications of magnetic fields are separations, ranging from steel and mineral manufacturing through wastewater treatment, down to protein or cell isolation in biopharmaceutical processing [45]. Separations of paramagnetic or ferromagnetic minerals are carried out on the industrial scale, and Figure 4.10 presents an example of a continuous high-gradient magnetic separator developed by METSO Corporation. It consists of several magnetic heads and a ring with matrix cassettes. The feed and rinse water enter through the boxes located on top of each magnet head. Rinse water and nonmagnetic solids are collected in the boxes below the matrix ring while the magnetic product is flushed from above the slots by low-pressure water and vacuum.

Magnetic field-assisted selective separations in environmental and biopharmaceutical processes usually take place via *magnetic carriers* or *magnetic tags* [36]. The difference between those two is schematically shown in Figure 4.11. Magnetic carriers are usually 10–1000 times larger than the target species and usually have an engineered surface, to which colloidal, macromolecular, or ionic species attach [46]. The tags in turn are usually (much) smaller than the particles to be separated and they attach to those particles by various mechanisms, such as antibody–antigen interaction or electrostatic adsorption.

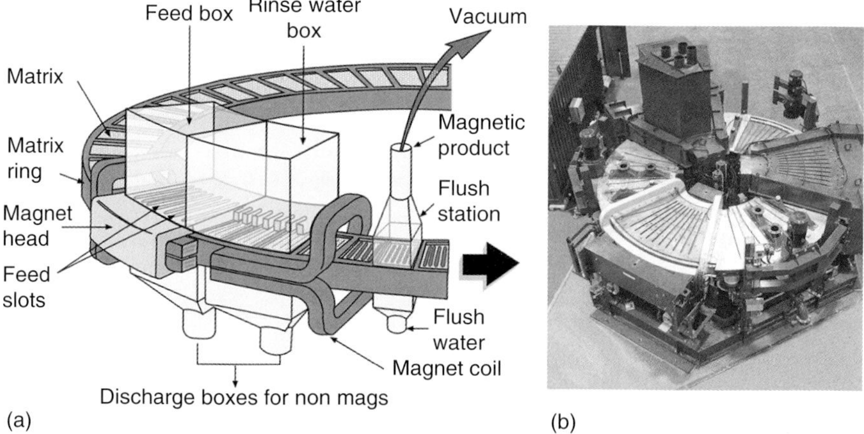

Figure 4.10 High-field magnetic separator for mineral processing Source: Courtesy of METSO Corporation, Finland, www.metso.com.

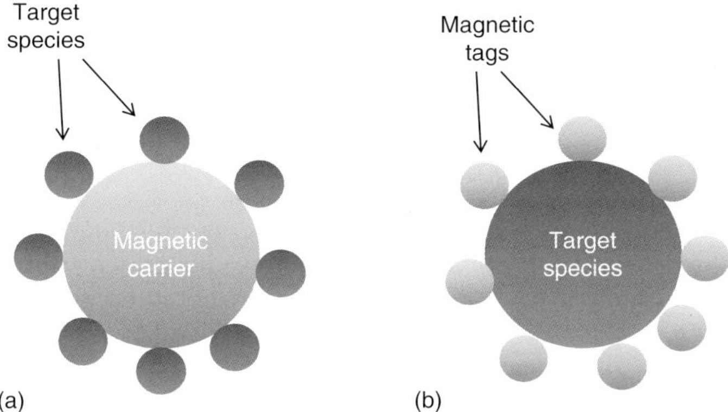

Figure 4.11 Selective separation via magnetic carrier (a) and magnetic tags (b).

A relatively new and interesting area related to the magnetic field-assisted processing is *ferrofluids* (or magnetite nanofluids). Ferrofluids present suspensions of magnetic nanoparticles (usually 10–15 nm diameter) in nonmagnetic carrier fluids. Experimental data from different groups show that utilization of ferrofluids in magnetic fields leads to a significant increase of the heat transfer rates and up to fourfold enhancements are reported [47]. The postulated mechanisms for such enhancements include accumulation of ferroparticles near the magnets leading to locally increased thermal conductivity and the formation of aggregates that enhance momentum and energy transfer in the flow. Also, mass transfer enhancement effects have been reported for systems using nanofluids under magnetic fields. Samadi et al. [48] observed a significant increase in mass flux and mass transfer coefficients (22% and 59%, respectively) in the CO_2 absorption in nanofluids, in a wetted-wall column with the magnetic field directed downward parallel to the flow. Finally, magnetic particles and magnetic fields find applications in microfluidic devices. More information can be found in several good review papers on that topic [49–51].

4.4 Electromagnetic Fields

4.4.1 Microwaves

Microwave energy is a form of electromagnetic radiation between 300 MHz and 300 GHz (wavelengths 1 m to 1 mm, respectively, in free space). Legislation imposes that the usable frequencies for chemical processes are 915 MHz, 2.45 GHz, and 5.85 GHz, the so-called ISM (Industrial Scientific and Medical) frequency bands. In liquid systems, microwave heating requires the presence of polar molecules or ions in the liquid solution. In the presence of a rapidly alternating electric field, polar molecules with a permanent dipole moment try to align themselves with the field direction and ions are subjected to rapid translational movements (Figure 4.12). The result of this rotational or translational motion is internal friction and eventually heating of the polar medium. It follows

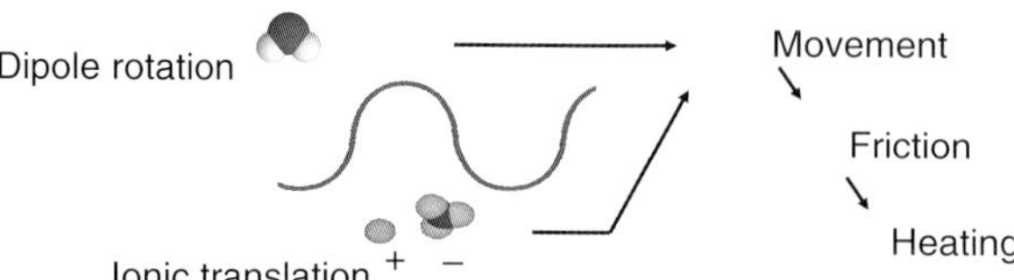

Figure 4.12 Mechanisms of liquid heating by microwaves.

that microwave heating of liquids is volumetric in nature and fundamentally different from conduction-based heating, which requires the presence of a heat transfer surface. As such, it is attractive from the process intensification point of view because it can intensify heat transfer in viscous and low thermal conductivity media.

Solid materials can be divided into three categories with regard to their interaction with microwaves. Conductors, such as metals or graphite, reflect microwaves from their surface. In insulators, such as polypropylene or quartz glass, microwaves penetrate through the material, whereas in the so-called "*dielectric lossy materials*," such as silicon carbide, microwaves are absorbed causing heating. The latter are electron-rich solids, which have no freely rotatable dipoles, such as carbon-based solids. In this case, it is the motion of electrons that generates heat through joule heating or the formation of arcs at the phase boundaries [52].

The properties of any material with respect to microwave radiation are described by its *complex permittivity, ε^**, and *complex permeability, μ^**.

$$\varepsilon^* = \varepsilon' - j\varepsilon'' = \varepsilon_0(\varepsilon_r' - j\varepsilon''_{eff}) \tag{4.1}$$

$$\mu^* = \mu' - j\mu'' \tag{4.2}$$

The real part (ε') of the relative permittivity is called the *dielectric constant* and characterizes the ability of the material to store electrical energy. The imaginary part (ε'') is called the *loss factor* that reflects the ability of the material to dissipate electrical energy. Similarly, the real part of complex permeability (μ') represents the amount of magnetic energy stored within the material, whereas the imaginary part (μ'') represents the amount of magnetic energy, which can be converted into thermal energy. The ratios between the imaginary and the real parts of the permittivity and permeability are called loss tangent ($\tan \delta$) and magnetic loss tangent ($\tan \delta_\mu$), respectively.

$$\tan \delta = \frac{\varepsilon''}{\varepsilon'} \tag{4.3}$$

$$\tan \delta_\mu = \frac{\mu''}{\mu'} \tag{4.4}$$

Microwaves can penetrate into the material up to a certain depth, called the *penetration depth – D_p*. This is usually described as the depth where the microwave power drops to about 37% of the initial value. For materials where $\varepsilon''/\varepsilon' < 1$ (lossy dielectrics), D_p is defined as

$$D_p = \frac{\lambda}{2\pi} \frac{\sqrt{\varepsilon'}}{\varepsilon''} \tag{4.5}$$

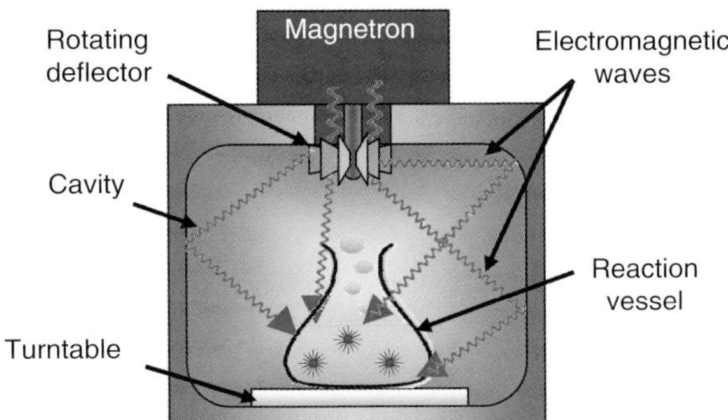

Figure 4.13 Multimode microwave cavity.

where λ is the wavelength.

The average power dissipated in a material of volume V can be expressed as

$$P = \omega \varepsilon_0 \varepsilon'' E_{\mathrm{rms}}^2 V + \omega \mu_0 \mu'' H_{\mathrm{rms}}^2 V \tag{4.6}$$

where E_{rms} is the root mean square of the electric field intensity, H_{rms} is the root mean square of the magnetic field intensity, ω the angular frequency, and ε_0, μ_0 are the permittivity and permeability in free space, respectively.

The microwave equipment used in chemical processing can roughly be divided into two basic categories, namely cavity-based equipment, including multimode and monomode cavities, and noncavity-based equipment, such as the internal transmission line-based microwave reactors.

The multimode cavity type is the most celebrated equipment type. The applications range from domestic ovens up to large-scale industrial dryers. They usually have the form of a rectangular metal box (Figure 4.13). The three-cavity dimensions are longer than half of the wavelength. Inside the cavity, microwaves get reflected on the cavity walls, combine constructively and destructively, and form a large number of resonance modes. Because of the wave interference, the microwave field is highly inhomogeneous. To improve heating uniformity over the heated material, a mode stirrer and/or a rotating disc is used. The main advantage of multimode cavities is that they can be made large enough to enable microwave processing of large material volumes (tens of liters).

In monomode cavities, a single, well-defined field mode is developed, which is relatively more robust compared to the field in multimode cavities. This makes analysis and optimization of this type of cavity and of the process occurring in it easier to carry out. The irradiated material is placed in one of the maxima of the electromagnetic field (Figure 4.14). However, the volume of the material to be processed is limited in size. Monomode cavities usually process volumes up to 200 ml. This is their main limitation. Enlargement of the cavity volume will lead to wave interference and a resonant multimode field. On the other side, the advantage of monomode cavities is that higher electromagnetic field densities can be developed inside the cavity, and thus higher heating rates. Popular off-the-shelf multimode and monomode cavities for lab-scale processing concerning mainly reactions, but also separation processes, are provided by a number of technology providers [53–56].

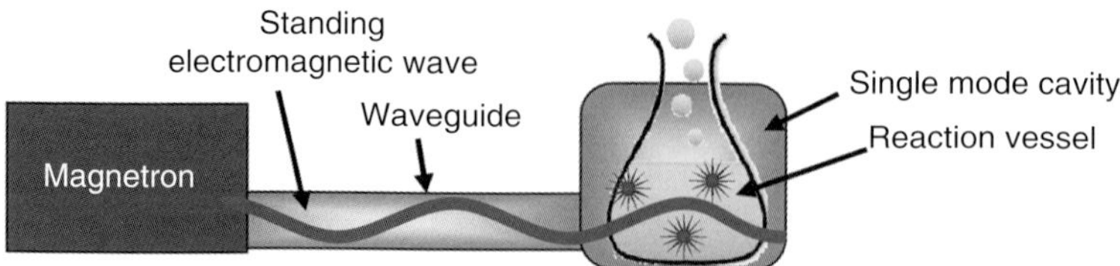

Figure 4.14 Monomode microwave cavity.

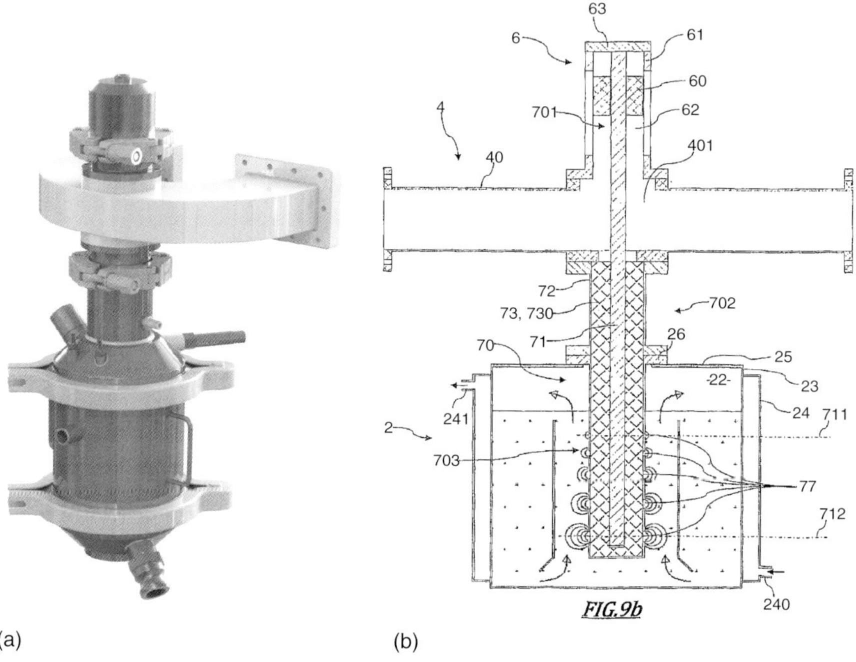

(a)   (b)

Figure 4.15 (a) Batch-jacketed reactor (2.2 l) mounted on a microwave processing system comprising a U-shaped waveguide and an internal transmission line. Source: Copyright SAIREM SAS. Reproduced with permission. (b) Schematic of the reactor internal showing the internal transmission line at the center of the reactor. Source: Patent: US 8759074 B2.

A different equipment concept, compared to the aforementioned designs, is based on an integrated reactor and microwave transmission system [57, 58]. These are systems specially designed to carry out microwave-assisted syntheses and extraction processes in batch or continuous flow both at laboratory and pilot scale. One type of relevant reactor/applicator system that has been demonstrated is shown in Figure 4.15. It is a stainless steel batch reactor with volumes of 1.5, 20, and 100 l mounted on a U-shaped waveguide. A cooling jacket is used to regulate the temperature and a mechanical stirrer to mix the fluids. Microwave energy is fed via the U-shaped waveguide and transmitted into the reactor via the internal transmission line that is placed in the middle of the reactor in direct contact with the reaction mixture. Overall, this type of equipment has a higher energy utilization efficiency compared to conventional multimode cavities and allows for high pressure processing because of the stainless steel wall and for precise temperature control because of the double-thermal actuation through the internal transmission line and the heating/cooling jacket.

Industrially, electromagnetic heating often involves *radio frequency heating* rather than microwave heating. Radio frequency fields involve longer wavelengths compared to microwave fields. This implies larger equipment and better field uniformity, as interference patterns need much more space to develop. In terms of equipment, traditional multimode cavities are impractical for these long wavelengths. Therefore, typical radio frequency field applicators are electrodes or coils, used, for example, in wood gluing presses and case hardening of steel, respectively [59]. In a more general context, scaling wavelength and equipment dimensions to approximately the same factor enables scale-up of microwave systems. For instance, lowering the operation frequency from the common 2.45 GHz band to the 915 MHz band enables scaling of (increase in) all dimensions by a factor of 2.68. This implies a 19-fold volumetric increase while retaining the same distribution of the microwave field in the domain.

Microwave (pre-)heating has traditional applications in several sectors of the process industry, such as rubber and plastic industry (preheating of resins and rubbers, vulcanization, bonding, curing, welding, and shrinking), food industry (drying, cooking, tempering, pasteurization, sterilization, and thawing), pharmaceutical industry (vacuum drying), ceramics (drying, joining, and sintering), paper and wood (drying and gluing), and textile (drying, dye fixation, and control of moisture content). The equipment used in those applications includes primarily large-scale multimode microwave chamber furnaces equipped with belt or roller conveyors (continuous operation) or racks (batch operation). The most important advantages of microwave heating in these applications are cost and energy savings, shorter processing times, space savings, and better product quality. In Table 4.2, some selected publications on industrial microwave applications are listed.

Other applications of microwaves investigated in the literature include liquid-phase and gas-phase reactions, extraction, adsorption, crystallization, distillation, and membrane separations.

4.4.1.1 Liquid-Phase Organic Synthesis Reactions

The literature concerning intensification of chemical reactions in homogeneous liquid-phase systems is rich with several good reviews on this subject [71–77]. Authors generally agree that under certain operating conditions, microwave heating can induce multifold acceleration of chemical reactions and in certain cases improve product yield as well. Some selected relevant examples are presented in Table 4.3.

Experiments on application of microwaves to continuous-flow microreactors have been published as well. He et al. [81] and Comer and Organ [82] have reported that Suzuki cross-coupling reactions in capillary reactors can be carried out at a higher rate and product yield when microwave-heated. Aside from possible process intensification effects, the combination of microwaves with micro- or millireactor devices can circumvent scalability issues of common bench-scale microwave oven designs. In particular, millireactors can be effectively irradiated

Table 4.2 Selected publications on traditional applications of microwaves.

Publication	Publication type	Applications discussed
Osepchuk 2002 [60]	Review	Various, including preheating and vulcanizing of rubber, sintering and synthesis of ceramics, pasteurization, sterilization, waste remediation
Regier et al. 2016 [61]	Book	Various, from baking and drying to blanching, thawing, and tempering. Covers the key area of process measurement and control to ensure more uniform heating of food products
Vadivambal and Jayas 2010 [62]	Review	Heating of food
Li et al. 2011 [63]	Review	Drying
Datta and Anantheswaran 2001 [64]	Book	Food processing
Ahmed and Ramaswamy 2004 [65]	Review	Food pasteurization and sterilization
Haghi 2005 [66]	Review	Textile processing
Kudra and Mujumdar 2009 [67]	Book	Drying
Duan et al. 2010 [68]	Review	Microwave-assisted freeze-drying of foods
Zhang et al. 2006 [69]	Review	Drying of fruits and vegetables

Source: Adapted from Stefanidis et al. 2014 [70].

Table 4.3 Effect of microwave heating on reaction time and product yield for several exemplary reactions.

Reaction	Reaction time		Product yield	
	Conventional	Microwave (min)	Conventional (%)	Microwave (%)
Hydrolysis of benzamide to benzoic acid	1 h	10	90	99
Oxidation of toluene to benzoic acid	25 min	5	40	40
Esterification of benzoic acid with methanol	8 h	5	74	76
S_N2 reaction of 4-cyanophenoxide ion with benzyl chloride	16 h	4	89	93
Heck arylation of olefines	20 h	3	68	68
Ruthenium-catalyzed reaction of endo-norborneneimide with dimethyl acetylenedicarboxylate (cycloaddition)	60 min	2	84	93

Source: Adapted from Gedye et al. 1988 [78], Larhed et al. 2002 [79], and Johnstone et al. 2010 [80]. Also published in [70].

by microwaves; then, a reactor numbering-up approach can bring throughput to the required scale.

The origin of some of the intensification effects discussed earlier has been the subject of a longstanding debate in the literature. Earlier works advocated the presence of nonthermal microwave effects such as microscopic hot spots, molecular agitation, and improved transport properties of molecules [83]. Further, phenomena, such as positioning of the transition states or decrease in the activation energy, have been put forward too [84–86]. Today, most authors agree that chemistry acceleration and subsequently better product yields and purities in some cases result from pure thermal effects [71, 87]. Stuerga and Gaillard [88, 89] studied theoretically the existence of nonthermal microwave effects from a thermodynamic point of view and concluded that electric fields cannot have any molecular effects. In simple terms, coupling of microwave energy with liquid mixtures in sealed vials can rapidly (within second or tens of seconds) raise temperature and pressure to very high levels (routinely up to 300 °C and 30 bar). In these small "autoclaves," chemistry can be significantly accelerated and produce super-equilibrium yields in comparison to those obtained in conventional open-reflux systems where temperature is limited by the boiling point (e.g. [90, 91]).

4.4.1.2 Gas-Phase Catalytic Reactions

In gas–solid reaction systems, macroscopic hot spot formation and selective heating of the solid catalytic phase are primarily considered to be the mechanisms behind the acceleration of the apparent reaction rates observed. Hot spots are places inside the catalytic bed where temperature is considerably higher (100–200 K) than the average temperature and hence reaction occurs at a much higher rate [92–95]. Very often, this is due to inhomogeneous distribution of the electromagnetic field inside the catalytic bed. Selective heating refers to selective dissipation of the microwave energy on the catalyst particles while the gas phase remains at lower temperature. This form of selective heating was proposed in [96] as the probable reason of the observed enhanced selectivity in methane coupling toward higher hydrocarbons in a microwave field. Recent works on selective heating of catalytic layers coated on monolithic reactor walls leading to notable temperature differences between the solid and bulk gas phase have been reported in [97, 98]. Finally, formation of microplasma (microarcs) among very "lossy" catalyst particles may intensify reaction rate and conversion [99, 100]. Some important works in the area of microwave-assisted heterogeneous catalysis are presented in Table 4.4.

4.4.1.3 Solid–Liquid Extraction

With regard to solid–liquid extraction, the techniques for extraction of valuable compounds using microwaves are mainly based on the modifications of steam distillation and solvent extraction. Thus, reported experiments and studies use solvents, water, and steam to extract the volatiles retained in the structure of plants [70].

In microwave extraction using steam as the carrying medium, microwaves are mainly applied to extract volatile fractions from plants and include, among

Table 4.4 Observed effects on gas–solid catalytic reactions carried out under the presence of microwave irradiation and their possible explanation.

Application	Observed effect	Explanation of observed effects	References
Methane decomposition	Higher selectivity	Nonuniform temperature conditions	Ioffe et al. 1995 [101]
	Higher conversion, higher selectivity	Plasma formation, arcing	Zhang et al. 2003 [102]
	Higher conversion	Microplasma formation	Fidalgo et al. 2008 [103]
	Reduction of reaction temperature	Hot spot formation	Bond et al. 1993 [92], Chen et al. 1995 [93], Chen et al. 1997 [104]
Methane dry reforming	Higher conversion, higher selectivity	Hot spot formation	Zhang et al. 2003 [105]
	Higher conversion	Microplasmas formation	Fidalgo et al. 2008 [99], Fidalgo et al. 2011 [100]
HCN synthesis	Higher selectivity	Selective catalyst heating	Koch et al. 1997 [106]
H_2S decomposition	Higher conversion	Hot spots formation	Zhang et al. 1999 [107]
SO_2 reduction	Higher conversion, phase transition of catalyst	Hot spots formation	Zhang et al. 2001 [94]
Propane oxidation	Higher conversion, reduction of reaction temperature	Hot spot formation	Beckers et al. 2006 [108]
Ethane dehydrogenation	Higher conversion[a]	Phase transition of catalyst	Sinev et al. 2009 [109]
o-xylene/toluene oxidation	Higher conversion, higher selectivity	Hot spots formation, more homogeneous dispersion of metallic particle	Liu et al.1999 [110]
2-Methylopentene isomerization	Higher conversion, higher selectivity	Changed dispersion of metallic particles[b]	Seyfried et al. 1994 [111], Roussy et al. 1997 [112]
Methanol steam reforming	Reduction of the reaction temperature	Homogeneous energy dissipation	Perry et al. 2002 [113]
	Higher conversion	Double absorption of microwaves by both the reagent and the catalyst	Chen and Lin 2010 [114]
	Higher conversion and significant spatial temperature gradients	Hot spots formation	Durka et al. 2011 [115]

a) Effect observed only for specific catalysts,
b) Effect observed only when the catalyst was pretreated under microwave conditions.
Source: Adapted from Stefanidis et al. 2014 [70].

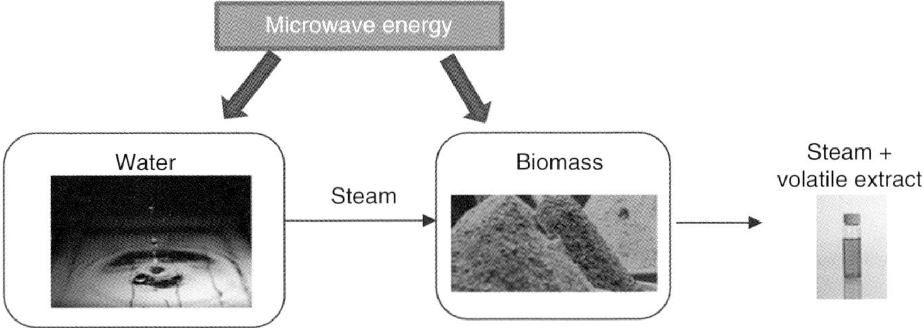

Figure 4.16 Representation of the extraction of essential oils with microwaves when steam is the carrying medium. Source: Taken from Stefanidis et al. 2014 [70].

others, solvent-free microwave extraction (SFME), improved microwave steam distillation (MSD), modified hydrodistillation, microwave-accelerated steam distillation (MASD), and combination of microwave hydrodiffusion and gravity (MHG). In these processes, steam can be produced inside and outside the bed of plants (Figure 4.16), or just applied to a moist plant bed.

Table 4.5 presents some applications of this kind of extraction. The core idea is the creation of steam and thereby overpressure inside the plant matrix, eventually leading to cell wall rupture and intensified mass transfer of the volatile fractions from inside the plant matrix toward the carrying medium (also steam).

In microwave extraction using liquid solvents as carrying media, soluble natural compounds such as antioxidants (e.g. polyphenols) [121–123] are extracted to water and organic solvents. Table 4.6 presents some more relevant applications. In the event, the solvent is a good absorber of microwaves (e.g. water), increased extraction rates result from the rapid solvent overheating [127]. In the case of organic solvents that are poor absorbers of microwave energy, microwave energy dissipation occurs primarily inside the plant structure. As mentioned earlier, this

Table 4.5 Microwave extraction with steam as the carrying medium.

Extraction technique	Plant materials	Mass of plant (g)	Moisture (%)	Microwave power (W)	Time (min)	References
SFME	Basil, garden mint, thyme	250	90, 95, and 80	500	30	Lucchesi et al. 2004 [116]
SFME	Lavandin super	100	50	600	10	Mato et al. 2008 [117]
SFME	Cardamom	100	67	390	75	Lucchesi et al. 2007 [118]
MHG	Citrus fruits	500	NR	500	15	Bousbia et al. 2009 [119]
MSD	Lavender flowers	20	Ambient dried	200	6	Sahraoui et al. 2008 [120]

NR, Not reported.
Source: Adapted from Stefanidis et al. 2014 [70].

Table 4.6 Microwave extraction with solvents as the carrying medium.

Product	Plant materials	Solvent	Conditions	Microwave power (W)	Time (min)	References
Polyphenols	Green tea	Ethanol/ water	1 bar, 85–90C	700	5	Pan et al. 2003 [121]
Microalgal pigments	Microalgae	Acetone	Vacuum and atmospheric pressure	25–100	3–13	Pasquet et al. 2011 [124]
Vegetable oils	Soybean and rice bran	Ethanol	Continuous system, at 73 C. 0.6 and 1.0 l/min	Up to 5 kW	21 (res. time)	Terigar et al. 2011 [125]
Flavonol glycosides	*Ginkgo biloba*	Ionic liquid	1 bar	120 W	15	Yao et al. 2012 [126]

Source: Adapted from Stefanidis et al. 2014 [70].

leads to rupture of the plant walls and fast extraction of secondary metabolites (e.g. oleic acid) toward the surrounding solvent. In turn, this enables reduction in process time and in energy consumption together with higher yields [122, 128].

4.4.1.4 Adsorbents Regeneration

Conventional thermal dehydration and regeneration of adsorbents, mainly zeolites and activated carbons, used for the adsorption of volatile organic compounds (VOCs) and hazardous air pollutants (HAPs), are time-consuming and energy-expensive processes in chemical industry [70]. Use of microwaves can induce fast volumetric heat transfer directly to the solid matrix and selective interaction with adsorbed polar compounds, thereby reducing the process time with concomitant energy savings. Yuen and Hameed [129] and Cherbański and Molga [130] have written extensive reviews in the field. Cherbanski et al. [131] presented a comparison of microwave swing regeneration and temperature swing regeneration of 13X molecular sieves in terms of kinetics of desorption of a polar (acetone) and a nonpolar (toluene) component. It has been reported that microwave swing regeneration is much faster compared to temperature swing regeneration, with the effect being more pronounced for the polar adsorbate (acetone). Chowdhury et al. [132] studied desorption of ethylene/ethane and carbon dioxide/methane mixtures from Na-ETS-10 molecular sieves under microwave and conductive heating. They reported higher gas recovery at three times lower process time and 1 order of magnitude lower specific energy consumption under microwave heating without a negative effect on the capacity of Na-ETS-10 after several successive adsorption–regeneration cycles. In the same context, Foo and Hameed presented studies on microwave-assisted desorption of methylene blue from activated carbons prepared from different materials [133, 134].

4.4.1.5 Crystallization

In the field of crystallization, Pinard and Aslan [135] performed microwave-assisted evaporative crystallization experiments with a system of silver metal nanoparticle-coated glass surface in a glycine water mixture. The metal nanostructures function as selective nucleation sites. The created temperature gradient between the metal nanostructures and the aqueous phase accelerates the mass transfer of glycine molecules from the solution to the nanoparticles and thereby intensifies nucleation. In addition, microwaves can speed up evaporation. In total, microwaves induced multifold decrease in the total crystallization time compared to conventional evaporative crystallization at room temperature. Similar work using L-alanine and L-arginine has been reported by the same group in Refs. [136, 137], respectively. Radacsi et al. [138] studied the effect of evaporation rate on crystal size. In particular, niflumic acid was crystallized from ethanol solutions with microwaves and conventional heating using a heating plate. The final product size was reduced when the evaporation rate was increased under microwave irradiation that couples directly with polar molecules, such as ethanol, and is not limited by the low thermal conductivity. Overall, higher solvent evaporation rates under microwaves result in shorter crystallization times and thereby in smaller crystals because of higher supersaturations (Figure 4.17).

4.4.1.6 Distillation

In the field of vapor–liquid equilibrium (VLE) equilibrium and distillation, Gao et al. [139] performed experiments with ethanol/benzene and

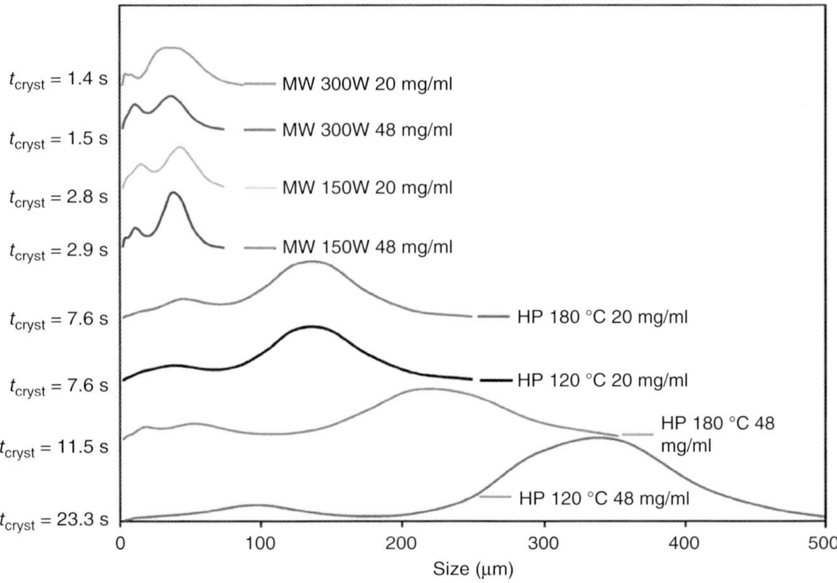

Figure 4.17 Crystal size (volume-based mean diameter) and crystal size distribution width decrease with decreasing crystallization time (t_{cryst}). Source: Radacsi et al. 2013 [138]. Reproduced with permission of American Chemical Society.

iso-octanol/bis(2-ethyhexyl)phthalate (DOP) binary mixtures. They reported that, in the case of ethanol/benzene binary, microwaves could shift the VLE line obtained under conventional heating. The effect becomes more pronounced with increasing field strength up to a plateau level. This was attributed to the selective microwave-ethanol interaction resulting in higher rate of microwave absorption by ethanol as compared to the heat transfer rate from ethanol to benzene. No microwave effect was reported for the other binary mixture (iso-octanol/DOP system) in which both components have similar dielectric constant values.

4.4.1.7 Membrane Processes

Microwave energy has been used both for the synthesis of various membrane materials [140–142] and for membrane-based separation processes, such as gas permeation and membrane distillation. In the latter case, it has been reported that the selective interaction of microwaves with the polar functional groups of polymeric membranes and/or with the feed mixture components can accelerate diffusivity and permeability through the membrane. As an example, Nakai et al. [143] have reported that gas permeability coefficients through hydroxypropyl cellulose, cellulose triacetate, and poly(methylmethacrylate) membranes were higher under microwave irradiation compared to conventional heating. The postulated reasoning was the accelerated molecular motion of —OH and —COO— groups inside the membrane matrix owing to their interaction with microwaves. Further, Ji et al. [144] reported intensified mass transfer under microwaves in experiments of vacuum membrane distillation experiments with polyvinylidene fluoride hollow fiber hydrophobic membranes comprising nonpolar functional groups. This was attributed to lower temperature polarization.

4.4.2 Plasmas

Plasma, the so-called "fourth state of matter" after gas, liquid, and solid media, is a partially ionized gas with balanced charges of electrons and ions. It is generated when an electric field, or electromagnetic field, applied to a gas medium is strong enough to generate and sustain ionization of the gas at a given pressure and density. Plasmas can generate a large number of chemically active species, such as electrons, ions, atoms, and radicals, with variable concentration depending on the operating conditions. The high reactivity of plasma is a unique property of the medium that is taken advantage of to intensify chemical processes for materials treatment and energy, environmental, and health applications. Plasmas are frequently characterized as thermal or nonthermal plasmas.

In *thermal plasmas,* all generated species are in thermal equilibrium and thus have the same temperature. Local thermal equilibrium is reached when the mean electron temperature ($T_{electron}$) is equal to the neutral species temperature ($T_{neutral}$). The absolute temperature value is very high in this case ($\sim 10^4$ K). At such temperatures, the active species are produced by both thermal activation and electron collisions. A characteristic type of thermal plasmas is arc discharges, which have been extensively used in applications, where high temperatures are required (e.g. metallurgy, acetylene synthesis, and waste pyrolysis) [145].

In *nonthermal plasmas*, the different plasma species have widely different temperatures. Electric energy is preferentially channeled to the electrons, which subsequently transfer it to the heavier species through electron–molecule collisions [145]. The electron temperature in nonthermal plasmas is much higher than the ion temperature that is in turn higher than the temperature of the neutral species ($T_{electron} \gg T_{ions} > T_{neutral}$). Dielectric barrier discharges (DBD), corona (DC, AC, and pulsed), and nanosecond-pulsed discharges are common examples of nonthermal plasmas.

It should be borne in mind that the classification of plasma media as thermal or nonthermal ones is a rough one. For example, some arc plasmas do not reach local thermal equilibrium. Further, in many nonthermal plasmas, the neutral species have relatively high temperature (e.g. 10^3 K). Consequently, nonthermal plasma models cannot describe chemistry accurately and thermal activation should be considered too. Such nonequilibrium discharges in which thermal activation is also important were recently characterized as *warm* plasmas [146]. Microwave, radio frequency, glow, and spark discharges belong to this group.

Depending on the way plasma is ignited and sustained as well as on the operating conditions, the different nonthermal or warm discharge types mentioned above can be generated. They cover different ranges in the domains of electron energy and density and neutrals temperature. DBP is the electrical discharge occurring between two electrodes (planar or cylindrical) separated by an insulating material (dielectric barrier). Alternating voltage (AC) is required for the operation of DBD discharges [145]. A *gliding arc* is initiated between two coplanar diverging electrodes, and the gas flow propels a thin plasma arc along the edges of those electrodes. When the length of the gliding arc exceeds a critical value, the heat losses surpass the energy supplied by the source and it is not possible to sustain the plasma in thermodynamic equilibrium. As a result, a fast transition from thermal into nonthermal plasma occurs [147]. *Corona discharge* is a partial discharge initiated around an active electrode. The restricted ionization regions imposed by the electrode geometry allow high energetic electrons to collide with gas molecules, promoting electron impact reactions. Corona discharges are scalable and can be operated under atmospheric pressure [148]. When the input voltage is raised, transitions from corona to *glow* and finally to *spark* take place [149]. *Glow discharge* is an unstable regime; therefore, corona turns almost immediately to spark [150].

Microwave (MW) discharges are initiated and sustained by electromagnetic waves with high frequency (typically 0.915 or 2.45 GHz). In most configurations, microwaves travel through a rectangular waveguide with a reduced height section followed by tapered sections. The plasma is generated inside a cylindrical tube, which is placed inside the waveguide [151]. MW discharges are widely used for generation of warm and nearly thermal plasmas for different applications. They have significant advantages over other types of discharges, such as simplicity of plasma ignition, wide region of operating pressures, electrodeless operation, and scalability. More detailed description of the different types of plasma sources can be found in ref. [152].

An enormous breadth of applications of plasma across different industries exists. Examples include

- *decomposition and synthesis of gas phase inorganic materials* (e.g. decomposition of CO_2, NH_3, SO_2, and N_2O; synthesis of nitrogen oxides; NH_3; and nitrides of organic compounds);
- *synthesis, treatment, and processing of inorganic materials* (e.g. reduction of oxides of metals and other elements; production of metals by decomposition of their oxides; plasma chemical synthesis of nitrides, carbides, hydrides, borides, and carbonyls of inorganic materials; plasma spraying; chemical etching; chemical vapor deposition; ion implantation; oxidation coating; nanoparticle formation; synthesis of nanoparticles; and carbon nanotubes);
- *organic and polymer chemistry* (e.g. hydrocarbon pyrolysis; acetylene synthesis from methane; synthesis and conversion of nitrogen fluorine and chlorine compounds; synthesis of aldehydes, alcohols, and organic acids; hydrocarbon polymerization; and treatment of polymer surfaces);
- *conversion of (oxygenated) hydrocarbons to hydrogen and syngas* (e.g. pyrolysis, partial oxidation, steam and dry reforming of light hydrocarbons; pyrolysis and gasification of coal, biomass and organic waste material); and
- *applications in biology and medicine* (e.g. nonthermal plasma sterilization of surfaces, air streams and water, tissue engineering and regeneration, blood coagulation, wound healing, and treatment of skin diseases).

An excellent review of many of the above processes and additional ones can be found in ref. [152]. In the rest of this section, three selected plasma-assisted chemical processes from the aforementioned process groups (i.e. acetylene synthesis from methane, low-grade coal gasification to synthesis gas, and CO_2 dissociation) are discussed in more detail to illustrate the process intensification potential of the plasma technology.

4.4.2.1 Plasma-Assisted Methane Coupling to Acetylene (Huels Process)

The Huels process was developed in Germany in 1925 to crack light hydrocarbons into acetylene for the synthesis of butadiene. The feedstock may be gaseous (Figure 4.18a), liquid, or solid (Figure 4.18b). In the second case, the process occurs in two stages: first, an arc is produced in hydrogen atmosphere and then the feed is added to the end of the discharge. In addition, the plant includes an electric arc furnace and a low- and a high-pressure purification system. The arc furnace consists of a cathode, a vortex chamber, and an anode. Both electrodes are water-jacketed carbon steel tubes, with a length of 0.8 and 1.5 m, respectively, and an inner diameter of 150 and 100 mm, respectively. The discharge (electric arc) is created by application of high voltage (7.1 kV in DC mode) between the electrodes and has a length of 1.2 m and typical current of 1200 A, for a total electric power input of ~8.5 MW. The gas phase (natural gas, other light hydrocarbons, or hydrogen) is injected into the chamber tangentially to stabilize the discharge. The temperature in the center of the arc can reach up to 20 000 °C, while it sharply decreases coaxially to 600 °C on the wall of the electrodes, limiting the thermal losses. The residence time of the gas in the arc region is a few milliseconds. During this time, hydrocarbons are cracked into acetylene, ethylene, hydrogen, and soot. The temperature is still as high as 1800 °C at the end of the arc. As acetylene rapidly decomposes into hydrogen and soot at such a high

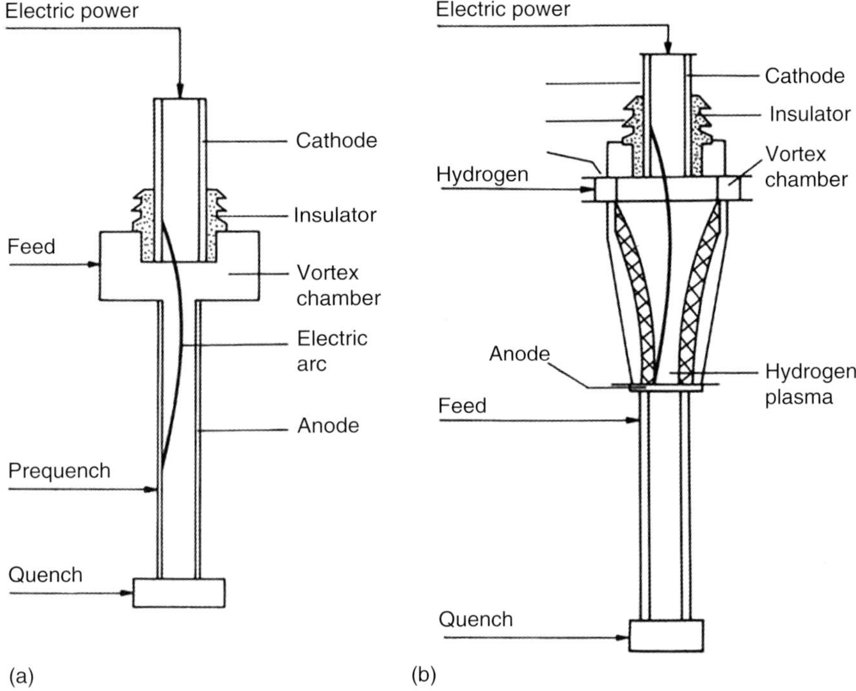

Figure 4.18 Huels electric-arc furnace: (a) one-step process (for gaseous feed) and (b) two-step process (for liquid and solid feed). Source: Pässler et al. 2008 [153]. Reproduced with permission of John Wiley and Sons Inc.

temperature, the gas is quenched immediately with water or other hydrocarbons down to 200 °C, with an extremely high cooling rate of about 10^6 °C/s.

Considering methane as feedstock, the thermal decomposition can be described as a sequence of consecutive reactions, presented in a simplified reaction scheme by [154]. The reaction starts with the formation of CH_2 radicals that react with methane to form ethane.

$$CH_4 \rightarrow CH_2 + H_2 \tag{4.7}$$

$$CH_4 + CH_2 + \rightarrow C_2H_6 \tag{4.8}$$

Then, we can write the following global reaction:

$$2\,CH_4 \rightarrow C_2H_6 + H_2 \tag{4.9}$$

with $\Delta H = 0.7\,eV/mol$, $k = 4.5 \times 10^{13} \exp(-46\,000\,K/T)$ 1/s [155].

As further dehydrogenations gradually convert ethane into ethylene (usually formed after 10^{-6} to 10^{-5} s), acetylene (formed after 10^{-4} to 10^{-3} s), and finally soot (see reaction scheme below [154, 155] and Figure 4.19), fast quenching is applied to avoid significant soot formation.

$$C_2H_6 \rightarrow C_2H_4 + H_2 \tag{4.10}$$

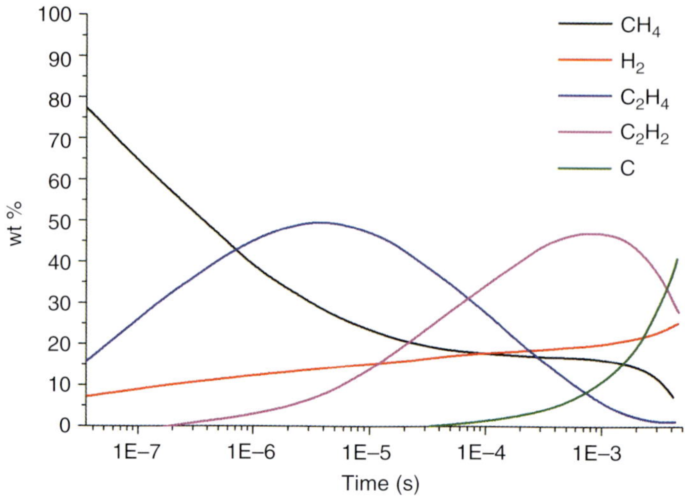

Figure 4.19 Kinetics of methane decomposition in a plasma jet. Source: Adapted from Polak 1966 [156].

with $\Delta H = 1.4\,\text{eV/mol}$, $k = 9.0 \times 10^{13} \exp(-35\,000\,\text{K/T})\ 1/\text{s}$

$$C_2H_4 \rightarrow C_2H_2 + H_2 \tag{4.11}$$

with $\Delta H = 1.8\,\text{eV/mol}$, $k = 2.6 \times 10^{8} \exp(-20\,500\,\text{K/T})\ 1/\text{s}$

$$C_2H_2 \rightarrow 2C_{(s)} + H_2 \tag{4.12}$$

with $\Delta H = -2.34\,\text{eV/mol}$, $k = 1.7 \times 10^{6} \exp(-15\,500\,\text{K/T})\ 1/\text{s}$

Overall, the combination of arc plasma and fast quenching enabled an efficient commercial process for acetylene production from natural gas at a feed rate of 2344 Nm^3/hour with 70% methane conversion and 51% acetylene yield at an energy cost of 12 kWh/kgC$_2$H$_2$. On the downside, a considerable amount of soot is formed [157].

4.4.2.2 Plasma-Assisted Coal Gasification for Synthesis Gas Production

Thermal (mainly arc) plasma has been used at different scales (from lab to commercial) for conversion of a wide range of organic materials (biomass, solid waste, and coal) to synthesis gas (mixture of hydrogen and carbon monoxide), an important building block for added value products and fuels. Thermal plasma technology essentially enables a very chemically reactive and high-temperature medium (10^3–$10^4\,°$C) that can robustly decompose organic feedstocks with large spatiotemporal composition variations to hydrogen and carbon monoxide [158–160]. Possible inorganic fraction in the feedstock is vitrified as slug and easily separated from the gaseous product. In addition, the inherent flexibility of plasma technology with regard to the choice of the plasma agent (e.g. nitrogen, oxygen, air, steam, or mixtures thereof) can be taken advantage of to "tune" product composition (i.e. H$_2$/CO ratio) according to downstream specifications. If the process is well designed, very high energy and conversion efficiencies can be attained, not only at large industrial scale (e.g. integrated gasification

Figure 4.20 A photograph of a swirl-type gasifier with one microwave plasma system installed at the upper part and the other one installed at the lower part. The inner configuration of the gasifier is a cylinder with a diameter of 90 cm and a height of 180 cm. The volume is 1145 l. Source: Uhm et al. 2014 [161]. Reproduced with permission of Elsevier.

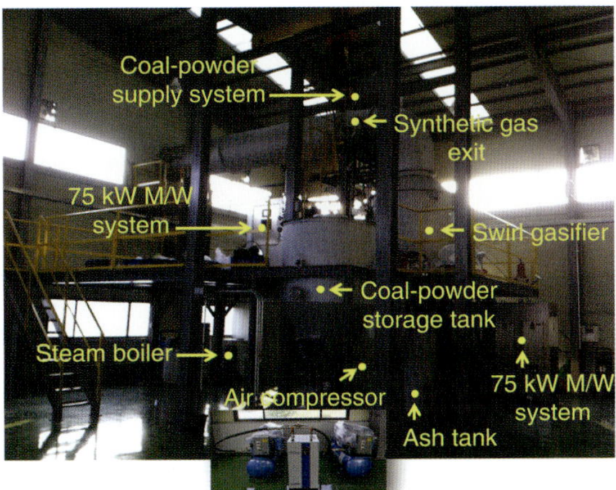

combined cycle [IGCC] plant) but also in moderately sized power plants that can be used for distributed production of energy and chemicals.

A very successful plasma-assisted coal gasification case has been published by Uhm et al. [161] and is explicitly discussed herein as an exemplary work in the field. The authors have constructed a pilot-scale microwave swirl steam plasma gasifier (cylindrical shape: 180 cm height, 90 cm diameter, Figure 4.20) powered by two 75 kW magnetrons operated at a frequency of 915 MHz. The gasifier processes at very high temperatures (1600–1700 °C) low-grade Indonesian coal in fine particle (34.30 wt% fixed carbon, 32.53 wt% volatile matter, 10.71 wt% water, 22.46 wt% ash) at a feed rate of 90 kg/h in plasma environment created by a mixture of steam, air, and oxygen plasma agents. Hydrogen-rich syngas product at a flow rate of 3299 l/minutes and a calorific value of ∼500 kW is obtained. To minimize heat losses, the reactor wall, made of a fire-resistant ceramic material, has been properly insulated by a cement layer. Overall, nearly 100% conversion of the feed is achieved at a very high cold gas efficiency of 84% (ratio of calorific value of syngas product over the calorific value of coal feed and the power consumption by the magnetron). This value is higher than that of most commercial gasifiers (70–80%). A photograph of the microwave plasma gasifier is shown in Figure 4.20 and a representative plot of the product composition is shown in Figure 4.21.

4.4.2.3 Plasma-Assisted CO_2 Dissociation

Solar fuels are fuels that are directly, or indirectly, produced from solar energy through photochemical, electrochemical, or thermochemical routes. They are considered as an alternative source of energy that can partially replace fossil fuels of which their reserves are finite and gradually depleting, they have an environmental impact, and their exploration, extraction, and utilization frequently engage geopolitical risks. The most investigated solar fuel is hydrogen and products from CO_2 reduction. CO_2 can be dissociated in the presence of H_2O to form syngas or methane. Solid oxide electrolysis is a suitable technology for efficient CO_2 dissociation. Plasmolysis is an alternative process that has gained attention because of the following advantages: fast response to renewable electricity fluctuation, no need for use of scarce materials, and high overall

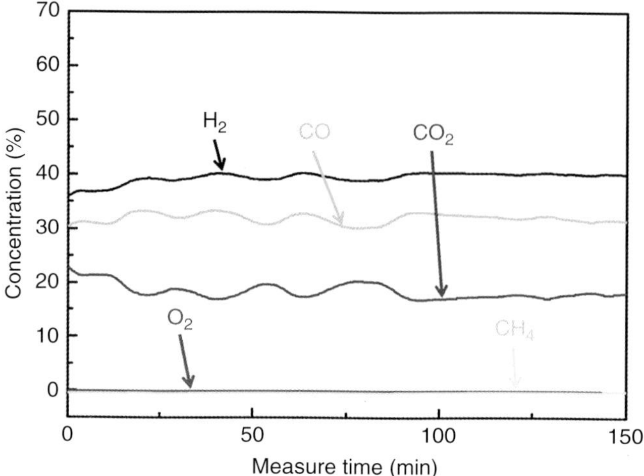

Figure 4.21 Plots of the relative concentrations of synthesized gas species during the operational period of last two and half hours. Source: Uhm et al. 2014 [161]. Reproduced with permission of Elsevier.

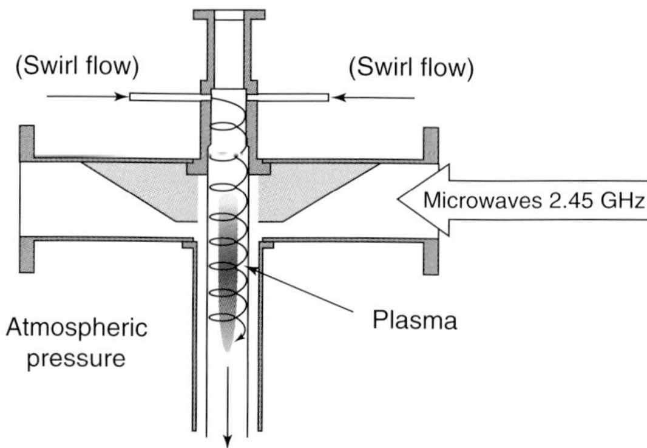

Figure 4.22 Schematic design of a vortex-stabilized microwave plasma reactor. Source: Adapted from Hrycak et al. 2015 [162].

energy efficiency, as the operation can be carried out at low temperature. A vortex-stabilized microwave plasma reactor can be employed for efficient dissociation of CO_2. The basic parts of a vortex-stabilized microwave reactor are presented in Figure 4.22.

The electron-induced dissociation reaction inside the plasma zone is

$$CO_2 \rightarrow CO + O \quad \Delta H = 5.5 \text{ eV} \tag{4.13}$$

The atomic oxygen (O) that is formed can initiate reaction with another CO_2 molecule:

$$CO_2 + O \rightarrow CO + O_2 \quad \Delta H = 0.3 \text{ eV} \tag{4.14}$$

The global reaction (considering both the reactions above) can be written as

$$CO_2 \rightarrow CO + \tfrac{1}{2}O_2 \quad \Delta H = 2.9 \text{ eV} \tag{4.15}$$

Two possible mechanisms can initiate CO_2 dissociation in the plasma zone: (i) thermal dissociation, which is induced by thermal heating at gas temperatures above 2000 K; in this case, fast nonequilibrium cooling is required to retain the products after CO_2 dissociation. (ii) Electron-driven dissociation, which is induced by vibrational excitation. In nonequilibrium plasma, the vibrational temperature (T_{vib}) is high, while the CO_2 gas (bulk) temperature is rather low; thus, high energy efficiency is achieved. The energy efficiency (η) of CO_2 dissociation is given by the formula

$$\eta = \alpha\, \Delta H \cdot \mathrm{SEI} \tag{4.16}$$

where α represents the fractional CO_2 conversion, ΔH the dissociation enthalpy for a CO_2 molecule (2.9 eV/molecule), and SEI the specific energy input (eV/molecule).

The energy efficiency of the CO_2 dissociation is crucial for the viability and economic feasibility of the process. This was studied by Bongers et al. [163], who focused on the energy efficiency maximization by activating mainly vibrational excitation channels using low-pressure microwave plasma. They investigated the effects of microwave power (expressed as specific energy input, SEI), residence time, rapid quenching of the products, and different vortex plasma configurations in two different microwave plasma setups (Figure 4.23). A high energy efficiency of 47% was obtained with the reactor configuration C of the second setup at SEI 0.9 eV/molecule, pressure 200 mbar, and quenching of the outlet to prevent backward reactions.

4.4.3 Photochemical and Photocatalytic Reactors (Artificial Light)

To activate chemical bonds, light is an attractive form of energy. The use of light, either artificial or solar, for carrying out chemical and biochemical reactions may render the process more sustainable for two reasons [164, 165]:

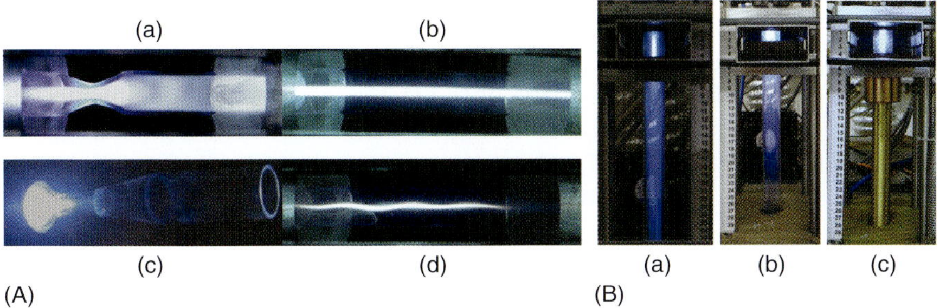

Figure 4.23 (A) First plasma reactor setup (operating at 915 MHz): (a) supersonic expansion of the plasma in the microwave cavity at 1 mbar gas pressure, (b) plasma in the microwave cavity (power 5 kW; flow rate 11 slm; pressure 200 mbar), (c) subsequent quenching of the plasma after the nozzle into the vacuum vessel (1 mbar), and (d) plasma in the microwave cavity (power 3.1 kW; flow rate 75 slm; pressure 200 mbar). (B) Second plasma reactor setup (1 kW, 2.45 GHz) in configurations (a), (b), and (c). The CO_2 gas is tangentially injected from the top just above the waveguide cavity to create a vortex gas flow separating the visible part of the plasma from the quartz tube. Photographs were recorded using pressures of 130 mbar, 150 mbar, and 130 mbar for (a), (b), and (c), respectively. Source: Bongers et al. 2017 [163]. Reproduced with permission of John Wiley and Sons.

- The process selectivity to the required products can drastically increase (e.g. because of a different chemistry, or low/ambient process temperature); and
- The energy consumption in the process can drastically decrease (e.g. because of the low-temperature processing or use of solar light).

In the field of synthetic organic photochemistry, light of high enough energy (deep UV) is used to directly activate molecules [166]. Inorganic photochemistry is applied for inducing oxidation or reduction of metals, often with the aim of separation. An example is the photochemical separation of europium from yttrium in the leachate of a red lamp phosphor [167, 168]. Photochemical conversion can also be induced at longer wavelengths with the aid of a catalyst, that is, by photocatalysis [169]. Here, light is used to activate a catalyst, typically a semiconductor, rather than the reactant. Photocatalysis is applicable for a large variety of reactions [170, 171]. By far, the dominant research area of heterogeneous photocatalysis is the photodegradation of organic compounds in either air or water [172]. TiO_2 is the most investigated photocatalyst in these applications, which converts contaminants to CO_2 and H_2O, and, if applicable, harmless inorganic ions. The superior performance of TiO_2 is attributed to the ability to form a relatively large concentration of oxidizing holes and hydroxyl radicals [171]. Hydroxyl radicals are considered as the reactive oxidant converting the contaminants [173]. Heterogeneous photocatalysis is less explored in the field of organic synthesis [174]. However, the possibility to induce selective, synthetically useful redox transformations has become increasingly more attractive and promising. Also, the combination of a photocatalyst and light allows the conversion of solar energy into chemical energy, by, for example, water splitting and/or CO_2 activation. An increasing intensity in research efforts can be observed in this field [164].

Although in the laboratory photochemistry is a popular research field, a relatively low number of industrial photon-induced processes exist. There are commercial photocatalytic reactors for water and air decontamination [173, 175], but applications in chemical synthesis are extremely rare. Notable exceptions are the photooxidation process, turning cyclohexane into cyclohexanone oxime, operated by Toray in Japan, and the photooxidation of citronellol to rose oxide applied by Dragoco in Germany. One of the factors responsible for this limited industrial implementation is the often suboptimal reactor design, hindering photon transfer in particular [176]. Intensification of photon transfer in chemical reactors has been studied via either improving the illumination efficiency of the light source into the reactor or by altering the light source altogether.

The main development in improving the illumination efficiency in the reactor has been the introduction of *optical fibers*. In these optical fibers, light is propagated along the fiber length by reflection on the fiber wall (Figure 4.24). Depending on the refractive index of the fiber wall, a portion of the light intensity is not reflected but refracted. In an optical fiber reactor, the catalyst is typically coated on the stripped fibers. The refracted light is absorbed by the catalyst that is subsequently activated. There has been an enormous amount of work put in optical fiber reactors by Marinangeli and Ollis [178].

Although this development is an interesting route, major problems still persist, such as the exponential decay of light along the axial direction of the coated fiber,

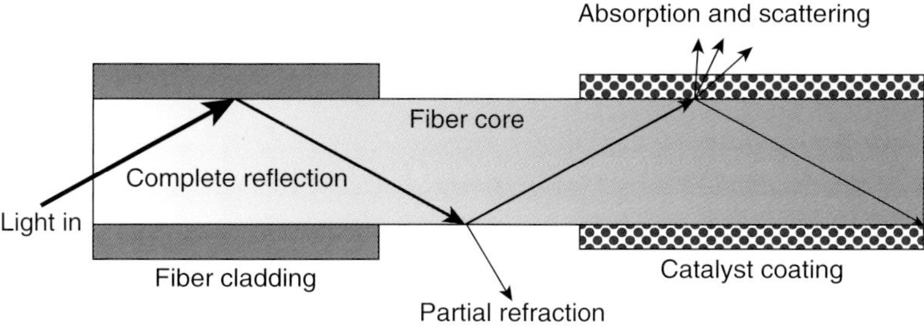

Figure 4.24 Propagation of light through optical fibers. Source: Adapted from Wang and Ku 2003 [177].

thus leading to light extinction after a distance in the order of centimeters; the phenomenon of back-irradiation where the charge carriers are generated at the fiber–catalyst interface, far from the catalyst–liquid interface, leading to important recombination losses; and the non-negligible volume of fibers in the reactor volume, thus decreasing the flow rate and increasing the pressure drop.

In spite of these drawbacks, the use of coated optical fibers has improved efficiency, in particular by applying them in monolith reactors, which provide a high surface-to-volume ratio (10–100 times more than plate or bead substrates with the same outer dimensions), allow high flow rates with low pressure drop, and are easy to scale up. The optical fiber monolith reactor, developed by Lin and Valsaraj [179], gives a 10-fold increase on the illuminated catalyst surface per unit of reactor volume compared to an annular reactor, and a 100 times increase of the apparent quantum efficiency. An improved variation of the design of Lin and Valsaraj is the one where side-light-emitting fibers are placed inside the channels of a ceramic monolith, but where the TiO_2 photocatalyst is coated on the wall of each individual channel, instead of on the fibers, in order to maximize light emission from the fibers and avoiding back-irradiation (Figure 4.25) [180]. An advanced type of optical fibers, the so-called photonic crystal fibers (PCF), have recently been reported to overcome the problems of both limited light propagation and back-irradiation [181]. The fiber itself consists of a waveguide core, surrounded by a cladding structure: a periodic array of holes running along its entire length. The major advantage is that by the appropriate design of the cladding structure, the light guided in the fiber core can strongly interact with tiny amounts of sample introduced in the micrometer-scale holes of the fiber. PCF offers much longer effective interaction lengths and much higher light intensities for a given amount of optical power (because of the small core area).

Another approach to intensify photon transport is to move away from the conventional Hg lamps altogether. Incorporation of annular Hg lamps is typically applied in industrial photoreactors but limits the design flexibility, although Hg lamps are costly, have a limited lifetime, and produce a significant amount of heat requiring extensive cooling [176, 182]. Alternative light sources include light emitting diodes *LEDs and lasers*. Solar illumination is discussed in the next section. A general overview is given in Figure 4.26.

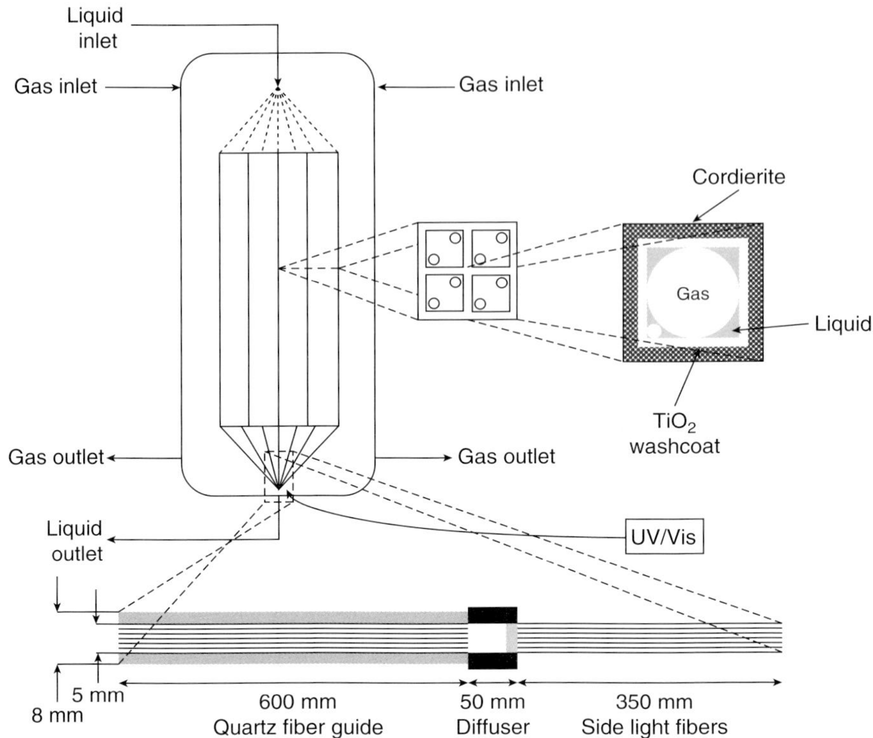

Figure 4.25 Internally illuminated monolith reactor scheme with a detail of the cross section of the monolith channels and the fiber optic bundle. Source: Du et al. 2008 [180]. Reproduced with permission of Elsevier.

Since the new millenium, a large number of designs have been reported where *LEDs* are integrated with microreactors (Figure 4.27). The earliest examples are those of Lu et al. [185], Gorges et al. [186], and Barthe et al. [187]. A recent review on continuous flow photochemistry has been compiled by Noël and coworkers [183]. Advantages of LEDs include their quasi-monochromatic output and small size (miniaturization of equipment) and that they are robust and long-lasting (hundred thousands of hours compared to thousands of hours in the case of classical lamps). The use of microscale illumination in microreactors provides both a large catalyst surface area per unit of reactor volume (ten thousands to a hundred thousands of m^2/m^3), well-defined control of the gas–liquid or liquid–liquid distributions and of the reactants residence times, and a high illumination efficiency. Without being exhaustive, microreactors have been successfully applied for various photochemical conversions, including cycloadditions, cyanation of aromatic compounds, photooxygenation of alkenes (dienes), and photochlorination. Compared to photocatalytic slurry reactors with the same selectivity and yields, conversion rates in microreactors have been reported to be 50–100 times higher (e.g. Li et al. [188] and Takei et al. [189]). Application of microreactors in combination with photochemistry appears quite feasible for the pharmaceutical industry.

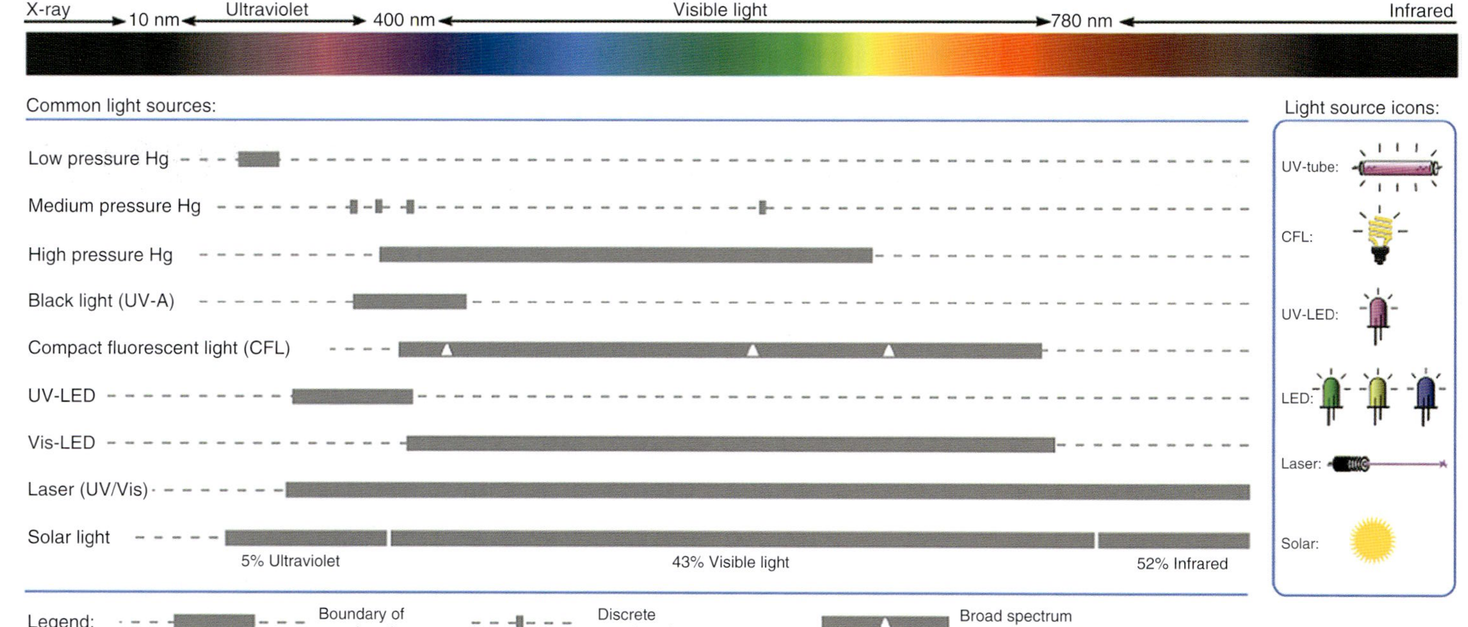

Figure 4.26 Overview of common light sources used for photochemical applications and their emission spectra. Source: Cambié et al. 2016 [183]. Reproduced with permission of American Chemical Society.

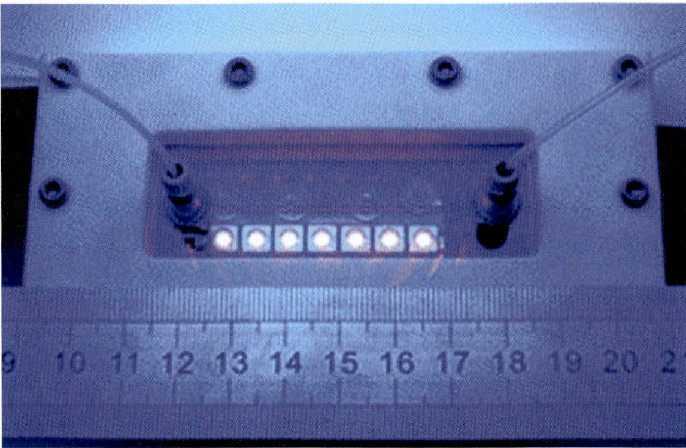

Figure 4.27 Quartz microchip design with 365 nm/500 mV UV–LED array. Source: Matsushita et al. 2008 [184]. Reproduced with permission of Elsevier.

The most important hurdle in photocatalytic reactors is their low energetic efficiency because of light absorption and dissipation between the source and the catalytic site. The development of optical fiber monolith reactors and LED microreactors aim to address this issue. The ultimate solution would be to generate the light directly in the catalyst or in the solvent containing the reactants, a concept captured in the term nanoillumination. So far, the topic of in situ light generation is underdeveloped and craves for more attention. Gole et al. [190] suggest the introduction of nitrogen-doped titania nanostructures into the pores of porous silicon (PS) in order to develop a device to produce visible light by electro-luminescence of PS, thus activating the photocatalyst particles. This device could then be incorporated in a microreactor. A scheme of this photoreactor principle is shown in Figure 4.28. Porous silicon emits visible light, hence the necessity of using nitrogen-doped TiO_2 samples to shift the absorption spectrum from UV to the visible spectrum. The integration of light source and catalyst surface enhances the options for other reactor types (e.g. monolithic reactor and spinning disc reactor, SDR), which are otherwise difficult to combine with microscale illumination. It remains to be verified whether nanoscale UV sources can provide sufficient energy (i.e. the required wavelength and intensity) for the desired reaction to occur.

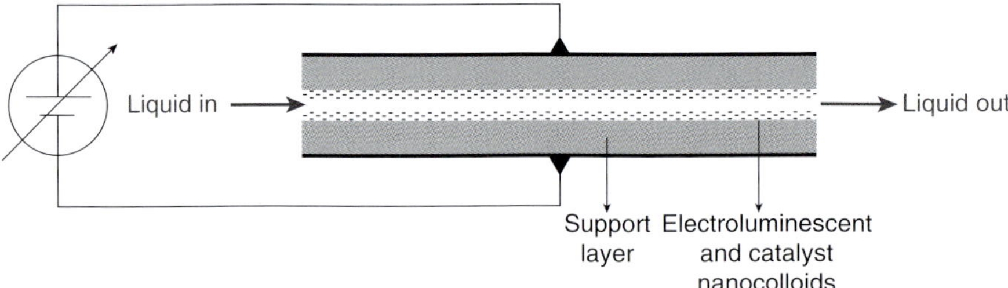

Figure 4.28 Nanoscale illumination reactor. Source: Van Gerven et al. 2007 [176]. Reproduced with permission of Elsevier.

Another example of nanoscale illumination is the use of phosphorescent solids, which are irradiated before their injection in the reactant stream [191, 192]. The presence of phosphors in a gas–solid photocatalytic fluidized bed reactor allowed to perform selective partial oxidation of ethanol to acetaldehyde with a constant linear increase of the ethanol consumption rate as a function of initial alcohol concentration up to the maximum concentration studied, i.e. 2.5 vol%. In the absence of phosphors, the ethanol consumption rate initially grows linearly with initial alcohol concentration, but then bends toward an asymptotic value for initial ethanol concentration higher than 0.5 vol%. The results showed that the presence of phosphors allowed improved photon transfer, increasing the apparent quantum yield from 2% to 30% together with a high photoreactivity.

An alternative to LED light is the *laser field*, which presents a fundamentally proven method to (selectively) excite molecules by addressing individual bonds, thereby significantly increasing their reactivity as reviewed by Richard Zare [193]. Laser radiation has five distinct features, making it unique with respect to all other light sources, such as LEDs or discharge lamps [5]. Laser light is monochromatic, directional, and coherent; it can be generated at high power; and it may be pulsed. Monochromatic means that the light emitted is extremely pure and has a very narrow line width. For comparison, in the UV region, the typical line width of the laser is 0.1 nm, whereas in the case of LED, it is 10 nm. This very narrow line width of laser light is crucial in view of selective molecular bond cleavage.

The concept of the vibrational excitation of molecules was proposed for the first time in 1972 by Polanyi [194] and since then has been shown to deliver spectacular effects, for example, with respect to CH_4 and H_2O [195]. In the Stanford group of Zare, chlorination of methane was investigated in experiments where a molecular beam was locally (spot c. 1 mm) subjected to the laser pulses [196, 197]. The resulting stretch of the C—H bonds in the methane molecule led to an increased number of stripping collisions because of which the reactivity had been enhanced by factors of more than 100 (see also Figure 2.5). On the other hand, Crim and coworkers showed the possibility of selective stretching and breaking either O—D or O—H bond in a singly deuterated water molecule, depending on the laser color used [198, 199]. Later works also reported laser-based bond dissociation and rearrangement in polyatomic molecules, such as acetone or acetophenone [200].

The application of these new light distributors and light sources offers unprecedented possibilities to revisit photoreactor designs that have been proposed over the last decades and to study whether their efficiencies can be increased further by intensified photon transfer. Besides the already mentioned monolith reactors and microreactors, these include fluidized-bed designs, spinning disc reactors (SDRs), and membrane reactors. In the *fluidized bed reactor*, for example, the catalyst is placed on supporting beads. These beads form the bed material and are fluidized by an air stream. Although the fluidized bed reactor is a dispersed-phase reactor with all its associated problems with respect to illumination, the generation of bubbles in the fluidized bed reactor will enhance light penetration compared to the conventional slurry reactor [201]. LEDs have recently been applied to irradiate the fluidized bed in photooxidative dehydrogenations, with a fourfold increase in apparent quantum yield of transformation of cyclohexane to benzene with UV-A LEDs compared to conventional mercury lamps [202]. Another type of dispersed phase reactor, nebulizing aerosols in an air stream, has also been combined with LEDs for the successful photooxidation of citronellol [203].

A SDR (see Section 4.7.2) has attractive features such as limited mass transfer limitations because of the combination of thin film and turbulent flow with subsequent high mass transfer coefficient, increase of conversion, better and more reliable product quality, introduction of the reaction liquid in a gentle nonoscillating motion, thus diminishing the occurrence of the phenomenon of catalyst attrition (see also the section on centrifugal fields). It is also easy to scale up, either by increasing the disc diameter (although this is technically not an infinite option) or by installing multiple discs (although costs would increase simultaneously). Problems with a uniform illumination of the discs were reported by Dionysiou et al. [204].

Membrane photoreactors have come into the picture c. 15 years ago. The idea is to immobilize the catalyst in the membrane structure in order to confine the photocatalyst in the reaction environment by means of the membrane, thus avoiding the need for catalyst recovery, to control the residence time of molecules in the reactor, and to realize a continuous process with simultaneous catalyst and product separation from the reaction environment [205]. Most reactor designs focus on the integration of the photocatalyst with the membrane and use conventional mercury lamps for irradiation [206, 207]. Only recently, some publications have appeared where LEDs [208, 209] or optical fibers [207] are integrated in the membrane system.

In order to assess the efficiency of photoreactors, the *photochemical space–time yield* (PSTY) can be used [210]. This new metric calculates the reactor space–time yield normalized to the power of the light source. Inherent to this assessment tool, intensified light emission and distribution will reflect in a high space–time yield to required lamp power ratio, and therefore, this tool can be used as a design tool for specific applications. In the comparison of 12 reactor designs for photocatalytic wastewater treatment, the highest scoring reactors are the pilot-scale slurry reactors. This is no surprise as they are also applied designs. The benchmark also shows that high area-to-volume geometries such as the microreactor technology is the most effective approach to introduce immobilized catalyst reactors to wastewater treatment processes. PSTY shows the weak points in a reactor concept as well. It is clear that the microreactor concept does not have a throughput problem, but an illumination scale-up limitation, although LED technology, as described earlier, can help address this issue. Provided effective scale up, the benchmark also shows that the microreactor concept can perform significantly better than the slurry reactor. The benchmark can also be applied to other domains, such as photochemical synthesis reactors.

4.4.4 Solar Reactors

In this section, we will review the *direct* (not solar cell-based) use of solar energy in chemical processes. We will focus on two groups of processes: those utilizing

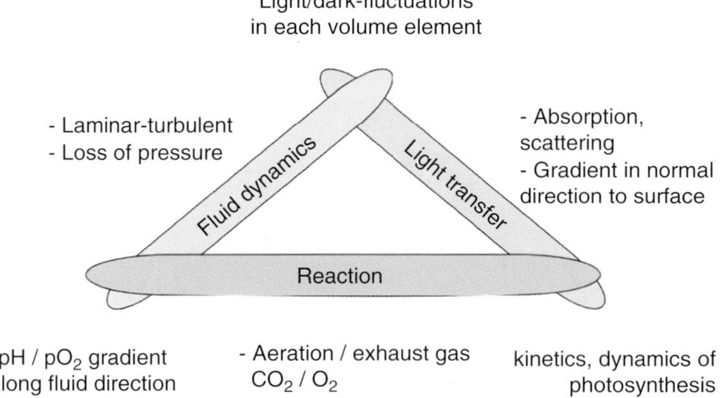

Figure 4.29 Three interacting elements in microalgae-based processes. Source: Posten 2009 [215]. Reproduced with permission of John Wiley and Sons.

solar photons to carry out chemical reactions (e.g. to manufacture synthetic fuels, specialty products, or for waste management) and those utilizing solar-derived heat for the production of hydrogen or synthesis gas via high-temperature, thermochemical water, carbon dioxide, or methane splitting. The enormous potential of all those technologies in terms of fossil energy saving and, in many cases, in terms of innovative equipment qualifies them all to be discussed in this book.

Perhaps, the least "intensive" in terms of equipment design, yet very relevant in terms of sustainable manufacturing, are *microalgae-based processes.* Microalgae are cultivated on the industrial scale to be used as protein-rich elements of animal and human nutrition or to manufacture biofuels (e.g. biodiesel, biosyngas, or biohydrogen) and biologically active compounds (e.g. cosmetics and pharmaceuticals) [211–214]. Three elements, light transfer, fluid dynamics, and bioreaction kinetics, interact with each other and affect the growth and productivity of microalgae – see Figure 4.29. These three elements must therefore be well understood and properly addressed in the microalgae reactor design. They translate directly to such design parameters as surface-to-volume ratio, light intensity, cell density, mixing time, etc.

In general, microalgae reactors can be divided into *open-air and closed systems.* The open-air systems are by far the simplest ones, yet the least efficient. Although quite widespread in terms of industrial-scale applications, they can only be applied to a limited range of algae species and in processes that do not require sterile conditions.

Most of the open-air systems are the *raceway ponds* shown in Figure 4.30. The movement of the microalgae suspension in the loop recirculation channels in those reactors is produced by a paddlewheel. The formation of dead zones presents one of the most important problems in raceway ponds and can be improved by the proper modification of pond geometry. The modeling study by Hadiyanto et al. [217] showed that the dead zones in a raceway pond can be reduced even by a factor of c. 6 relative to the standard design at 40% lower power consumption.

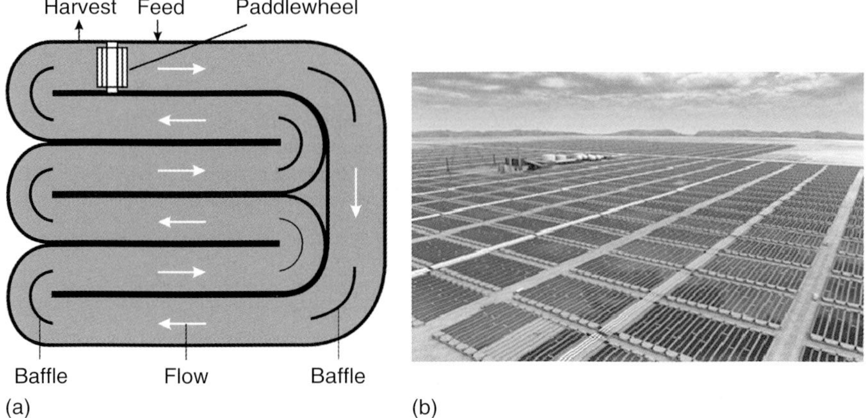

Figure 4.30 (a) Scheme of a microalgae raceway pond. Source: Bahadar and Bilal Khan 2013 [216]. Reproduced with permission of Elsevier. (b) An aerial view of an industrial location in Arizona. Source: Courtesy of Dr. Joel Cuello, University of Arizona.

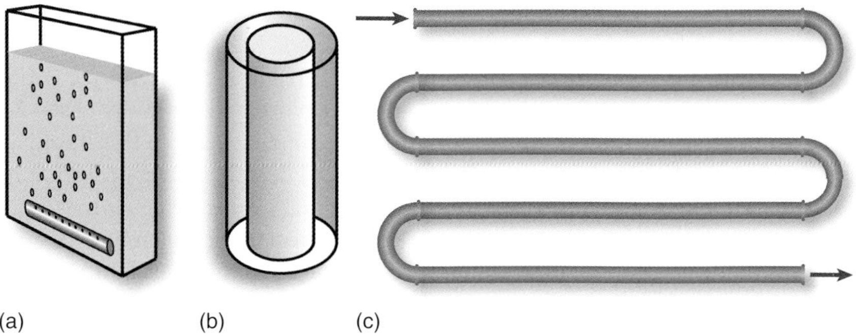

Figure 4.31 Most common types of closed microalgae reactors: flat panel (a), annular bubble column (b), and tubular (c). Source: Adapted from Posten 2009 [215].

Among the closed systems flat-plate reactor, (annular) bubble columns and tubular reactors (Figure 4.31) are the most common designs. In *flat-panel designs*, the movement of the liquid is generated using the airlift principle and baffles in the panels produce mixing and serve also as light-conducting structures. Also in *annular bubble columns*, the airlift principle is often employed. *Tubular microalgae reactors*, in turn, consist of a multitude of narrow, transparent tubes, with lengths varying between 100 m and 500 km [215]. Table 4.7 summarizes basic features of the open and closed microalgae reactor types.

Current activities aiming at intensification of photobioreactors for microalgae-based processes focus on better gas exchange (CO_2 supply and oxygen removal, e.g. via hollow fiber membranes) and on better light distribution inside the reactor volume, e.g. via side-emitting glass fibers. Also, intensified methods are being developed for the extraction of valuable products from the algal biomass. Those include, among other things, ultrasonic-assisted organic solvent extraction and microwave-assisted organic solvent extraction [216].

Table 4.7 Comparison of basic microalgae reactor types.

	Raceway pond	Flat panel	Annular bubble column	Tubular
Light utilization	Low	Excellent	Good	Very good
Mixing	Fair	Good	Fair–good	Good
Area/volume ratio	Low	Very high	Reasonable	High
Footprint	Large	Small	Small	Medium
Temperature control	Difficult	Excellent	Good	Excellent
Sterility	Impossible	Achievable	Achievable	Achievable
Stress on algae	Very low	Low–high	Low	Low–high
Scale-up	Very difficult	Difficult	Difficult	Reasonable

Source: Adapted from Borowitzka [211].

Obviously, algae-based production belongs to low-temperature processes. In order to carry out reactions at higher temperatures, say above 150 °C, concentration of sunlight is needed. For the medium temperature range, solar collectors have been developed that concentrate sunlight between 5 and 50 times. Those collectors need to track the sun, as schematically shown in Figure 4.32a [218]. Here, a parabolic collector concentrates solar rays on a transparent reactor tube that lies in the focus of the parabolic mirror. These types of systems can be used, for instance, in the area of water purification. Figure 4.32b presents such a system operated at Almería, Spain.

Another way of utilizing the solar energy is via a thermochemical cycle. Such a cycle is basically an engine that converts heat into work in the form of stored chemical energy. Efficiency gains are possible as initial conversion to mechanical work and electricity are avoided [219]. Reactors used for high-temperature solar thermochemical processes look differently than the ones described above. They are developed for different purposes, which include metal oxide to metal reduction, carbonaceous matter gasification, methane reforming, and (mostly) solar production of fuels via carbon dioxide and water splitting. A very good review on the latter topic has recently been published [220]. Practically, all those reactors operate via a thermal reduction/oxidation cycle of a metal oxide catalyst (e.g. ZnO, Fe_3O_4, Mn_3O_4, and Co_3O_4) involving the following generic steps:

Thermal reduction (solar)

$$\frac{1}{\delta}MO_x \rightarrow \frac{1}{\delta}MO_{x-\delta} + \frac{1}{2}O_2 \tag{4.17}$$

Water splitting (nonsolar)

$$\frac{1}{\delta}MO_{x-\delta} + H_2O \rightarrow \frac{1}{\delta}MO_x + H_2 \tag{4.18}$$

CO_2 splitting (nonsolar)

$$\frac{1}{\delta}MO_{x-\delta} + CO_2 \rightarrow \frac{1}{\delta}MO_x + CO \tag{4.19}$$

The majority of the thermochemical cycle reactors can be roughly divided into three categories:

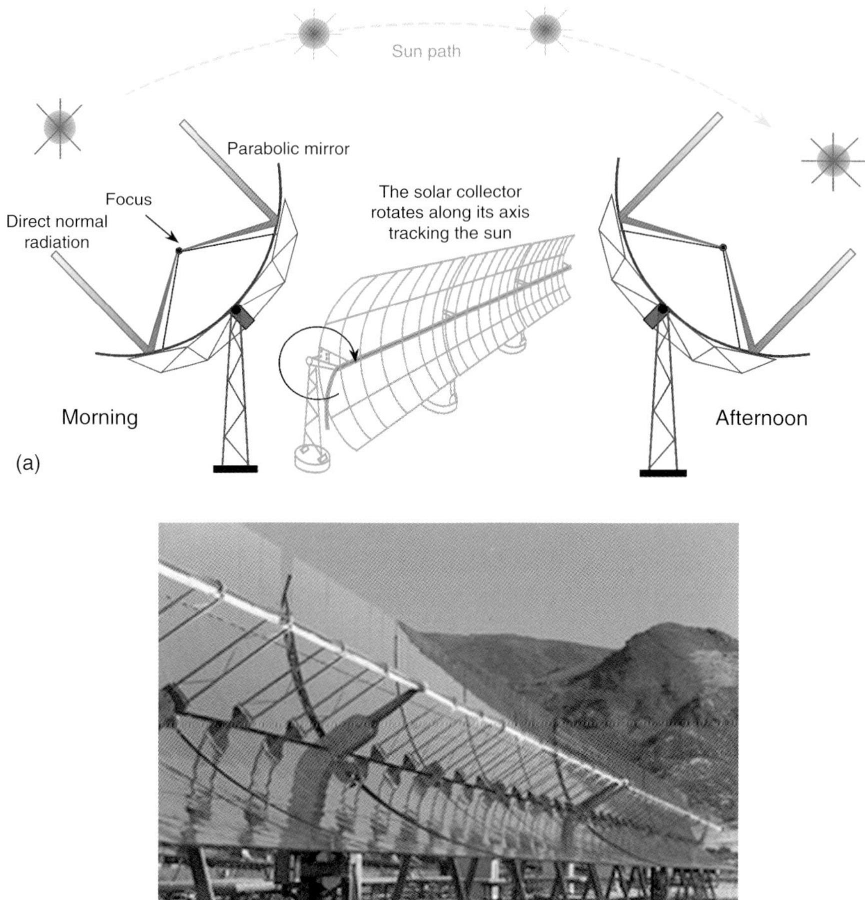

Figure 4.32 (a) Tracking the sun with a parabolic sunlight collector © UNESCO-Encyclopedia of Life Support Systems (EOLSS) Source: From Solar Photochemistry Technology, Blanco Gálvez and Malato Rodríguez Solar Energy Conversion and Photoenergy System, Vol. II, with permission of UNESCO-Encyclopedia of Life Support Systems (EOLSS) [218] and (b) solar photochemical reactor with a parabolic collector for medium temperature range applications developed and installed at Platforma Solar de Almería, Spain. Source: Courtesy of PSA, Spain, www.psa.es.

- cavity-type reactors with fixed catalytic/ceramic elements (lining blocks, tiles, monoliths, and foams), in which the solar radiation enters the cavity axially;
- rotor-type reactors with rotating catalytic/ceramic elements, in which solar radiation is introduced radially through the circumference; and
- reactors in which the catalytic/ceramic material is in the form of moving particles.

Figure 4.33 presents a reactor belonging to the first category. It is a pilot-scale 100 kW solar reactor for thermal dissociation of ZnO developed in the Steinfeld group at Paul Scherrer Institute of ETH, as an intermediate step toward the

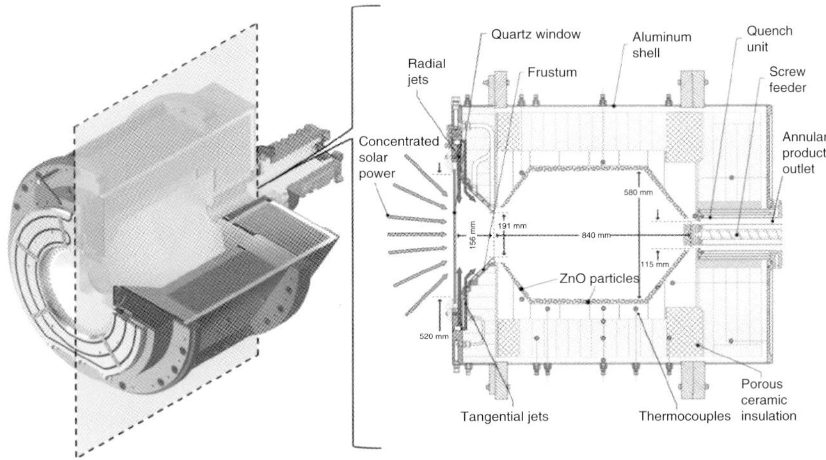

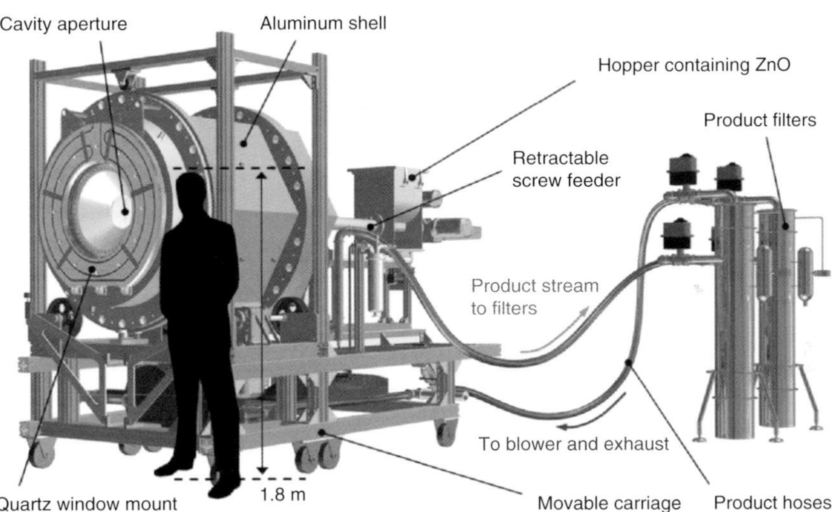

Figure 4.33 Pilot reactor for thermal dissociation of ZnO developed at Paul Scherrer Institute of ETH. Source: Koepf et al. 2016 [221]. Reproduced with permission of American Institute of Physics.

integrated solar production of zinc and hydrogen (https://www.psi.ch/lst/bfe-solarzinc-pandd). It consists of a rotating cylindrical cavity receiver with a quartz window and an aperture for the access of concentrated solar radiation. The cavity wall is built of Al_2O_3 and lined with ZnO particles. With this arrangement, ZnO serves simultaneously the functions of radiant absorber, chemical reactant, and thermal insulator. During the operation, ZnO particles are added batchwise with a dynamic screw feeder. Product gases, Zn(g) and O_2, are rapidly quenched with an inert gas to prevent the recombination reaction. Operation temperatures in the reactor reach as high as 2064 K.

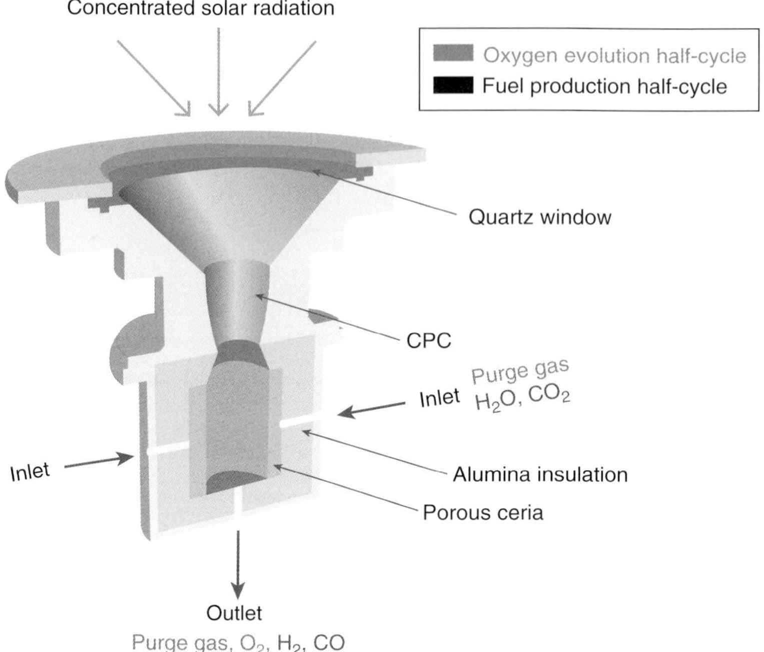

Figure 4.34 Thermochemical solar reactor based on porous Ceria as catalysts. Source: Adapted from Chueh et al. 2010 [222].

Another reactor in this category codeveloped by ETH and CALTECH is shown in Figure 4.34 [222]. Here, the cavity is lined with porous ceria. The reactor is not rotating but works in a cyclic mode. In the fuel production, half-cycle H$_2$ or CO is generated; in the oxygen evolution, half-cycle, a purge gas is being used to remove the oxygen from the cavity. Other reactors belonging to the same category include rotary kilns for production of lime [223] and methane reformers [224, 225].

A representative of the second category of thermochemical solar reactors is the pilot-scale CR5 water/CO$_2$ splitter developed at Sandia National Laboratory [226, 227]. The reactor is shown in Figure 4.35. It is based on the FeO/Fe$_3$O$_4$ cycle and consists of an assembly of ceramic discs rotating in opposite directions. The reduction step for this catalyst requires temperatures above 1600 K. Conceptual studies indicated the possibility of utilizing the CR5 reactors as the first step in the new processes, leading from CO$_2$ and water to methanol or Fischer–Tropsch fuel [228]. Another rotor-type thermochemical solar reactor for water splitting has been developed by Kaneko et al. [229]. The reactor uses reactive ceramics of CeO$_2$ and Ni,Mn-ferrite.

Finally, in the third category, various reactor types have been proposed. One of them, developed at Sandia National Laboratory, is presented in Figure 4.36 [230]. Here, a packed bed of catalyst particles is moving through the reactor cavity by means of a stationary screw elevator with rotating casing. The particle bed fills the entire reactor but is shown only selectively in the figure, to preserve the clarity of the schematics.

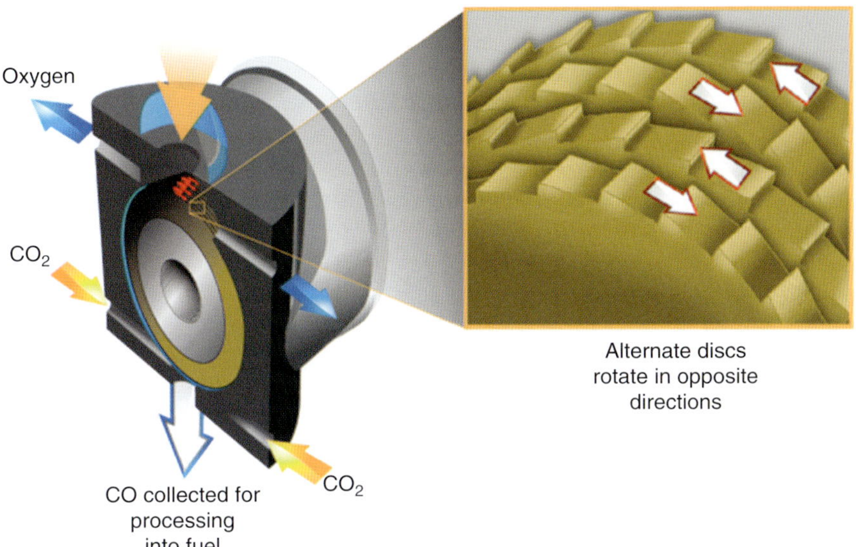

Figure 4.35 Water/CO_2 splitter CR5 developed at Sandia National Laboratory © 2012 Sandia Corporation. All Rights Reserved. Source: Reproduced with permission of Sandia Corporation.

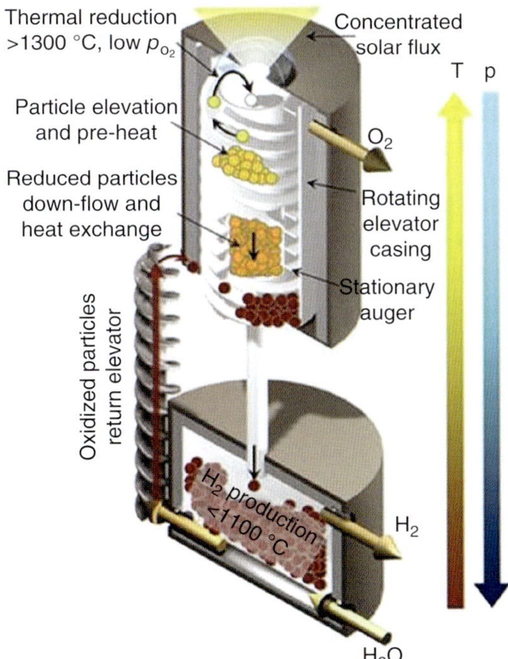

Figure 4.36 Solar thermochemical reactor with moving bed of catalyst particles. Source: Ermanoski and Siegel 2014 [230]. Reproduced with permission of Elsevier.

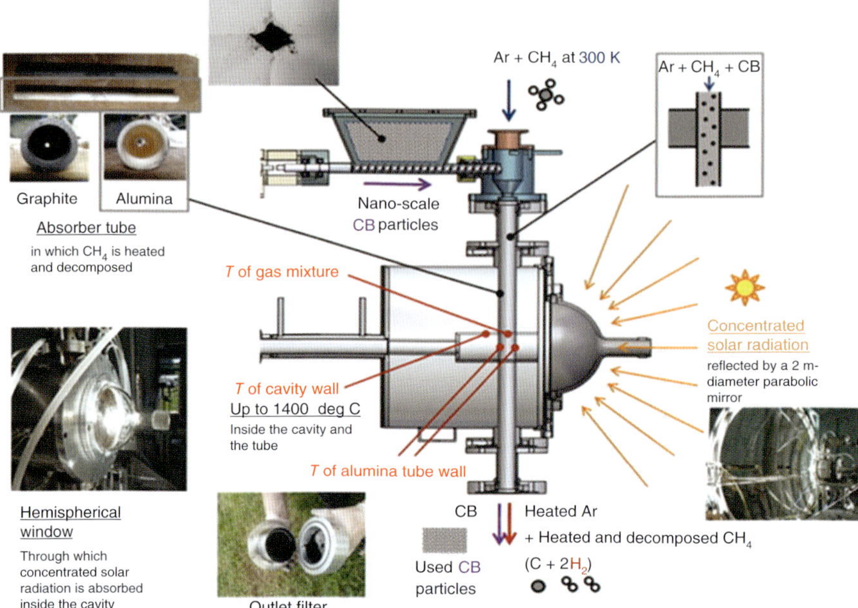

Figure 4.37 Particle-entrained flow solar reactor for hydrogen production from methane. Source: Abanades et al. 2015 [231]. Reproduced with permission of Elsevier.

Recently, an interesting particle-entrained flow reactor has been designed and tested for CO_2-free hydrogen production from thermal methane decomposition (Figure 4.37, [231]). In the reactor, nanoscale carbon-black particles are co-currently fed with methane stream to a solar absorber tube, in which CH_4 is heated and decomposed.

The key issue for cost and scalability of solar thermochemical reactors is their efficiency. Development of new constructional and catalytic materials will play crucial role in the route to commercial-scale applications.

4.4.5   Induction Heating

Induction heating goes back in history to Michael Faraday's law of induction formulated in 1831. It presents a very rapid method of heating electrically conducting objects by electromagnetic induction. It is based on the so-called eddy currents (sometimes also called Foucault currents) that are generated by rapidly alternating magnetic field inside the bulk of the object (Figure 4.38). The flow of those currents through the material resistance leads to Joule heating.

Induction heating is extremely powerful in terms of power transmission (see Table 4.8) and therefore finds diverse applications, starting from home cooking and hyperthermia in cancer treatment, through welding, bonding, brazing, melting, various forms of heat treatment of metal items, and plastic processing. The latter primarily concerns thermoplastic welding, thermoset curing, and injection molding processes [234, 235].

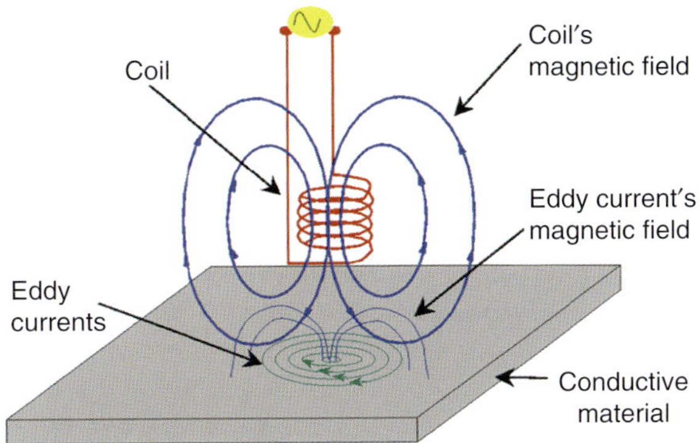

Figure 4.38 Formation of eddy currents in a conducting object placed in a magnetic field. Source: From http://www.microwavesoft.com/eddycurrent.html.

Table 4.8 Power transmission of different heating processes.

Type of heating	Power transmission (W/m^2)
Convection	0.5
Irradiation	8
Heat conduction	20
Flame	1 000
Inductive heating	30 000

Source: Benkowsky 1990 [232], and Kirschning et al. 2012 [233].

Because of the very high heating rates achievable with the induction heating, the technique has attracted the attention of researchers working in the area of chemical reactor design. The first publication related to this topic dates from 1952 and comes from Shell [236]. The researchers investigated the possibility of using the induction heating for very fast preheating of petroleum fractions entering a catalytic cracking unit. Such a fast preheating to very high temperatures (700 °C or more) should help reduce thermal cracking and coke formation on the catalyst. Accordingly, an induction-heated cracking unit was designed and investigated. In the unit shown in Figure 4.39, the catalyst bed is preceded by a preheater consisting of an induction coil and a bundle of gold wires. In the preheater, petroleum fractions were heated up from 400 to 700 °C in about 10 ms. As a result, carbon formation in the novel reactor was less than 1/10th of that in a conventional system.

Also, *induction-heated stirred tank reactors* have been investigated [237]. In those reactors, the cooling is realized by an external jacket ("envelope") system, as presented in Figure 4.40. *Induction-heated tubular reactors* have been proposed for fast pyrolysis of different types of biomass, including rice straw, sugarcane bagasse, coconut shell, napier grass, pinewood sawdust, and sewage sludge [238–241]. The heating rates in those reactors were as high as 500 °C/minutes. Among other things, higher quality of the bio-oils obtained and lower carbon deposition were observed in the induction-heated units (Figure 4.41, [241]).

More recently, inductive heating has been investigated in relation to organic synthesis in microreactors [233, 242]. The heating in the reactor channels was

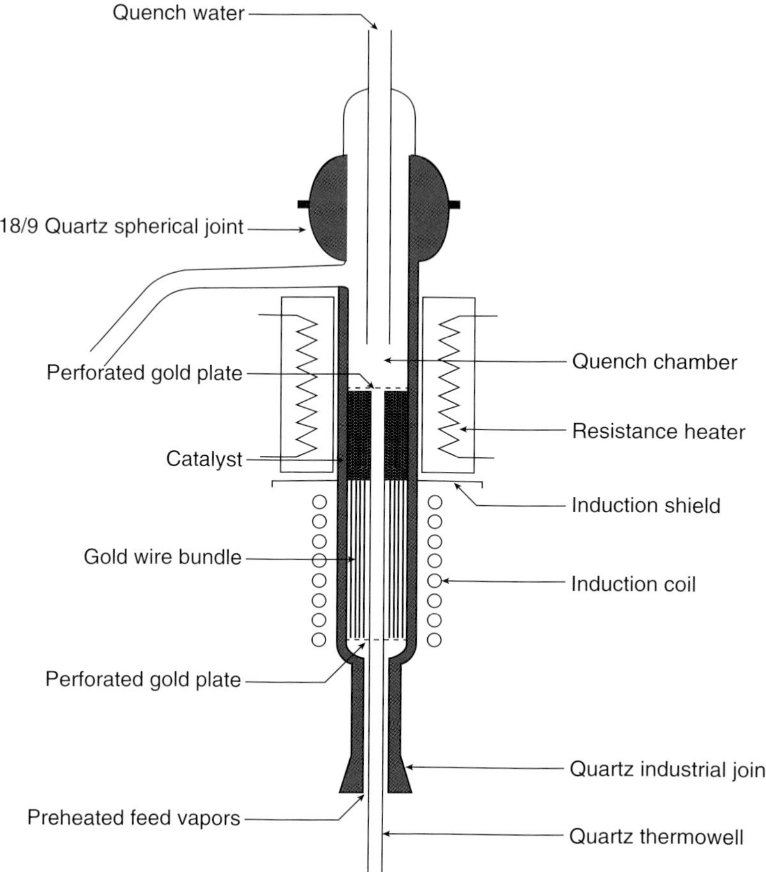

Figure 4.39 Induction-heated cracking unit developed at Shell. Source: Archibald et al. 1952 [236]. Reproduced with permission of American Chemical Society.

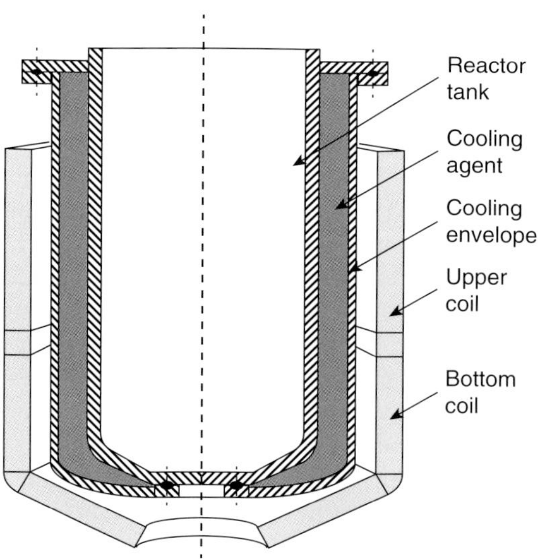

Figure 4.40 Induction-heated tank reactor. Source: Fireteanu et al. 2005 [237]. Reproduced with permission of Emerald Group Publishing.

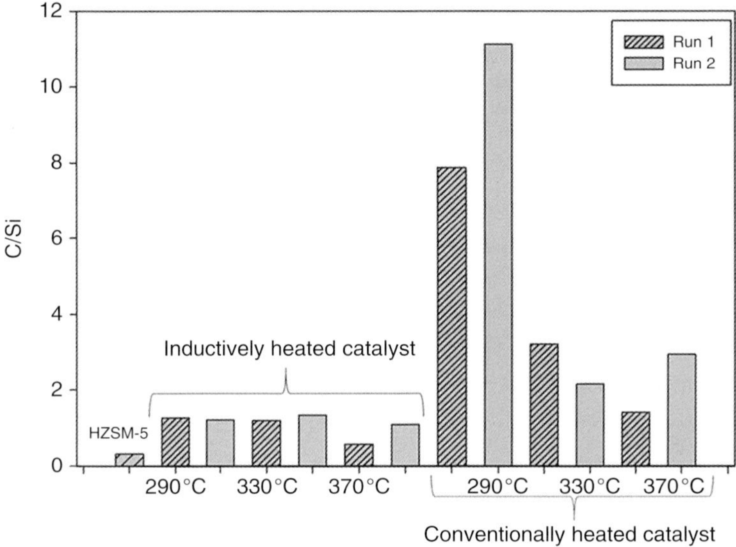

Figure 4.41 Carbon-to-silicate ratio as a measure of coke formation in the inductively and conventionally heated catalysts. Source: Muley et al. 2015 [241]. Reproduced with permission of American Chemical Society.

realized via superparamagnetic core–shell nanoparticles, e.g. silica-coated manganese ferrite particles. This led to significantly shorter reaction times and higher yields than in the corresponding batch reactors.

Finally, Canadian researchers from École Polytechnique de Montréal and Western University developed novel induction-heated micro- and mini-fluidized-bed reactors for screening tests of solid feedstocks and catalysts [243, 244]. Stainless steel rods and Inconel wires were used as heating elements.

4.5 Acoustic Fields

The research of the acoustic energy effects in the chemical processing has a long history and ultrasonic devices have already found a number of commercial applications, e.g. in cleaning and decontamination or for dye dispersion and fixation in the textile industry [165]. First reports on chemical and biological rate enhancement by acoustic fields were published in the late 1920s. Since early 1980s, the field of sonochemistry has become a very popular area of (mostly) chemical research [245–248].

Ultrasound is a mechanical sound wave with a frequency above 20 kHz, exceeding the capability of human hearing and reaching up to several GHz [249]. Depending on the frequency and power, it is further classified into two different categories, displayed in Figure 4.42 [249, 251, 252]:

- *Diagnostic ultrasound*: These ultrasonic waves have a frequency between 5 and 10 MHz and operate with a power in the milliwatts range. This frequency range

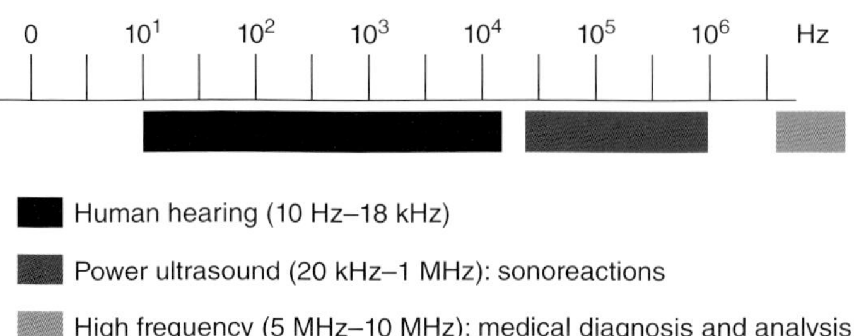

Figure 4.42 Classification of ultrasound into separate classes depending on the frequency. Source: Cravotto and Cintas 2011 [250]. Reproduced with permission of Taylor and Francis Group.

is mostly used for medical and analytical applications as it does not alter the medium through which it propagates.

- *Power ultrasound*: This range is most exploited in sonochemical applications and usually ranges from 20 to 100 kHz, with a maximum operating power of several thousand watts. However, advancements in the construction of ultrasound generators extended the frequency range to the MHz region. A large amount of acoustic energy can be generated to induce cavitation bubbles in the liquid. The effects originating from these cavitation bubbles are mostly used to intensify chemical processes and will be discussed further on in this section.

Acoustic frequencies lower than 20 kHz can also be used to intensify transport processes [165]. Already in 1957, the effects of sonic energy at frequencies between 100 and 700 Hz on gas–solid mass transfer were studied with enhancements of up to about 120% for a bed of spheres and up to about 220% for a single sphere. More recently, Krishna and Ellenberger [253] used low-frequency (50–200 Hz) vibrations to improve the performance of a bubble column and observed 50–100% increase of the volumetric mass transfer coefficient, depending on the flow rate. All in all, the applications of acoustic fields in the human hearing frequency domain remain relatively scarce.

When applying ultrasound in the power frequency range to a liquid, the acoustic wave generates alternate compression and expansion (i.e. rarefaction) zones (Figure 4.43). At low intensity, the ultrasonic wave generates vortices and mixing within the fluid that improve both mass and heat transfer. This effect is commonly referred to as *acoustic streaming*, although different types can be distinguished [254]. At higher intensity, the negative pressure during the rarefaction phase can exceed the intermolecular van der Waals forces locally and small cavities or gas-filled microbubbles are formed. Above a critical acoustic pressure of about 1010 Pa, voids, called *cavitation bubbles*, will be created in water [255, 256]. In practice, however, impurities or microbubbles of dissolved gasses will be present in the liquid. These bubbles act as nuclei and lower the required pressure for the creation of cavitation bubbles [257].

Mainly two types of cavitation bubbles can be distinguished, namely stable and transient bubbles [255, 258]. In general, the bubbles will oscillate around an equilibrium radius for hundreds of acoustic cycles leading to a relatively long lifetime

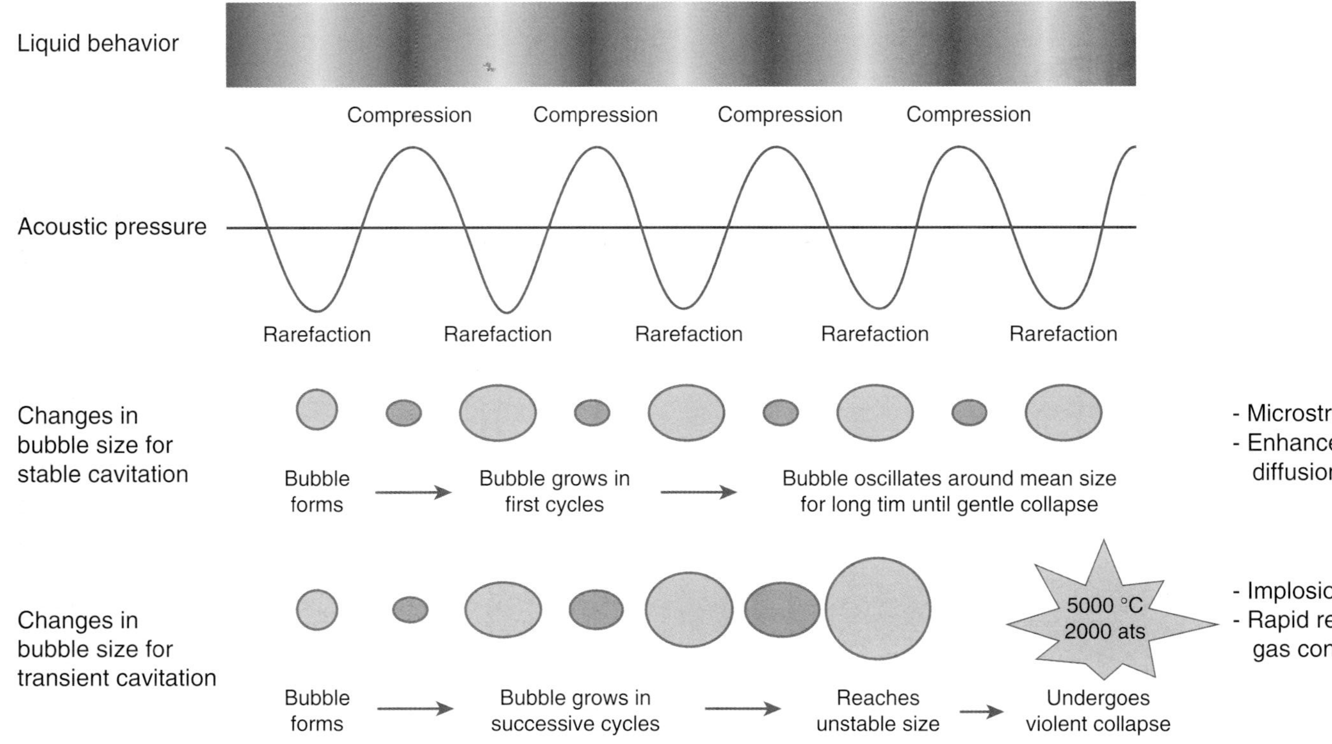

Figure 4.43 Schematic overview of two types of acoustic cavitation. Stable cavitation arises at low amplitude and is characterized by a periodic oscillation around an equilibrium radius. Transient cavitation dominates at high amplitudes, with bubbles rapidly growing to at least twice their initial size and imploding violently afterward. Source: Adapted from http://www.sonochemistry.info/introdution.htm.

in the order of 10 ms [259]. This cavitation type is known as *stable cavitation* [260–262]. As a result, gasses will be able to migrate through the bubble–liquid interface and enter the cavitation bubble. These gas molecules will absorb energy upon collapse and cushion the implosion. The bubbles that are oscillating in the sound field experience a net influx of gas because more gas diffuses into the bubble during the rarefaction than diffuses outward during the compression phase. This process, called rectified diffusion, can be attributed to the available surface area of the bubbles in both phases. As the bubbles are larger during the rarefaction, they exhibit a larger area that is available for gas diffusion [263, 264].

If a high acoustic power is applied, the negative pressure during the rarefaction can even overcome the intermolecular forces of the liquid molecules, generating *transient cavitation* bubbles. In this case, the bubbles rapidly expand during the rarefaction and can even grow to twice their initial size. Because of the short time frame in which this happens, the dissolved gasses have insufficient time to cross the bubble–liquid interface and accumulate inside the bubble [265]. The resulting lack of gas cushioning causes transient bubbles to implode very violently under the influence of the static pressure and surface tension, creating extremely high local temperature (up to 5000 K) and pressure (up to 50 000 atm has been suggested). Overall, these transient bubbles only exist for one or a few acoustic cycles [260–262].

Transient bubbles are primarily present at low ultrasonic frequencies, whereas stable bubbles are predominantly present at higher frequencies. The emitter design, the applied power, and frequency define which cavitation type is present in the reactor. Transient bubbles will only emerge when the initial bubble radius is between certain upper and lower limits [255]. When the bubble radius becomes too large, the bubbles will coalesce and not be sonochemically active. When the bubble radius is too small, stable cavitation bubbles will be present. The bubble size is impacted by the frequency and acoustic pressure. The lower the frequency and the higher the acoustic pressure, the larger the cavitation bubbles become. Therefore, transient cavitation bubbles are primarily present at low ultrasound frequencies (around 20 kHz). At higher frequencies, the bubble radius becomes smaller, so that higher acoustic pressures are needed to exceed the minimum radius size. The higher the frequency, the more difficult it therefore becomes to create transient bubbles. In addition, also attenuation of the ultrasound field increases at higher frequencies. Higher frequencies increase the rate at which liquid molecules move past each other, thus generating more frictional heat. Furthermore, at high frequencies, vibrating liquid molecules have less time to revert to their equilibrium positions before the next disturbance. When the next ultrasonic pulse encounters the liquid molecules before they are fully relaxed, the molecules will move in the direction of the wave propagation and cause acoustic streaming. As a result, high acoustic pressures are more difficult to achieve and hence primarily stable bubbles are present at these frequencies [249, 255].

The implosions of both stable and transient bubbles create, among other effects, shockwaves, high pressures, microjets, and enhanced mixing [255, 266, 267]. These effects are in general classified as chemical or mechanical effects.

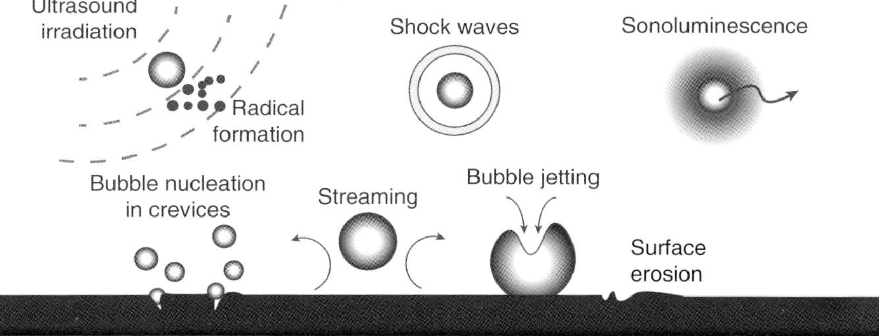

Figure 4.44 Schematic representation of the physical and chemical effects of ultrasound. Source: Adapted from Fernandez Rivas et al. 2012 [268].

Figure 4.44 gives a schematic overview of the possible effects induced by ultrasound [268].

Ultrasonic cavitation, i.e. the formation, growth, and collapse of micron-sized bubbles, results in the release of a large amount of energy. Upon implosion, intense local heating with temperatures of above 5000 K and pressures of about 10^5 kPa were reported. Additionally, these bubbles exhibit heating and cooling rates of 10^7–10^{10} K/s. As a result, cavitation bubbles are often considered as localized microreactors, which are also used as the base for the hot spot theory. This model allows to estimate the generated heat and pressure by considering the implosion as a nearly adiabatic compression of gas and vapor, resulting in a localized hot spot [251, 265, 269–271]. Equations (4.20) and (4.21) allow to calculate the final temperature T_f (K) and pressure p_f (Pa) using the initial temperature T_i and bubble radius R_i. Additionally, the vapor pressure p_V, the pressure in the bubble at the moment of collapse p_{bubble}, the final bubble radius R_f after implosion, and the adiabatic index γ are used in the calculations [270, 272].

$$T_f = T_i \left(\frac{p_{\text{bubble}}(\gamma - 1)}{p_v} \right) = T_i \left(\frac{R_i}{R_f} \right)^{3(\gamma-1)} \tag{4.20}$$

$$p_f = pv \left(\frac{p_{\text{bubble}}(\gamma - 1)}{p_v} \right)^{\frac{\gamma}{\gamma-1}} \tag{4.21}$$

These equations overestimate the temperature and pressure as heat transfer, and thermal conductivity of the gases and energy loss due to decomposition of gas or vapor from the surrounding liquid are neglected. Hence, depending on the temperature or vapor pressure of the solvent, a significant amount of solvent molecules can enter into the bubble, lowering the efficacy of the collapse and the maximum pressure and temperature attained. A more recent work by Pflieger et al. stated that bubble temperatures cannot be quantified by one single temperature. Instead, these authors consider the bubble core as a nonequilibrium plasma, characterized by an electron (T_e), vibrational (T_v), rotational (T_r), and translational (T_t) temperature that can be determined by spectral analysis [273–275].

The extreme conditions generated by the bubble implosion promote both *chemical and mechanical effects.* The former is often indicated by the term sonochemistry and is initiated by the thermal dissociation of water vapor into reactive hydroxyl (OH·) and hydrogen radicals (H·). Because of radical and recombination reactions, other species such as HO_2·, O·, and H_2O_2 may form in the bubble core. In turn, these products can participate in new reactions or diffuse toward the bulk medium in order to oxidize soluble compounds. In addition, volatile solutes can evaporate into the bubble and decompose mainly by means of pyrolysis. In general, three different reaction zones are distinguished in an aqueous sonochemical reaction process. In the core of the cavitation bubble, the most hydrophobic and volatile solutes will be present and take part in pyrolysis. In contrast, nonvolatile hydrophilic compounds remain in the bulk liquid, which is dominated by reactions with hydroxyl radicals, diffused from within the bubble. Finally, hydrophobic nonvolatile products are concentrated at the bubble–liquid interface, participating in predominantly radical reactions.

Apart from the chemical events, ultrasound will also induce mechanical effects in a sonicated medium. In homogeneous liquid systems, the shockwaves generated by the collapsing bubbles will induce improved mixing within the bulk liquid [276], and by that heat and mass transfer. In solid–liquid systems, depending on the localization of the solid particles, the collapse of the ultrasonic cavitation bubbles exhibits two different natures, which in turn determine the arising mechanical effects [277]. Figure 4.45 shows a schematic representation of the collapse types. In the absence of a solid surface close by, bubbles implode predominantly in a symmetric way, resulting in shockwaves and microturbulence that promote mass and heat transfer as well as (micro-)mixing [263, 267]. These shock waves travel at speeds of about 4000 m/s and reach pressures in the order of 10^5 kPa [267, 278, 279]. Within a slurry of brittle solid particles, these perturbations induce high-speed collisions that can lead to fragmentation and breakage. Moreover, in the vicinity of an extended solid boundary that is several times larger than the bubble radius, some bubbles can exhibit an asymmetrical collapse, generating a liquid jet that first penetrates the bubble wall before impacting the solid interface. This jet can develop a velocity of more than 100 m/s and significantly affect the solids, resulting in pit erosion of the surface, disruption of aggregates, and abrasion of surface coatings [280, 281]. This causes fragmentation of solids, removal of passivating surface layers, and exposure of new reaction surfaces [282, 283].

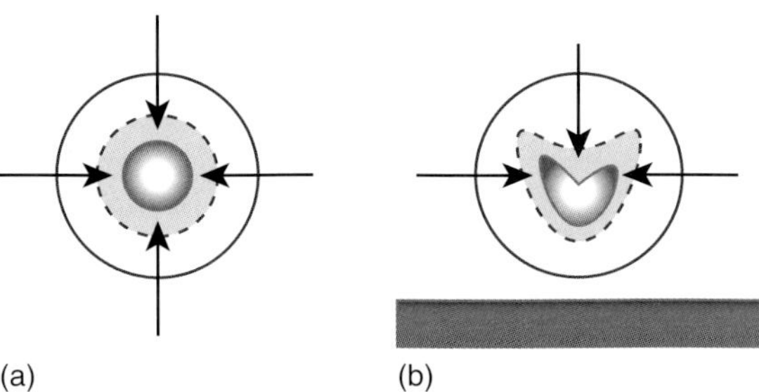

(a) (b)

Figure 4.45 Two different types of cavitation bubble implosions: (a) symmetric bubble collapse, (b) asymmetric bubble collapse in the vicinity of a solid surface resulting in a liquid jet toward the solid surface. Source: Adapted from Leighton 1994 [263].

In immiscible liquid–liquid systems, the shear forces result in a jet of one phase moving into the other, causing disruption of the interface, mixing, and the formation of fine emulsions [283–286]. The resulting emulsification can contribute to increased interfacial area per unit volume required for mass transfer. In biological systems, induced forces will be able to cause cell destruction [287], although less intense ultrasound has been shown to enhance metabolism and subsequent growth rate and biomass yield. Meticulously controlled ultrasonication is also able to extract extracellular polymer substances while avoiding cell lysis [288].

Ultrasound can be used to enhance *kinetics, flow, and mass and heat transfer*. The overall results are that organic synthetic reactions show increased rate (sometimes even from hours to minutes, up to 25 times faster) and/or increased yield (tens of percentages, sometimes even starting from 0% yield in nonsonicated conditions). In multiphase systems, gas–liquid and solid–liquid mass transfers have been observed to increase by 5- and 20-fold, respectively [165]. Membrane fluxes have been enhanced by up to a factor of 8 and membrane fouling has been reduced considerably [289]. The chemical effects are mainly applied in two domains. One is the degradation of organic compounds where ultrasound is considered to be an advanced oxidation process (AOP) [272, 290, 291]. Other AOPs are photocatalysis, Fenton reaction, ozone, and hydrogen peroxide oxidation. These degradation reactions occur because of the hydroxyl radicals formed by the collapse of the cavitation bubbles. The very extreme temperatures and pressures created upon bubble collapse in water create ·OH and ·H radicals, which subsequently cause radical reactions with the organic compound. In research labs, ultrasound is often combined with photocatalysis as sonophotocatalysis, although no industrial processes are known.

Another field is that of *organic synthesis*. A huge variety of organic synthesis reactions have been reported to improve by the use of ultrasound. Improvement is to be understood as the increase of reaction rate to achieve the same yield as without ultrasound enhancement, as the increase of yield within the same reaction time as without ultrasound, or both. Examples are also reported where the reaction path is switched, allowing to increase selectivity for the desired product or even to form products that are not attainable without ultrasound. Thompson and Doraiswamy [248], in their seminal work on science and engineering of sonochemistry, have reviewed the most important work till the end of the millennium, but the field is expanding exponentially. Table 4.9 illustrates the possibilities.

Quite some attention has been given to the use of the ultrasound-induced mechanical effects in the intensification of chemical processes. Originally, the first applications were in cleaning of solid surfaces. This is also a major application of ultrasound in industry, e.g. for the cleaning of silicon wafers for electronics. Many other applications have been described as well, particularly in the multiphase systems. Examples include ultrasound-assisted leaching, extraction, crystallization, acoustophoresis, and biotechnology. They are shortly discussed in the following paragraphs.

The domain of *ultrasound-assisted leaching* of metals is investigated in view of improved industrial leaching of ores and slags such as mine tailings, converter slag, zinc roasting slag, electrolysis sludge, and PCB waste sludge [292–296]. The

Table 4.9 Intensification of organic synthesis reactions by the use of ultrasound – a selection of examples.

Reaction	Reaction time		Product Yield	
	Conventional	Ultrasound	Conventional (%)	Ultrasound (%)
Diels–Alder cyclization	35 h	3.5 h	77.9	97.3
Oxidation of indane to indan-1-one	3 h	3 h	<27	73
Reduction of methoxyaminosilane	No reaction	3 h	0	100
Epoxidation of long-chain unsaturated fatty esters	2 h	15 min	48	92
Oxidation of arylalkanes	4 h	4 h	12	80
Michael addition of nitroalkanes to monosubstituted α,β-unsaturated esters	2 d	2 h	85	90
Permanganate oxidation of 2-octanol	5 h	5 h	3	93
Synthesis of chalcones by Claisen–Schmidt condensation	60 min	10 min	5	76
Ullmann coupling of 2-iodonitrobenzene	2 h	2 h	<1.5	70.4
Reformatsky reaction	12 h	30 min	50	98

overall conclusion is that ultrasound does increase leaching of metals, but it does not do so in a selective manner, although the group of Fengchun Xie has reported the possibility of selective leaching of Cu, Ni, and Zn without significant leaching of Fe and Cr, from two types of waste (printed circuit board waste sludge and electroplating sludge) [295, 297–299]. The authors suggest that ultrasound is able to enhance the dissolution and diffusion of Cu into the liquid while at the same time enhancing precipitation of $Fe(OH)_3$, possibly aided by oxidation of Fe(II) by hydroxyl radicals formed during cavitation.

Ultrasound-assisted extraction (UAE) is a very popular research field in the domain of food processing and API (active pharmaceutical ingredient) and oil extraction from natural products [300–302]. This includes the extraction of herbal, oil, protein, and bioactive compounds from plant and animal materials (e.g. polyphenolics, anthocyanins, aromatic compounds, polysaccharides, and functional compounds) with an increased yield of extracted components and/or increased rate of extraction, thus achieving a reduction in extraction time and solvent consumption, and a higher processing throughput and quality of the extracts. The integration with microwave-assisted extraction is often studied as well [303], but industrial implementation of this hybrid technology is even further away.

Sonocrystallization, i.e. the use of ultrasound in combination with crystallization, is investigated because of its ability to reduce the metastable zone width (MSZW), alter the particle size distribution, reduce agglomeration, improve reproducibility, control polymorphism, and induce nucleation without the use of the seed material [304]. Although applications usually require sonication during the crystallization process, ultrasound can also be used as an after-treatment for fragmentation or particle rounding [266, 267]. The optimal moment and duration of ultrasound irradiation during crystallization is further studied in the work of Gielen [305]. Application of ultrasound in crystallization was first reported in 1927 by Richards and Loomis, but because of a lack of consistent results and proper ultrasonic equipment, follow-up research was discarded for many years [306]. Only in the 1950s and 1960s, the benefits of sonocrystallization were acknowledged in the literature worldwide, encouraging scientists to investigate this technology in more detail [307]. As a result, sonication is now identified as an efficient tool for the promotion of crystallization and enhancement of the crystalline product quality in various systems including inorganic and pharmaceutical compounds [267, 308]. In general, the use of ultrasound during crystallization can result in the following improvements [304, 308–326]:

- *Faster primary nucleation*: This effect will result in a more narrow MSZW or a reduction in the induction time.
- *Improved crystal morphology*: Application of ultrasound during or after the crystallization process can alter the shape of the crystals or avoid the formation of agglomerates. In some cases, this causes improved handling characteristics, filtration capability, and bulk density.
- *Control over the polymorph formation*: Because of the effect of ultrasound on the induction time, the nucleation of different polymorph types can be altered, providing a tool to selectively produce a particular polymorph.
- *Increased reproducibility of the process*: Spontaneous crystallization is a stochastic process and therefore exhibits a broad distribution of the nucleation event. Spontaneous nucleation, often employed in sterile crystallization processes, can therefore benefit from ultrasound as it significantly reduces the time and variability of the nucleation kinetics.
- *Avoiding the use of seed crystals*: With seeding, the nucleation depends on the available surface area, which is strongly affected by the dispersion of the seeds and the reproducibility of the pretreatment step. Hence, any change in the properties of the seed material can result in batch-to-batch variations. The formation of cavitation bubbles and the accompanying effects that arise in the presence of ultrasound can induce nucleation in a reproducible way without the addition of the seed material. Sonication can therefore be successfully applied as an alternative seeding technique.

Overall, the main benefits of sonocrystallization can be attributed to the effect of ultrasound on crystal nucleation, crystal growth, and the interaction between cavitation bubbles and solid crystals in suspension affecting particle size and shape as well as (de-)agglomeration rate and extent. Ultrasound has the potential to control the final particle size and morphology, increase the reproducibility, and hence improve the final product quality. Therefore, sonocrystallization seems a

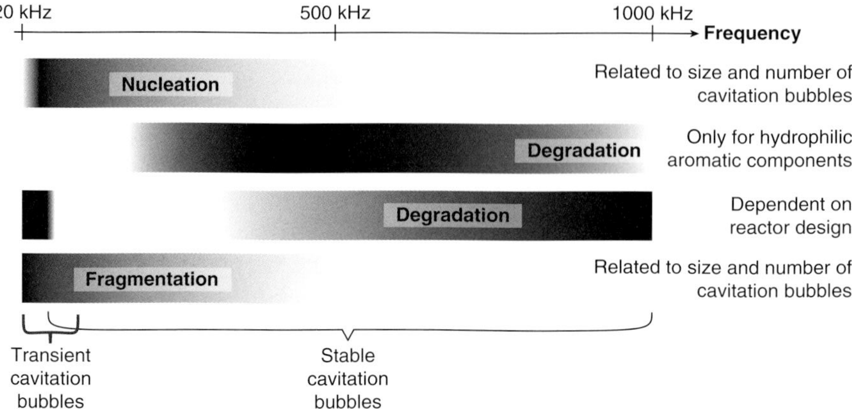

Figure 4.46 Overview of optimal frequencies for nucleation, degradation, micromixing, and fragmentation during the crystallization process. Darker colors indicate elevated levels of nucleation, degradation, micromixing, and fragmentation rates. Source: Jordens 2016 [327].

promising technique to intensify the crystallization process and thus to reduce the number of unit operations by avoiding expensive post-processing such as milling. Process conditions should be optimized toward the desired effect that is strived for. Figure 4.46 illustrates the optimal frequency selection in the crystallization process [327].

A phenomenon that has the potential to intensify separation processes is *acoustophoresis*, which relates to controlled movement of particles by an acoustic field [328–332]. The research group of Thomas Laurell has done extensive research in the field and showed that by inducing standing waves in a microchannel, particles can be directed toward certain directions and thus be separated from each other or form concentrated suspensions (Figure 4.47). The magnitude of the movement depends on many factors, such as the size of the particle, the acoustic pressure amplitude, and the frequency of the sound wave. The direction the particle is moved depends on the density and compressibility of the particle as well as of the liquid medium. Acoustophoresis can be applied for medical

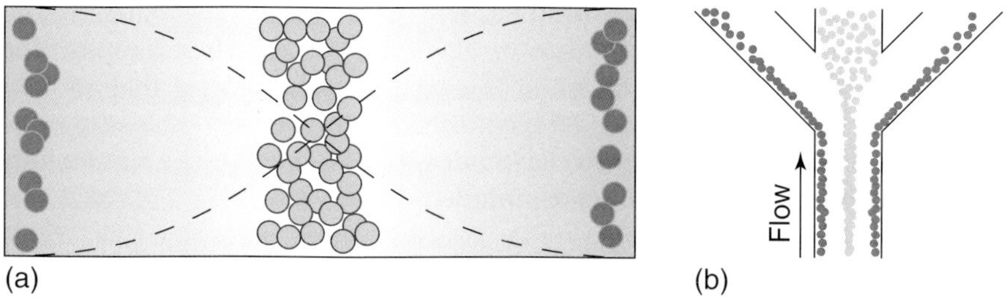

Figure 4.47 Separation of rigid particles from air bubbles in a liquid medium by an ultrasound standing wave (a) and application of this phenomenon to concentrate and separate particles in a flow system (b). Source: Reproduced with permission of Dr. Andreas Lenshof, Dept. Biomedical Engineering, Lund University, Sweden [333].

purposes (e.g. removal of lipid contamination in blood and separating plasma from blood samples) and in food processing (e.g. lipid removal from raw milk).

Similar to acoustophoresis in liquid, research has been performed on the removal of fine particles from the gas phase, e.g. in the case of industrial flue gases. When a sound source is applied to a dust-laden flue gas, the finer particles will experience the so-called *orthokinetic collisions*, which is the increased probability of collisions with coarser particles because the finer particles are selectively put into vibration by the ultrasonic pressure waves because of their lower inertia [334]. The result is an increased agglomeration, which improves the efficiency of particle matter removal technologies. It has been shown that a 50% reduction in PM2.5 (i.e. particle matter smaller than 2.5 µm) could be achieved by applying high power ultrasound (160 dB at 20 kHz) before a baghouse filter in an air pollution control process [335].

Finally, a last application is in the field of *biological processes* [287, 336–338]. The use of ultrasound has been reported to increase enzymatic activity and microorganism biomass production and to improve disintegration and dewatering of sludge. In bioleaching processes, metal extraction yields have been increased by a factor of 4 and metal extraction rates by 25% [339–342]. Also, the use of low-intensity ultrasound in biological wastewater appears promising [343].

Design and scale-up of *sonochemical reactors* is an ongoing point of attention. On the one hand, work is done on the design of the electromechanical transducers, i.e. the equipment to convert electricity into the mechanical sound wave. These transducers can be either of the piezoelectric type or of the magnetostrictive type [248]. *Piezoelectric transducers* are constructed using a piezoelectric material, such as quartz, which expands and contracts in an alternating *electric* field, thus producing sound waves from the electric signal. *Magnetostrictive transducers* are constructed from materials, such as nickel alloys, which expand and contract in an alternating *magnetic* field. Each transducer has its own advantages and disadvantages, but in general, piezoelectric transducers are the more efficient in energy conversion and are used with small-volume processes, while the more robust magnetostrictive transducers are less efficient and applied when large volumes and/or long, continuous reaction times are required. On the other hand, ultrasound reactor design is following the general transition of batch to continuous processing. Large-scale batch operation is replaced by small-scale flow operation, often via intermediate solutions such as loop reactors (Figure 4.48).

These two sides of equipment design (transducer and reactor) come together when developing an integrated reactor [245, 248, 344]. The transducer can be integrated with the reactor in different ways, the most common being the bath concept, the probe, and the planar transducer type. The bath concept is applied by the often used ultrasound baths in chemical laboratories for cleaning of glassware. Ultrasound baths have transducers attached to the bottom of the bath and the reaction vessel is submerged in the coupling liquid. This type of operation is rarely used for scale-up because of disadvantages such as the irradiation being dependent on the location of the vessel in the bath and the ultrasound power ending up in the reaction vessel being low. Temperature control of the coupling fluid, however, allows cooling and heating of the reaction vessel. The probe

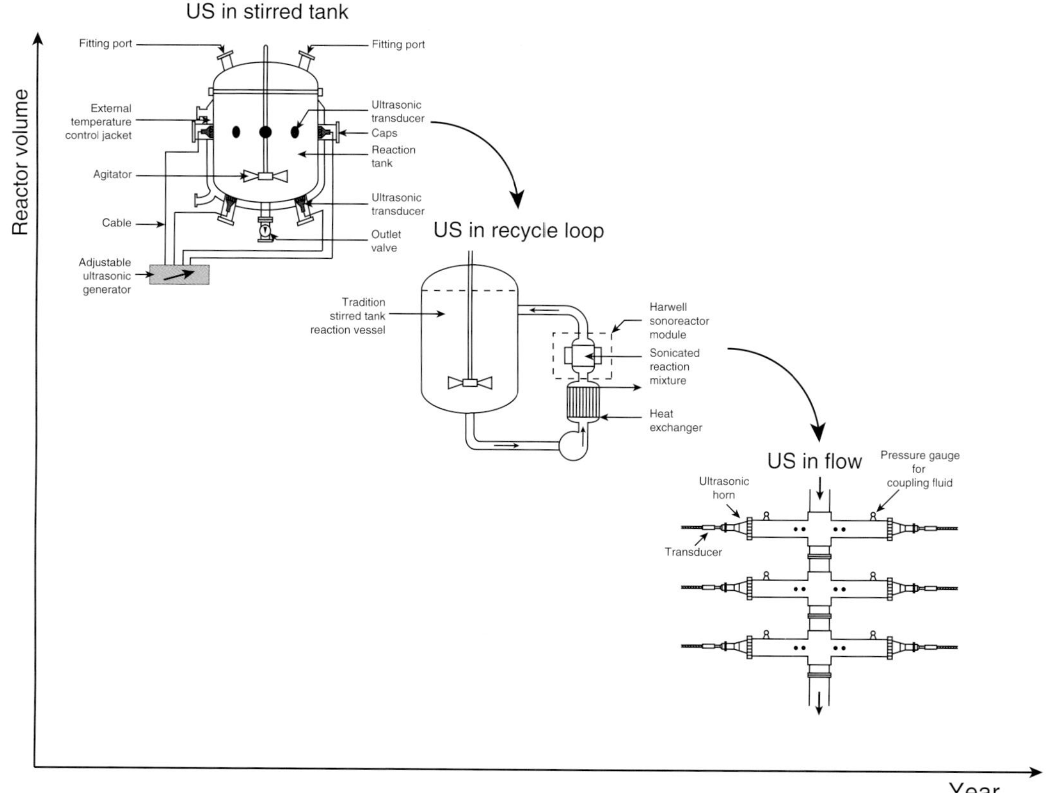

Figure 4.48 Research evolution of large-scale batch operation to small-scale flow systems in sonochemical process design. Source: Reactor schemes taken from Thompson and Doraiswamy 1999 [248]. Reproduced with permission of American Chemical Society.

system, where the ultrasound horn is inserted in the reaction vessel, is much more used, in particular in chemistry research. The efficiency of power transfer from the transducer to the reaction liquid is much higher compared to the bath system. Downsides, however, include the erosion of the probe tip and subsequent contamination of the reaction liquid with eroded and dissolved metals [345]. The third and last type is the planar transducer. Although slightly less efficient than the horn in transferring ultrasound into the reaction liquid, this transducer is glued or clamped on the exterior wall of the reaction vessel. No problems with contamination occur. In addition, these transducers can be manufactured and optimized toward a wide variety of frequencies, in contrast to horns where the optimal frequency is related to the length of the probe. Gogate and coworkers have investigated the use of transducers with multiple frequencies in flow cells and show that higher energy efficiencies can be attained because of a more uniform energy dissipation over a wider area [346, 347]. Further developments in design are oriented at matching the geometry of the transducer with the reactor with, e.g. cylindrical or cone-shaped transducers. The HIFU (high-intensity focused ultrasound) design, originating from therapeutical applications, is an example that is under investigation by the group of Hihn [348, 349]. Another approach is to couple transducers with microreactors with either dedicated piezoelectric elements (e.g. [350, 351]) or off-the-shelf transducers [352, 353] (Figure 4.49). The combination of ultrasound actuation and microreactors is attractive because conventional mechanical actuation of mixing is difficult in microstructured geometries. In recent years, the field has been growing tremendously (e.g. [268, 354, 355]).

Despite all the encouraging results described earlier, use of ultrasound in chemical industry is mainly limited to the fields of cleaning and decontamination [247]. One of the main barriers to industrial application of sonochemical processes is the control and scale-up of ultrasound concepts into operable processes. Therefore, a better understanding is required of the relation between a cavitation collapse and chemical reactivity, as well as a better understanding and

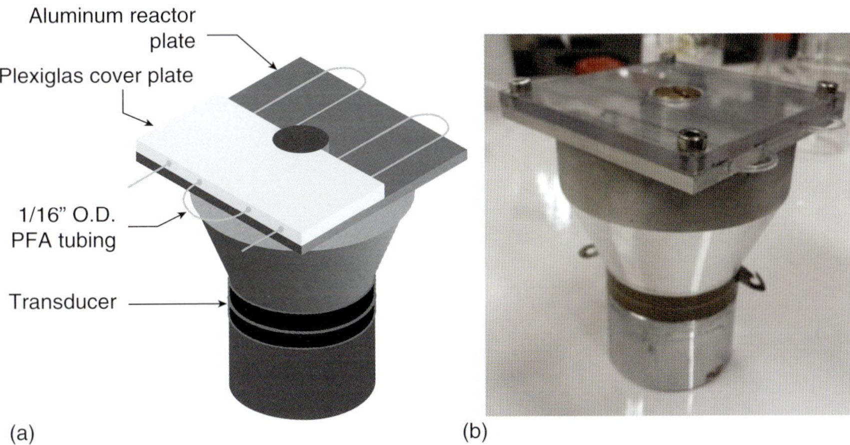

Figure 4.49 An ultrasound-actuated microreactor design with off-the-shelf available transducers. Source: John et al. 2016 [353]. Reproduced with permission of Elsevier.

reproducibility of the influence of various design and operational parameters on the cavitation process. Also, reliable mathematical models and scale-up procedures need to be developed [165, 246, 247].

4.6 Flow Fields

The kinetic energy of fluid flow can be advantageously utilized to intensify mixing, phase dispersion, and mass and heat transfer. Static mixers discussed in the chapter on STRUCTURE present an example of process intensification technology making use of the flow energy. More of such technologies and corresponding devices are presented below.

4.6.1 Hydrodynamic Cavitation

Acoustic fields, about which you could read in Section 4.5, are not the only method to generate cavitation in a liquid. Next to acoustic cavitation, we also know the optic cavitation (generated by high intensity photons, e.g. a laser beam), particle cavitation (produced by a beam of elementary particles, e.g. protons), and *hydrodynamic cavitation* [356]. That latter type of cavitation has the greatest importance in engineering and has been investigated for many decennia in connection to the erosion of ship propellers and rudders. In the early 1990s, researchers started studying hydrodynamic cavitation as another means for intensification of chemical reactions, and there are several good review papers on that topic (e.g. [356, 357]). There are several types of devices that can be used to generate cavitation in a flowing liquid. Contrary to the ultrasonic devices, they are pretty simple and easy to maintain. Three types of cavitation reactors most frequently studied in the literature include liquid whistle reactors, constriction-based reactors, and rotating reactors.

A *liquid whistle reactor*, shown in Figure 4.50, is based on a metal blade placed against the liquid stream. The blade vibrates as the liquid moves along, and the frequency of those vibrations is adjusted so that cavitation occurs. Liquid whistles can be used, among other things, for homogenization or production of fine emulsions [358]. They are simple and flexible. However, they suffer from low vibrational power, hence low intensity of cavitation. Also, blade erosion presents a problem in systems involving solid particles.

The presence of solid particles does not pose any specific problem in *constriction-based cavitation reactors* (Figure 4.51). These devices use the principle well known in the hydraulics of pipelines. Here, the pressure drop in the orifice or a venturing nozzle creates pressure lower than the vapor pressure of the fluid, as a result of which cavitation sites are created. The dimensionless number characterizing this process is called *cavitation number*, and is defined as

$$C_N = \frac{P_2 - P_v}{\frac{\rho v^2}{2}} \tag{4.22}$$

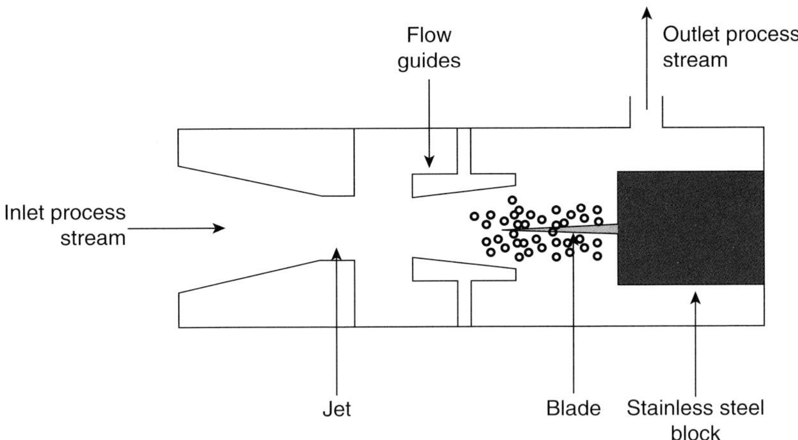

Figure 4.50 Liquid whistle reactor. Source: Thompson and Doraiswamy 1999 [248]. Reproduced with permission of American Chemical Society.

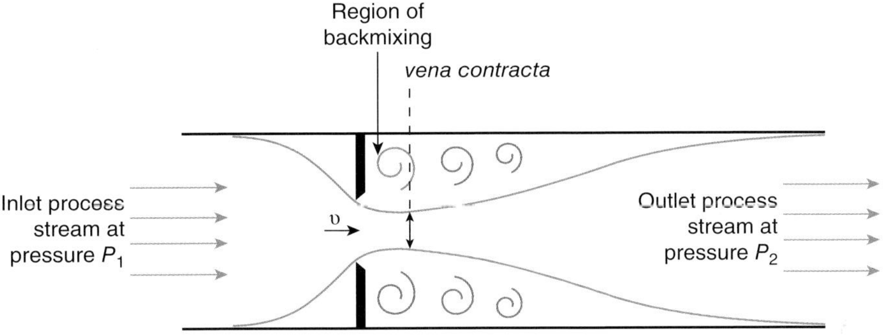

Figure 4.51 Constriction-based cavitation reactor. Source: Thompson and Doraiswamy 1999 [248]. Reproduced with permission of American Chemical Society.

where P_2 is the fluid pressure after orifice, P_v is the vapor pressure, ρ is the density, and v is the velocity of the fluid in the constriction. It has been shown that optimum cavitation conditions occur at C_N values slightly below one [356].

The constriction-based cavitation devices have been recently studied, among other things, in the production of water–oil emulsions [359], biodiesel synthesis [360], and pretreatment of lignocellulosic biomass [361]. Also, water treatment-related processes were investigated [362, 363].

Rotating cavitation reactors are rotor–stator devices (Figure 4.52), in which liquid flowing at high speeds between the rotor and the stator enters the indentations on the rotor surface. When leaving the indentation, a low pressure/vacuum is created, which results in cavitation. Processes investigated in the recent years in rotating cavitation reactors include wastewater treatment [366] and biomass pretreatment in relation to paper or biogas production [364, 365].

Comparison of hydrodynamic and acoustic cavitation reactors is usually done using a general parameter of *cavitational yield* defined as the ratio between the mass of the product and the amount of energy supplied to the unit [357]. Data

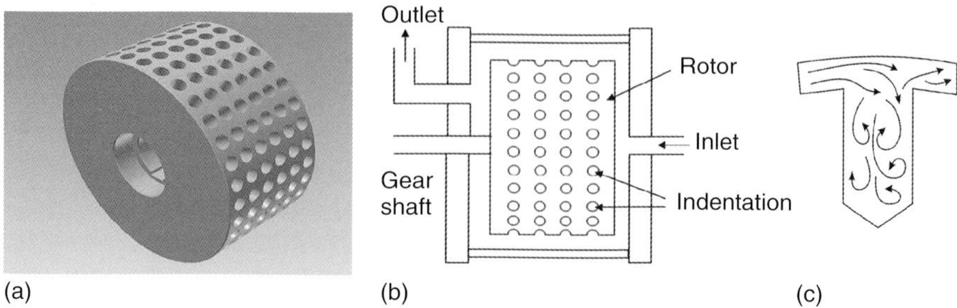

Figure 4.52 Rotating cavitation reactor (a, b) and flow pattern in an indentation (c). Source: (a) Patil et al. 2016 [364]. Reproduced with permission of Elsevier; (b, c) Badve et al. 2014 [365]. Reproduced with permission of Elsevier.

provided in the literature unanimously indicate that cavitational yields in hydrodynamic cavitation reactors are much (usually an order of magnitude!) higher than those in the acoustic cavitation reactors [357, 367].

4.6.2 Ejector-based Liquid Jet Reactors

In *ejector-based liquid jet reactors,* high-speed liquid flow through a nozzle is used to create a fine gas–liquid dispersion, hence high mass transfer rates. In those systems, the liquid is supplied through a nozzle and sucks the gas into the reactor chamber producing the so-called "mixing shock." The fundamentals of mixing shocks and liquid jets have been well described by Witte [368] and Cunningham and Dopkin [369]. A scheme of a typical ejector system is shown in Figure 4.53. The liquid passes through a Venturi-type nozzle (sometimes preceded by a swirl

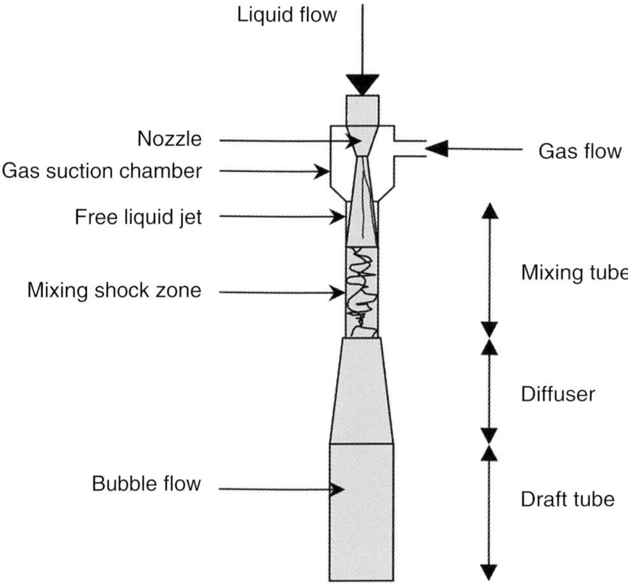

Figure 4.53 Scheme of a typical ejector system. Source: Tinge and Rodriguez Casado 2002 [370]. Reproduced with permission of Elsevier.

device to stabilize the liquid flow), which creates a high-velocity liquid jet. The jet creates the suction of the gas into the ejector. In the so-called mixing tube, a rapid dissipation of liquid kinetic energy takes place that results in the creation of a mixing shock zone where high turbulence generates a fine dispersion of the gas bubbles and a large specific interfacial area for the mass transfer. Gas bubbles created in that section are usually 30–70 μm in diameter while the volumetric mass transfer coefficients $k_L a$ can reach as high as 10/s [371]. Below the mixing tube, the turbulence fades away and common bubble flow occurs.

Reactor systems based on the above principle have been investigated since mid-1960s and both upflow and downflow configurations have been studied [372–380]. In the *upflow configurations* (Figure 4.54a), the ejector usually plays the role of a gas sparger in a bubble column or in a gas lift system with a riser-downcomer loop. In the *downflow configurations* (Figure 4.54b), the ejector is partially submerged in the liquid and the gas is sucked into the reaction chamber directly from the head space above the liquid surface. Both gas–liquid mass transfer coefficients and specific interfacial areas are higher in downflow configurations [382].

The history of industrial development of ejector-based liquid jet reactors is longer than the academic research in the area and goes back to mid-1950s when the Swiss company BUSS ChemTech AG developed the so-called BUSS Loop Reactor (BLR). The reactor was a downflow unit similar to the one shown in Figure 4.54. Its further modification is the Advanced BUSS Loop® Reactor (ABLR). The difference between this model, shown in Figure 4.55, and the previous one is that although in BLR, the gas was separated from the liquid in

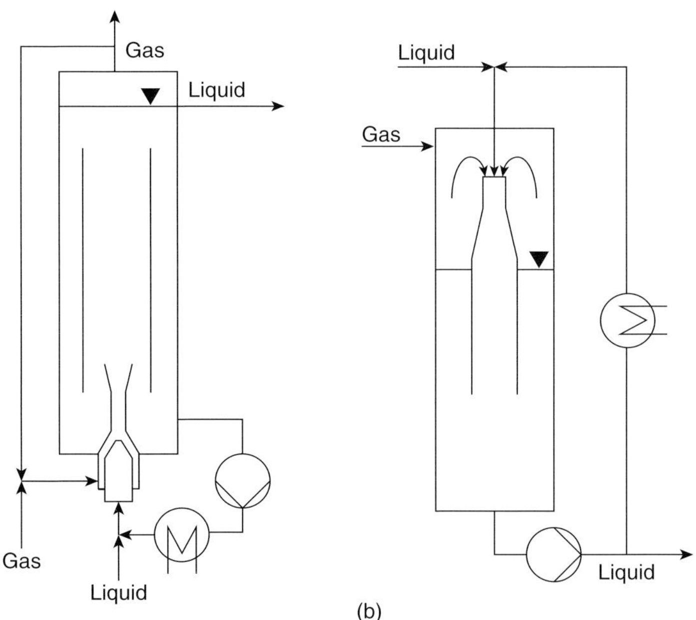

Figure 4.54 Ejector-based liquid jet reactors: (a) upflow configuration, (b) downflow configuration. Source: Zehner and Kraume 2000 [381]. Reproduced with permission of John Wiley and Sons.

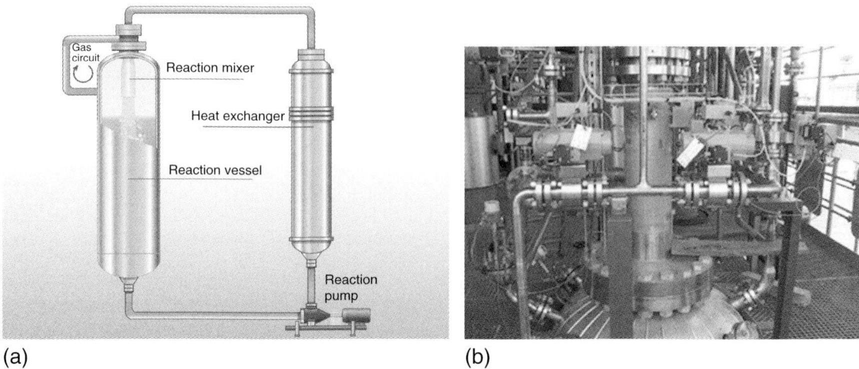

Figure 4.55 Scheme and the top view of an advanced BUSS Loop reactor. Source: Courtesy of Buss ChemTech AG, Switzerland, www.buss-ct.com.

the reaction vessel and brought back to the head space, the ABLR is equipped with a special pump able to handle gas loads of up to 30%, which ultimately results in a significantly improved performance of the system.

As might be expected, ABLR delivers very high gas–liquid volumetric mass transfer coefficients at relatively low power input, especially when compared to different types of stirred-tank systems – see Figure 4.56. Because of this fact, the

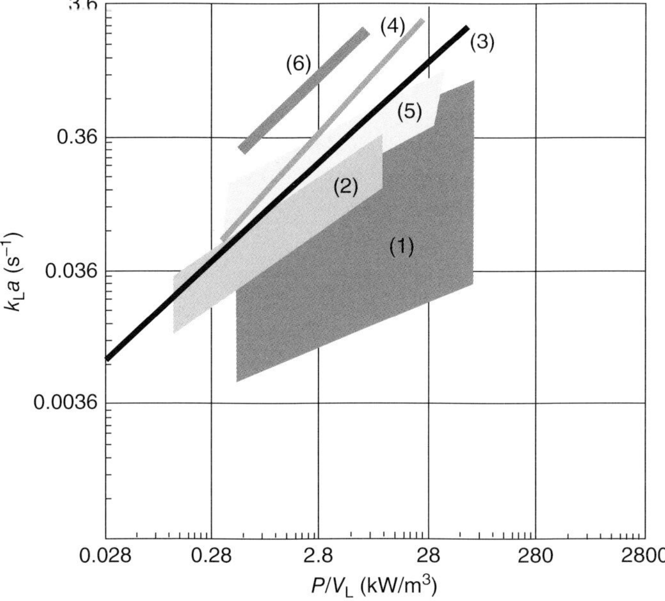

Figure 4.56 Volumetric mass transfer coefficients versus specific power input in various types of gas–liquid reactors: (1) stirred reactor with external cooling, (2) conventional vessel with six-blade turbine stirrer, (3) self-aspirating stirrer in a vessel, (4) self-aspirating stirrer in combination with EKATO phasejet in autoclave, (5) Buss Loop reactor (BLR); (6) advanced BUSS Loop reactor (ABLR). Source: Data from Buss ChemTech AG, Switzerland, www.buss-ct.com.

Table 4.10 Performance comparison of the advanced BUSS Loop reactor (BLR) and a conventional stirred-tank system (STR) in different types of reactions.

Reaction	Reaction time (min)		Catalyst load (wt%)		Yield (%)	
	STR	ABLR	STR	ABLR	STR	ABLR
Hydrogenation of nitrobenzene	80	45	0.09	0.03	95.7	97.0
Hydrogenation of aldehyde	360	100	6	0.6	90	95
Oxidation	600	42	0.6	0.4	—	—

Source: Data from Buss ChemTech AG, Switzerland, www.buss-ct.com.

device has a wide applicability range, particularly in fast gas–liquid reactions that are mass transfer limited in the conventional equipment, such as hydrogenations, oxidations, alkoxylations, or phosgenations. Very high mass transfer rates achievable in BUSS reactors result in significant shortening of the reaction times, increased yields/selectivities, and reduced catalyst loads. A comparison of the performance of the Advanced BUSS Loop Reactor and a conventional stirred-tank system in different types of reactions is presented in Table 4.10 based on the data published by BUSS ChemTech.

4.6.3 Supersonic Flow

Using the energy of the *supersonic shockwave* presents another promising alternative method for intensification of the phase contacting and transport processes. First publications in this field appeared in mid-1990s and, interestingly, the vast majority of them come from the industry. Mattick et al. [383] investigated a continuous-flow chemical reactor using shock waves for pyrolysis of hydrocarbons in the commercial manufacturing of olefins (Figure 4.57). In the reactor, heat is added to a carrier gas, which is cooled to sub-pyrolysis temperature by expansion to supersonic speed and is mixed with a supersonic flow of feedstock. This produced high pyrolysis temperatures at very short reaction times resulting in higher olefin yield than in conventional systems. The ethylene yield in the supersonic unit increased by 20% as compared to conventional technology, while the energy consumption dropped by 15%.

Company Praxair Technology Inc. has developed a *supersonic gas–liquid reactor* for carrying out fast processes [384]. The device is shown in Figure 4.58. The energy of the shockwave is used here to disperse gas into tiny microbubbles and by that to create an enhanced interfacial area for mass transfer. The oxygen transfer rates in Praxair reactor are up to c. 10 times higher than in a tee-mixer with gas–liquid mass transfer coefficients exceeding 2.0/s.

Messer Griesheim GmbH has patented and commercialized a *supersonic nozzle for fluidized-bed applications* [385, 386]. The nozzle was applied on the industrial scale in a fluidized-bed reactor for iron sulfate decomposition at Bayer AG. Supersonic injection of oxygen has increased the capacity of the reactor by 124% [387]. The same technology has been implemented in the sludge combustion reactors increasing the throughput by approximately 40% [388].

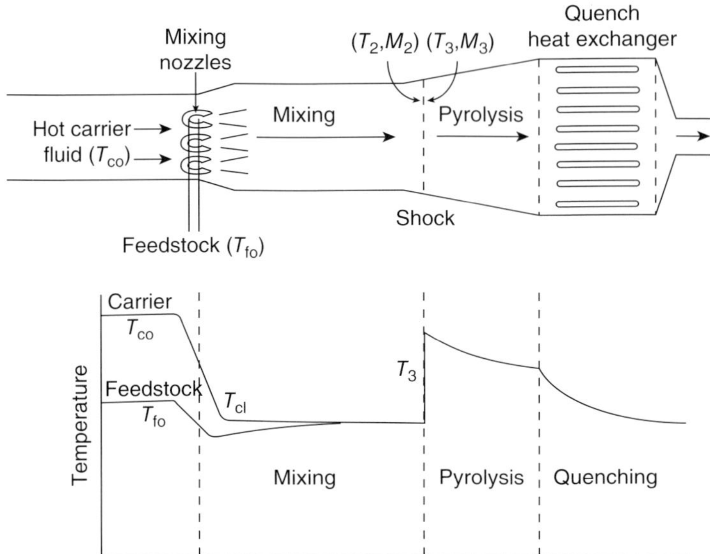

Figure 4.57 Shockwave reactor for pyrolysis of hydrocarbons and gas temperature profile. Source: Mattick et al. 1995 [383]. Reproduced with permission of Springer.

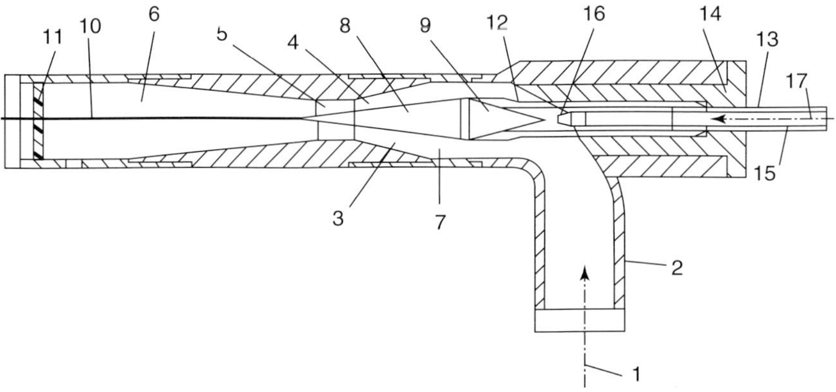

Figure 4.58 Praxair's supersonic gas–liquid reactor. Source: USA Patent #: US 5061406 A.

In the area of bioprocessing, a *transonic oxygen sparger* has been developed at DSM and applied in a large-scale yeast fermentation system (Figure 4.59). That supersonic injection system has doubled the yeast productivity of the fermenter [389]. More recently, supersonic flow reactors have also been patented in relation to the cracking of heavy hydrocarbons [390] and methane pyrolysis [391].

Finally, it needs to be noted that supersonic flow is not exclusively applied for carrying out chemical or biochemical reactions. It can also be used for intensification of separation operations, such as separation of water and heavy hydrocarbons from natural gas. The device developed for this purpose, Twister®

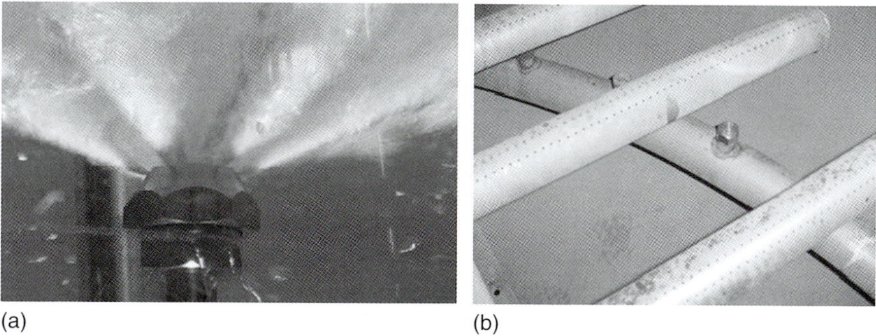

(a)   (b)

Figure 4.59 DSM's transonic sparger for oxygen injection in large-scale fermenter. Source: Courtesy of DSM N.V, The Netherlands, www.dsm.com.

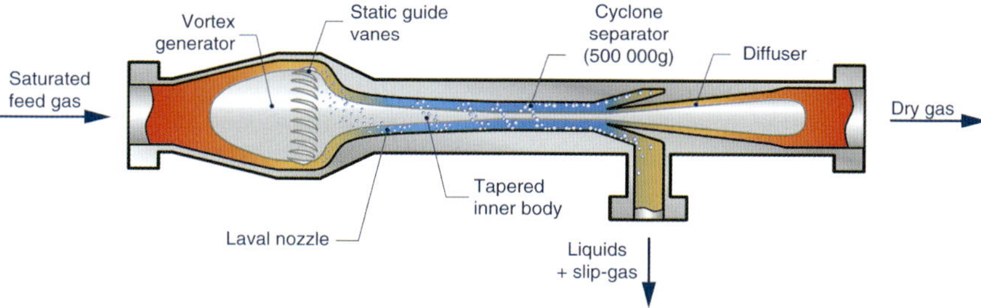

Figure 4.60 Scheme of Twister Supersonic Separator. Source: Courtesy of Twister BV, The Netherlands, www.twisterbv.com.

Supersonic Separator, has thermodynamics similar to a turboexpander, combining expansion, cyclonic gas/liquid separation, and recompression. According to Twister BV, the Dutch company that has developed and commercializes the supersonic separator, Twister operates via the sequence of the following steps (see Figure 4.60):

- Multiple inlet guide vanes generate a high vorticity, concentric swirl (up to $500\,000g$).
- A Laval nozzle is used to expand the saturated feed gas to supersonic velocity, which results in a low temperature and pressure.
- This leads to the formation of a mist of water and hydrocarbon condensation droplets.
- The high vorticity swirl centrifuges the droplets to the wall.
- The liquids are split from the gas using a cyclonic separator.
- The separated streams are slowed down in separate diffusers, typically recovering 70–75% of the initial pressure.
- The liquid stream contains slip-gas, which will be removed in a compact liquid degassing vessel and recombined with the dry gas stream.

As claimed by Twister BV, significantly lower life cycle cost, simplicity, reliability, small footprint, and lack of fouling/poisoning of the system present the most important advantages of their separator over conventional silica gel- and glycol-based systems. Its compactness and low maintenance needs make it feasible to apply the separator as a subsea processing tool.

4.6.4 Impinging-stream Reactors

The concept of *impinging-stream devices* is old and dates back to the end of the nineteenth century [392], but the more extended research in this area has started about 100 years later, and meanwhile two books on that topic have been published [393, 394]. At the base of this concept lies the energy of collision between two single-phase or multiphase streams moving along the same axis from opposite directions (Figure 4.61a). Multistream impinging devices with streams colliding perpendicularly or circumferentially are also known. The impinging event produces a zone of high shear and turbulence, in which excellent conditions for mixing, mass, and heat transfer are created.

A typical scheme of an impinging-stream reactor for gas–liquid systems is presented in Figure 4.61b. Here, the gas and liquid feeds are mixed with each other and accelerated toward the impingement zone by means of two nozzles. A part of the liquid circulates in the downcomer-riser system. Performance

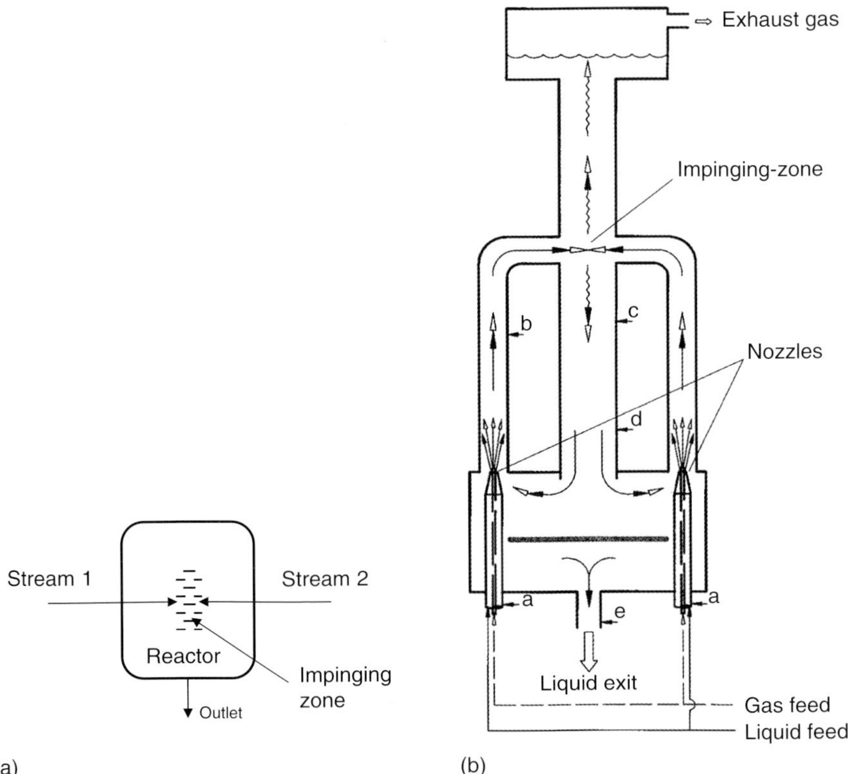

Figure 4.61 The impinging-stream concept (a) and gas–liquid impinging-stream reactor (b). Source: Sprehe et al. 1998 [395]. Reproduced with permission of John Wiley and Sons.

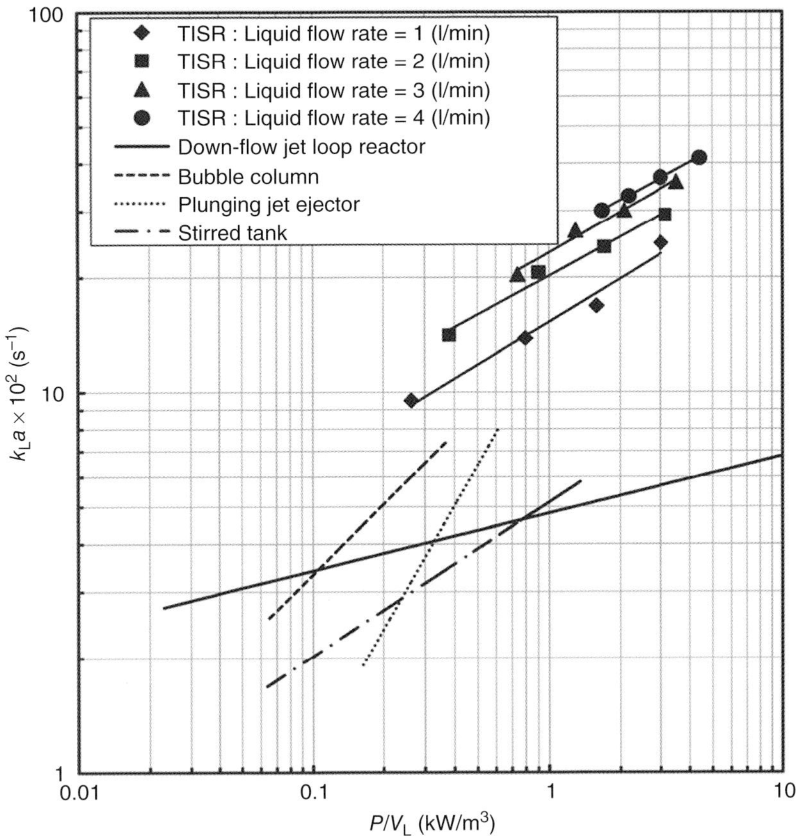

Figure 4.62 Volumetric mass transfer coefficients versus power input in various types of gas–liquid reactors. Source: Dehkordi and Savari 2011 [398]. Reproduced with permission of American Chemical Society.

of the gas–liquid impinging-stream reactors has been extensively studied in the literature [395–398], and indeed, very high mass transfer rates have been reported. Figure 4.62 presents a comparison of the volumetric mass transfer coefficients as a function of power input in various types of gas–liquid reactors [398]. As one can see, the two impinging-stream reactor clearly surpasses all other types, including the ejector reactor (see Section 4.6.2).

Impinging-stream reactors are used on the commercial scale for a variety of processes ranging from fine chemicals production, crystallization of nanoparticles, and oxidations in waste treatment to coal gasification. The reported effects include up to 85% reduction of the reactor volume and up to 85% reduction of energy. Figure 4.63 presents a two-step E-Gas™ technology for coal and petcoke conversion. The technology, currently owned by CB&I, utilizes the impinging-stream principle in the first stage, in which coal slurry is mixed with 95% pure oxygen and undergoes a highly exothermic oxidation process delivering synthesis gas. The second stage is used for fine-tuning of the syngas

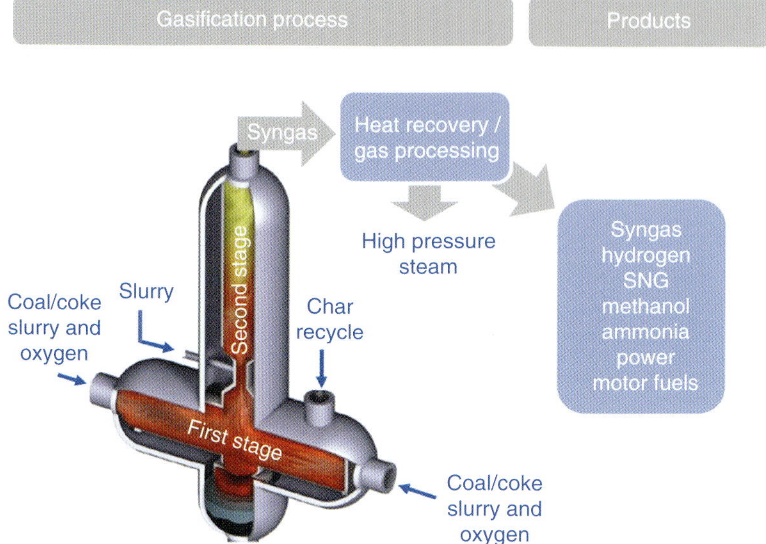

Figure 4.63 E-GasTM coal gasifier utilizing impinging-stream concept. Source: Courtesy of CB&I, www.cbi.com.

composition by bringing it in contact with more coal slurry that result in the generation of some methane in the product gas.

Another group of impinging-stream devices, the so-called "nonrotating vertical circulative flow impinging-stream reactor" (Figure 4.64a) and "nonrotating flow impinging-stream crystallizer" (Figure 4.64b) have been developed and commercially applied in China. In both designs, two drawing tubes equipped with propellers force the circulation of the liquid into the impingement zone. Both the reactor and the crystallizer have been successfully applied on the pilot and on the commercial scale for the production of various noncrystalline and crystalline products, including nanoparticles [399].

Other (potential) applications of the impinging-stream technology discussed in the literature include, among other things, coal liquefaction [400], flue gas desulfurization [401], solid–liquid enzyme reactions [402], methanol synthesis [403], photooxidation [404], extraction [405], and drying [406]. The concept has also been applied in microreactors, in the production of nanoparticles [407, 408].

4.7 High-Gravity and High-Shear Fields

In this section, we will discuss process intensification by means of high-gravity (centrifugal) and high-shear fields created in the rotating equipment. The use of rotating equipment is old and goes back to the beginning of industry. It included such devices as water wheels, pumps, compressors, or centrifuges for solid–liquid separations. However, the use of rotating equipment for intensification of mixing and heat and mass transfer in carrying out reactions or separations presents

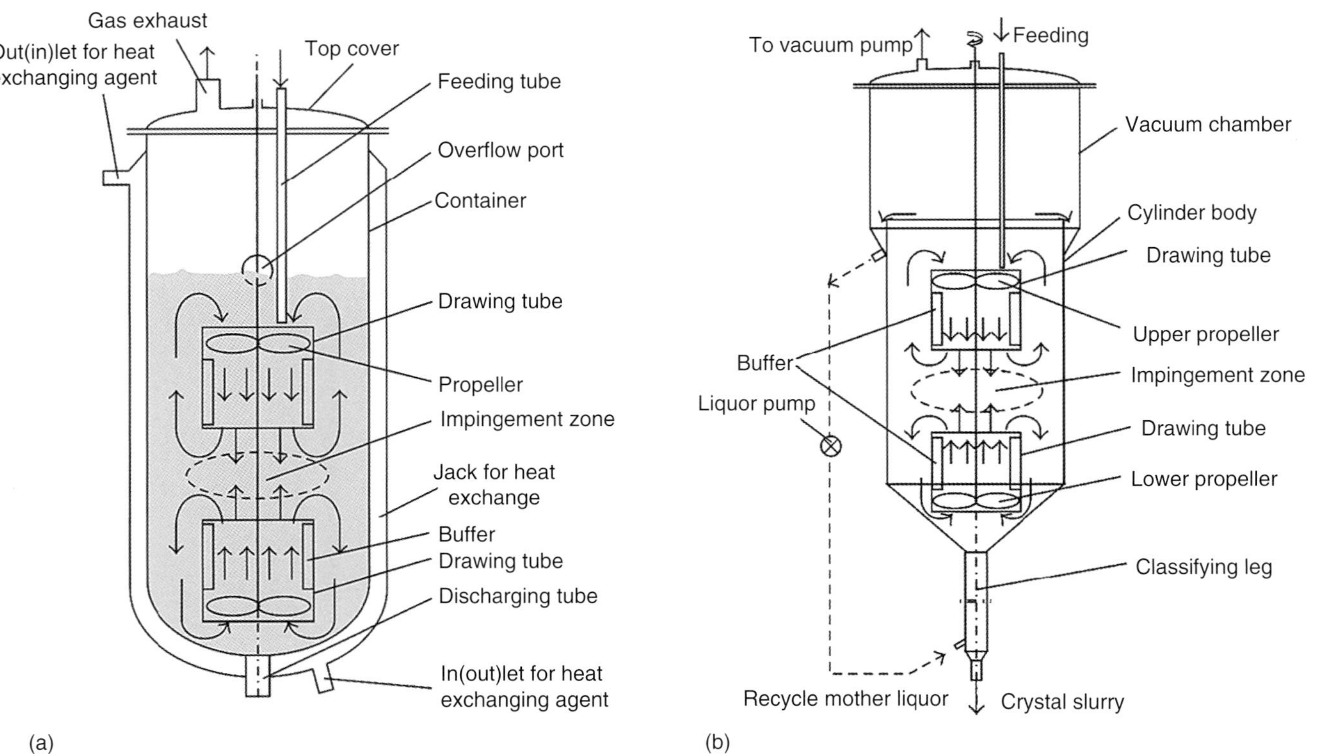

Figure 4.64 The nonrotating vertical circulative flow impinging-stream reactor (a) and nonrotating flow impinging-stream crystallizer (b). Source: Wu et al. 2010 [399]. Reproduced with permission of Hindawi Publishers.

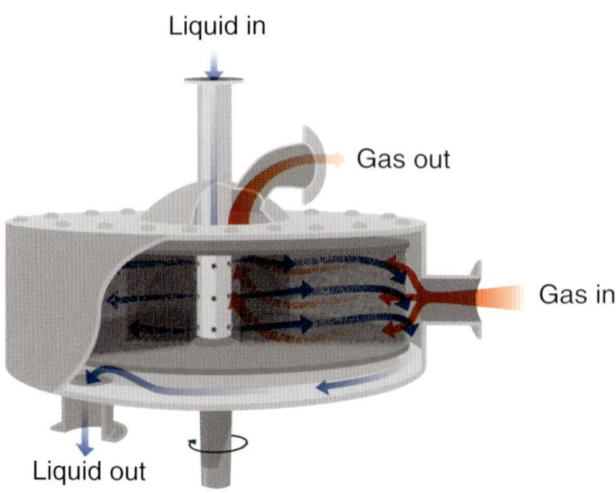

Figure 4.65 Rotating packed bed. Source: Górak and Stankiewicz 2011 [415]. Reproduced with permission of Annual Reviews.

a much younger development, an important milestone of which was the application of the Podbielniak extractor to the recovery of penicillin [409]. We will focus here on the four most important categories of high-gravity and high-shear equipment. We will explain the mechanisms of its operation, examine its advantages and limitations, and discuss its applicability areas.

4.7.1 Rotating Packed Beds

The research on *rotating packed beds* (RPBs) for process intensification was initiated by the ICI New Science Group who carried out studies on the application of high-gravity fields (so-called "HiGee") in distillation processes [410, 411]. Since then, the technology has been investigated at numerous groups and has been subject to several excellent reviews [412–414].

RPB in its basic form presents a pretty simple device (Figure 4.65) used for an intensive two-phase (usually gas–liquid, and sometimes liquid–liquid) contacting. It consists of a stationary casing, in which a structured packing rotates, usually with speeds between 500 and 2000 rpm. The packing can be made of foam, mesh, wire, etc., and has the specific surface area usually between 150 and 5000/m. The liquid is fed centrally, flows outward driven by the centrifugal force (typically 100–500 g), and comes in contact with the gas moving in the opposite direction. Usually, the gas forms the continuous phase, while liquid moves as a film flowing along the packing surface and as microdroplets filling in the void spaces in the packing. That second mechanism is important as it creates very high interfacial surface areas, which result in the intensified mass transfer. The residence times in RPBs are short, usually in the range of a fraction of second.

Indeed, excellent mass transfer presents the most important advantage of RPBs. Already in 1981, Ramshaw and Mallison [410] reported a 27- to 44-fold increase of the liquid-side mass transfer coefficient and four- to ninefold increase of the gas-side mass transfer coefficient, with respect to the stationary bed of Intalox saddles. Chong Zheng and coworkers found the height of the mass transfer unit for the liquid side to be as low as 2.5–4 cm [416]. Lin et al. reported HETP (height equivalent of theoretical plate) values of 3–9 cm, compared to 30–40 cm for the conventional structured packings [417]. Rao et al. point out

Table 4.11 Comparison between RPB-based stripping and vacuum desorption technologies for water deaeration in off-shore environment.

	Vacuum tower (one set 10 000 t/d)	HiGrav Deaerator (two sets 6000 t/d each)
Platform area (m^2)	30	2×10
Height (m)	14	3
Weight at operation (t)	130	2×10.5
Residual oxygen(ppm)	1 (summer) 2–3 (winter)	<0.05 <0.05
Investment (normalized)	1	0.8
Power consumed (kW)	155	2×160

Source: Zheng et al. 1997 [418]. Used with permission of BHR Group.

that up to c. 200 times increase of the $k_L a$ in the RPBs with respect to the conventional packed columns is possible [413]. This brings us to the conclusion that the use of RPBs can be particularly beneficial for fast and very fast, mass transfer-limited processes, especially those that do not require intensive heat exchange, as the latter is difficult in rotating systems.

Excellent mass transfer properties of the RPBs have led to a number of industrial applications. Initially, they were applied to water deaeration in oil fields in China [418]. A comparison between the conventional technology for water deaeration (vacuum desorption tower) and the HiGee technology designed for the off-shore application is presented in Table 4.11. As one can see, the RPB-based option delivers much smaller and lighter equipment (very important for on-platform operations), as well as 1 order of magnitude better removal efficiency of oxygen. On the other hand, the power consumption in the operation of a vacuum tower is about half of that in the equivalent set of two HiGee deaerators.

In 1999, Dow Chemical applied RPBs in their hypochlorous acid process [412]. The conventional process was based on two absorption-stripping columns and operated with product yields of c. 80%. The yields were limited by a secondary reaction, which could be avoided by a very fast product removal from the reaction environment. Application of reactive stripping in three RPB units has led a substantial yield increase (to c. 95%), a 50% decrease in stripping gas consumption, and a 40-fold (!) decrease in equipment size. The RPB-based plant in the foreground and the conventional plant in the background of Figure 4.66a have the same production capacity.

More recent applications of RPBs have been seen in the production of nanoparticles [419, 420]. According to Zhao et al. [414] in 2010, five enterprises in China were manufacturing $CaCO_3$ nanoparticles with an overall production capacity reaching 36 000 ton/year. In Figure 4.66b, one of such production lines with the capacity of 10 000 ton/year is shown.

A newer development in the field of HiGee, slightly similar to RPBs, is the *rotating zigzag bed* (RZB). The device, shown in Figure 4.67, consists of a stationary disc and a rotating disc, to which concentric baffles are fixed. In the RZB, gas is the continuous phase and flows a zigzag path, whereas liquid is the dispersed phase

(a) (b)

Figure 4.66 (a) RPB-based hypochlorous acid of Dow Chemical with the old absorption tower in the background. Source: Courtesy of Dow Chemicals, www.dow.com. (b) Chinese production line for $CaCO_3$ nanoparticles. Source: Courtesy of Research Center of the Ministry of Education for High Gravity Engineering & Technology, Beijing.

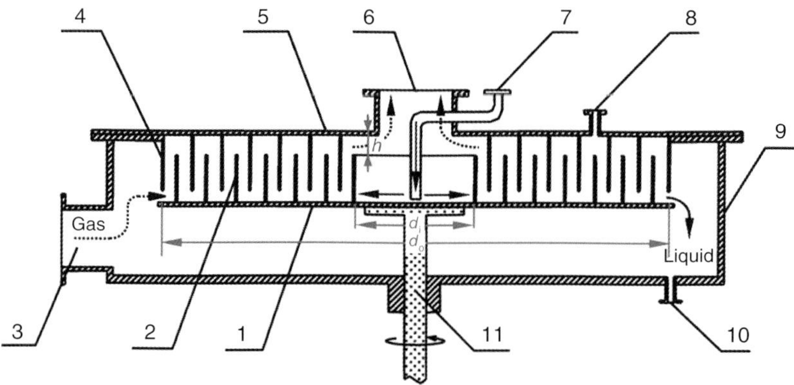

Figure 4.67 Rotating zigzag bed device: 1 – rotational disc, 2 – rotational baffle, 3 – gas inlet, 4 – stationary baffle, 5 – stationary disc, 6 – gas outlet, 7 – liquid inlet, 8 – intermediate feed, 9 – liquid outlet, 10 – rotor casing, and 11 – rotating shaft. Source: Wang et al. 2011 [421]. Reproduced with permission of Elsevier.

in fine droplets. The droplets are accumulated by the stationary baffles and form a liquid film falling on the rotational disk and flowing toward the next baffle, where it becomes dispersed again. This impingement and spray of liquid results in fast renewal and high surface areas available for mass transfer. The RZB can also be manufactured as a multirotor device. The advantages of RZB in comparison with RPB include lack of necessity for a liquid distributor, possibility for the intermediate feed, larger liquid holdup at constant gas velocity and longer, and adjustable residence times (up to c. 100 seconds). On the other hand, RZBs produce larger pressure drop and exhibit higher power consumption than RPBs.

Table 4.12 Dimensions of rotating zigzag beds and packed columns applied in selected distillation processes.

	Methanol–water		Ethanol–water	
	RZB	Packed column	RZB	Packed column
Diameter (m)	0.83	0.60	0.80	0.40
Total height (m)	0.80	11.0	0.55	9.0
Volume (m^3)	0.43	3.11	0.28	1.13

Source: Wang et al. 2011 [421]. Reproduced with permission of Elsevier.

Relatively long residence times that can be realized in rotating zigzag beds as well as the ease of having multiple rotors within one housing make these devices an attractive option for applications in continuous distillation. Wang et al. [421] reported in 2011 about 200 RZB units successfully commercialized in various distillation, absorption, and stripping processes. A comparison of the dimensions of RZBs and packed columns used for methanol–water and ethanol–water distillation is provided in Table 4.12. It can be seen that 1–2 orders of magnitude reduction in height and volume, respectively, could be attained by using the RZB units.

Finally, although not known as "rotating bed," the earlier mentioned *Podbielniak liquid–liquid extractor* can also be classified within this category. The "bed" in this device is formed by concentric cylindrical bands that fit into grooves on the rotor at one end and the endplate at the other (Figure 4.68). The bands are perforated, which enables intensive contacting between the heavy phase flowing outward and the light phase moving in the opposite direction. The Podbielniak extractor is therefore in fact a sieve-plate extraction column being rotated around its top. Podbielniak et al. extractor was first commercialized shortly after World War II in the extraction of penicillin [409] and is nowadays offered by Baker Perkins (B&P Process Equipment). Other types of multistage centrifugal extractors including those of Alfa-Laval, Lurgi-Westfalia, and Robatel have been reviewed by Gebauer et al. [422].

4.7.2 Spinning Disc Reactors

The SDR (Figure 4.69a), originally developed by Ramshaw at the University of Newcastle [423], makes use of thin, highly sheared, wavy film flow of the liquid on the surface of the rotating disc. The liquid enters the disk in the center and moves outward driven by the centrifugal force. The gas (in gas–liquid applications) moves in counter-current with respect to the liquid. The flow is intrinsically unstable and an array of spiral ripples is formed (Figure 4.69b). This provides excellent conditions for heat exchange with a heat transfer fluid flowing in the space under the disc surface.

Indeed, highly intensive heat and mass transfer are the most important advantages of the spinning disc reactor. Aoune and Ramshaw [424] investigated the heat and mass transfer characteristics of liquid films on rotating discs and found local heat transfer coefficients ranging from c. 10 000 to 30 000 W/m^2 K and local mass transfer coefficients between c. 4E-04 and 1E-03/ms, much higher than predicted by the Higbie model. The enhancement of the local mass transfer coefficients appears to be associated with the passage of the ripple [425]. Because of the above and the short residence times on the discs (usually in the order of 1–5 seconds), the SDR is a very good solution for performing any intrinsically rapid, highly exothermic transformation in a liquid, even if it is viscous. This has been confirmed in a number of demonstration projects, such as a phase-transfer-catalyzed Darzen's reaction for preparing a drug intermediate at GlaxoSmithKline. The SDR allowed here for a 1000-fold reduction in reaction time, 100-fold reduction of inventory, and 93% reduction of impurity level [426]. Other possible applications of the SDR investigated included polymerizations and polycondensations. In both cases, considerable shortening of the processing times was the most important effect. Figure 4.70 shows the conversion versus time curve for free radical polymerization of styrene [423]. The "steps" on the batch reaction profile indicate the savings in the processing time with respect to the batch operation because of an almost instantaneous increase of the conversion in the SDR.

Also, heterogeneously catalyzed reactions have been studied on SDRs. For instance, Vicevic et al. [427] investigated the rearrangement reaction of α-pinene oxide to campholenic aldehyde on a Zn-triflate/SiO_2 catalyst. The SDR-based continuous process showed clear advantages over conventional batch technology – see Table 4.13.

More recently, spinning disc reactors were studied in the context of nanoparticle manufacturing [428–430]. It has been found, among other things, that the particle size distribution can be controlled in a wide range by changing the flow rate and/or the rotational speed of the disc (Figure 4.71). SDRs have also been studied in connection to photocatalysis. In one study, a photocatalytic spinning disc reactor exhibited 1 order of magnitude higher reaction rates than the reference annular reactor [431].

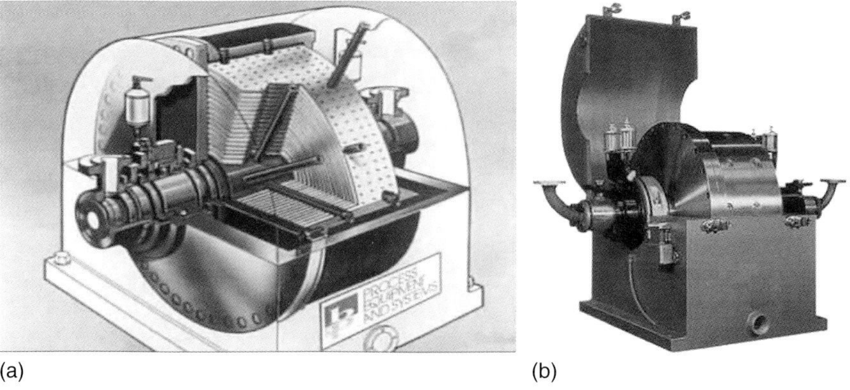

(a)   (b)

Figure 4.68 Podbielniak extractor. Source: Courtesy of B&P Littleford LLC, USA, www .bplittleford.com.

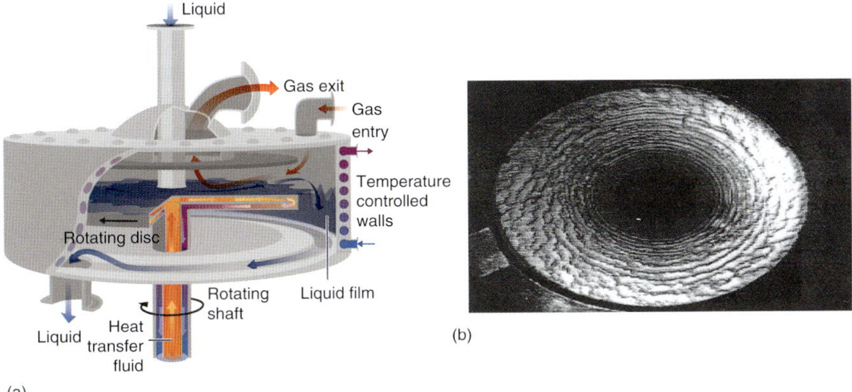

Figure 4.69 (a) Spinning disc reactor. Source: Górak and Stankiewicz 2011 [415]. Reproduced with permission of Annual Reviews. (b) Ripples formed on the disc surface. Source: Ramshaw 2004 [423]. Reproduced with permission of Taylor and Francis Group.

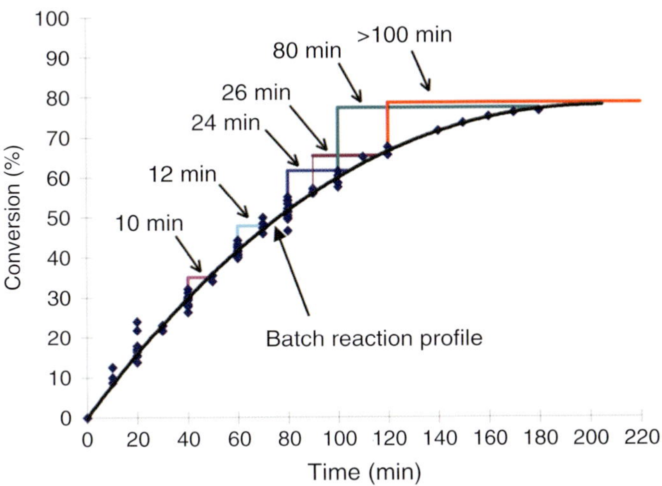

Figure 4.70 Time savings in free radical polymerization of styrene by introduction of the spinning disc reactors at various conversion levels. Source: Ramshaw 2004 [423]. Reproduced with permission of Taylor and Francis Group.

4.7.3 Rotor–Stator Devices

The HiGee devices discussed so far, whether rotating beds or spinning discs, were based on a structured rotor (packing, concentric baffles, and disc) rotating "freely" inside a housing. The rotor–stator devices make purposeful use of the highly intensive shearing between two solid surfaces, one rotating, and another stationary. This principle is used to enhance mixing, transfer processes, and/or phase dispersion and can be applied both in gas–liquid and liquid–liquid systems.

Rotor–stator spinning disc reactors (RSSDR) represent a new generation of the earlier described SDRs. Here, instead of a free turbulent liquid film flow on the rotating disc, a sheared flow between the rotating disc and the stator is executed. RSSDR can be applied in both gas–liquid and liquid–liquid systems. Figure 4.72a presents a single-stage RSSDR for gas–liquid contacting. Both phases are fed via the top of the reactor. First, the liquid flows as the film over the rotor with the gas-phase above it, similarly to what we see in classical SDRs. Then, at the rim of the rotor, gas bubbles are sheared off. From that point on, between the rotor and the stator (or rotors and stators, as the device can be multistage as well), the liquid flows as continuous phase with small gas bubbles dispersed in it. Figure 4.72b presents a multistage RSSDR for single-phase fluid processing, with the cooling liquid flowing on the stator side.

The rotor–stator spinning disc reactors exhibit superior properties as far as heat transfer and interfacial mass transfer are concerned. The reported values of heat and mass transfer coefficients are summarized in Table 4.14. A comparison between the gas–liquid mass transfer performance of an RSSDR and other types of contactors is shown in Figure 4.73.

Annular centrifugal contactors (ACC) present a group of rotor–stator-based devices that apply Taylor–Couette type flow to mix two immiscible fluids and then to separate them centrifugally. Initial design of the annular centrifugal contactor was done at the Argonne National Laboratory and has been employed in solvent extraction processes for metals valuable to the nuclear industry. Figure 4.74 presents a cross-sectional scheme of an annular centrifugal contactor of CINC Industries, Inc. In the highly sheared Taylor–Couette zone, intensive mixing and dispersion of heavy and light phase take place. Next, the mixture flows into the inner cylinder, where it is held stationary with respect to the cylinder by means of the vanes while the centrifugal force separates the phases. The ACC provides a single separation stage, but the devices can be combined in counter-current multistage systems. Schuur et al. [440] reported specific interfacial areas in the mixing zone of an ACC to be up to an order of magnitude larger than in stirred tanks.

As said earlier, first applications of ACCs took place in the nuclear waste processing industry, where large remote handling operations are very capital intensive. Here, the reduction of the equipment size, simpler design, and better reliability can bring large returns on investment. Yarbro and Schreiber [441]

Table 4.13 Comparison between SDR-based continuous process and conventional batch technology for conversion of α-pinene oxide to campholenic aldehyde.

	Batch	Continuous (SDR)
Process time (s)	300	1
Processing capacity (kg/h)	1.2	209
Conversion (%)	100	100
Selectivity (%)	64	62
Note	Necessity for catalyst separation from product stream	No loss of catalyst (catalyst is fixed on SDR surface)

Source: Vicevic et al. 2001 [427]. Presented and published at BHR Group's conference.

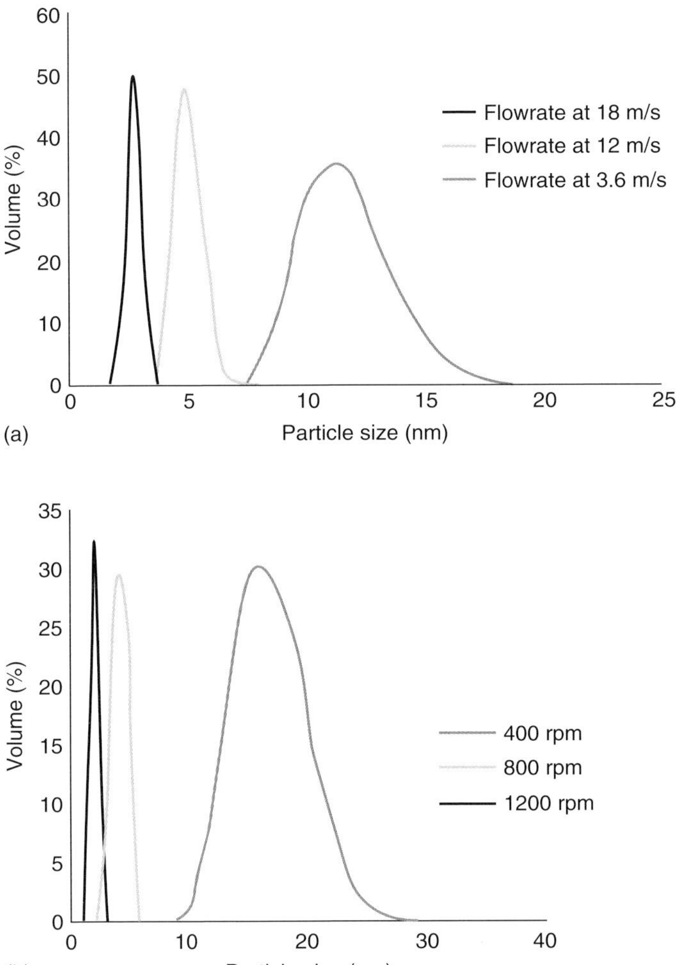

Figure 4.71 Particle size distribution of TiO$_2$ nanoparticles in a spinning disc reactor as a function of flow rate (a) and rotational speed (b). Source: Mohammadi et al. 2014 [430]. Reproduced with permission of Elsevier.

described the application of an annular centrifugal contactor to actinides purification at Los Alamos National Laboratory. The ratio between the flow rate of the processed liquids and the working volume of the apparatus was found for the ACC to be more than 16 times higher than for the packed column, 3.7 times higher than for a mixer–settler system and almost 1.4 times higher than for a pulsed sieve plate column. A comparison of sizes of several types of contactors for actinides processing is shown in Figure 4.75. More recent applications of ACCs studied in the literature include, among other things, the resolution of racemates [442, 443] and catalytic biodiesel synthesis [444].

The last group of rotor–stator devices that definitely needs mentioning in the context of process intensification is the *rotor–stator mixers* (RSMs) [445, 446].

A RSM (Figure 4.76) contains a high-speed rotor spinning close to a motionless stator. Fluid passes through the region where rotor and stator interact and experiences highly pulsating flow and shear. In-line RSMs resemble centrifugal pumps and may contribute significantly to pumping the fluid. They provide extremely high shear and energy dissipation rates, respectively, up to c. 100 000/s

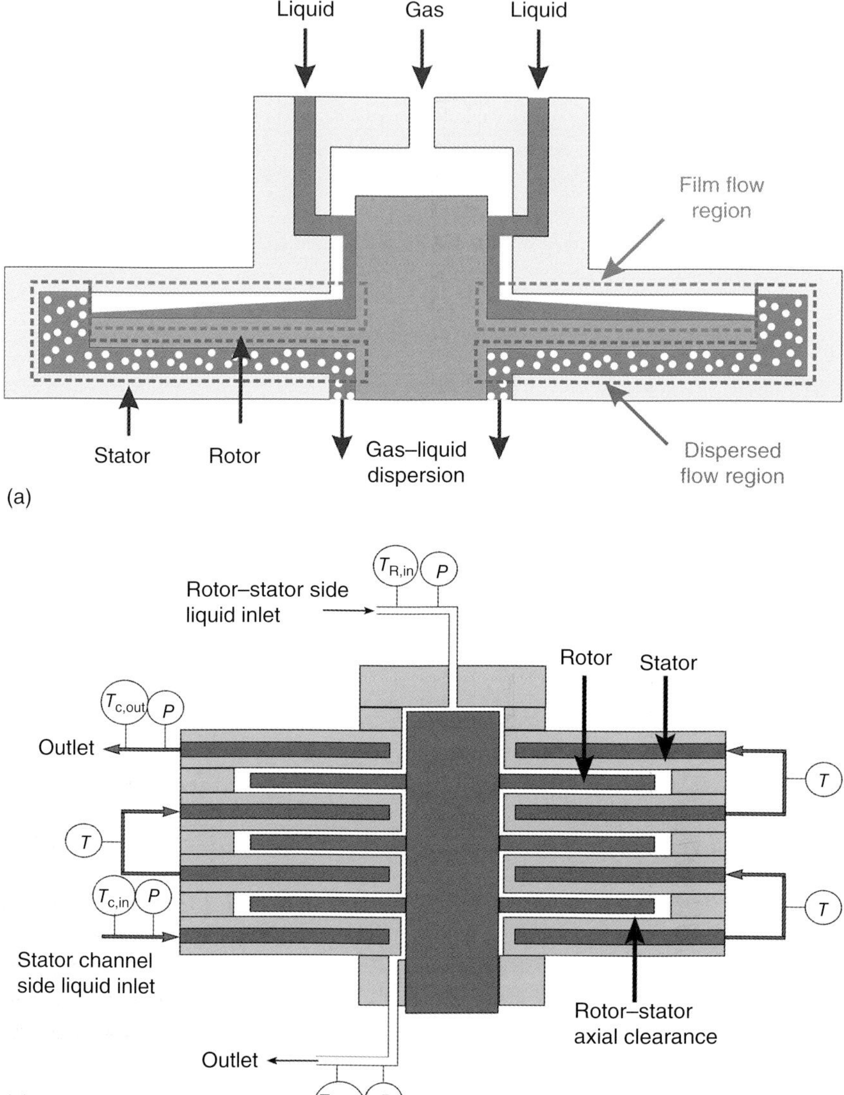

Figure 4.72 (a) Single-stage rotor–stator spinning disc reactor for gas–liquid contacting. Source: Meeuwse et al. 2012 [432]. Reproduced with permission of John Wiley and Sons. (b) Three-stage RSSDR for single-phase processing with cooling liquid on the stator side. Source: De Beer et al. 2014 [433]. Reproduced with permission of John Wiley and Sons.

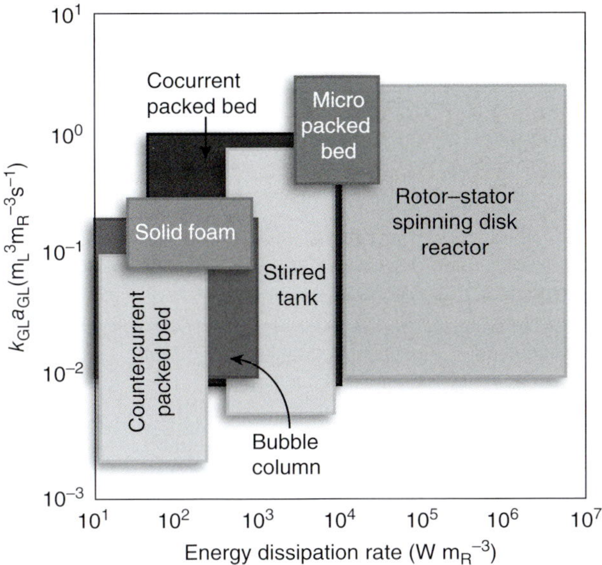

Figure 4.73 Volumetric gas–liquid mass transfer coefficients as a function of energy dissipation rate in various types of gas–liquid contactors. Source: Adapted from Meeuwse et al. 2012 [432], Stemmet et al. 2007 [438], and Trambouze and Euzen 1988 [439].

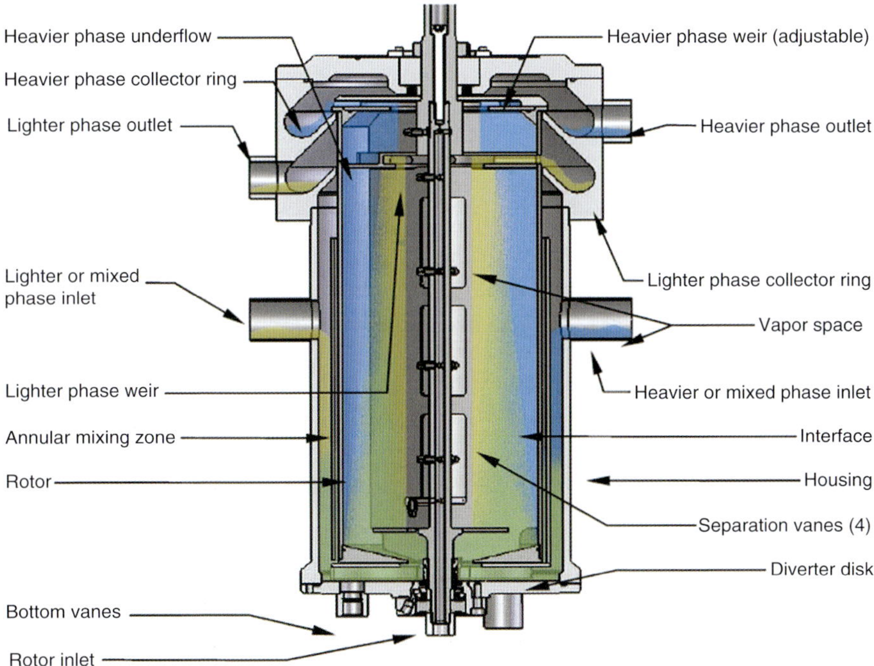

Figure 4.74 Annular centrifugal contactor offered by CINC Industries. Source: Courtesy of CINC Industries, USA, www.cincindustries.com.

Table 4.14 Heat and mass transfer performance of the rotor–stator spinning disc reactors.

System	Parameter	Value/magnitude	References
Single-phase liquid	Heat transfer coefficient	5× higher than in tubular reactors	De Beer et al. 2014 [433]
Single-phase liquid	Heat transfer coefficient	Up to 34 kW m^2/K	De Beer et al. 2015 [434]
Gas–liquid	Volumetric gas–liquid mass transfer coefficient	Up to 40× higher than in bubble columns	Meeuwse et al. 2010 [435]
Gas–liquid	Liquid–solid mass transfer coefficient	Up to 1 order of magnitude higher than in packed beds	Meeuwse et al. 2010 [436]
Liquid–liquid	Liquid–liquid mass transfer rate	At least 25× higher than in packed columns, up to 15× higher than in microchannels	Visscher et al. 2012 [437]

and 10^5 W/kg, which in turn translates to very short micro–/mesomixing times in the range of 10^{-4} s. RSMs are by far the most commercially widespread type of a rotor–stator device. They gained huge popularity in various industrial sectors, particularly in processes focusing on product properties/quality. Operations that use RSMs include emulsification, dispersion of powders, blending, homogenization, and cell disruption. Because of the very short mixing times, RSMs can also be used as reactors for carrying fast chemical reactions [448].

4.7.4 Process Intensification by Solids Moving in Centrifugal Fields

The basic working principle of all rotating systems discussed so far in this chapter was the movement of a liquid by the centrifugal force that led to intensive shearing, ripples, fine phase dispersions, etc. In the current section, we will focus on systems, which utilize the movement of solid particles in high-gravity fields. Such systems have been developed for both gas–solid and liquid–solid systems.

For gas–solid systems, *rotating fluidized beds* (RFBs) have been proposed [449–452]. Those devices are based on a porous cylindrical chamber, into which air of a process gas is injected (Figure 4.77). Upon rotation of the chamber, the bed on microparticles gets into the fluidized state and the intensity of the high-gravity field maintaining that fluidization depends on the rotational speed and cylinder diameter. Compared to conventional fluidized beds, the RFBs can operate in a wider operational range. Their advantages also include excellent gas–solid separation, increased slip velocities, hence transfer coefficients and a compact construction. Obvious disadvantages, however, are the use of the motor and moving geometry that may lead to mechanical vibrations or sealing problems as well as continuous feeding/removal of solid particles to the reactor chamber. To this end, the conventional RFBs usually start with a batch of particles preloaded in the chamber [453].

In order to address the above-mentioned disadvantages, RFBs in a static geometry (RFB-SG), also known as *gas–solid vortex reactors* (GSVR), have been proposed – see Dutta et al. [454] and De Wilde [455]. In those devices, rotating parts are eliminated. The process gas is injected tangentially in the vortex chamber (Figure 4.78) and the solid particles are entrained by the gas. The gas flows radially via the particle bed and leaves the chamber via a centrally positioned chimney. Particles are not able to follow the gas because of inertia forces and form a dense rotating particle bed near the wall. Increased slip velocities in GSVRs drastically increase the gas–solid heat transfer rates [456]. Because of this, GSVRs can be considered in processes where gas–solid heat transfer plays a predominant role, such as fluid catalytic cracking [457] or biomass pyrolysis [458].

In solid–liquid systems, a compact technology for adsorption and ion exchange called *centrifugal adsorption technology* (CAT) has been developed at the Delft University of Technology in the late 1990s [459]. The concept of CAT is based on counter-current flow between micrometer-range particles and a liquid in centrifugal field. The counter-current flow between liquid and adsorbent particles provides the driving force for intensive mass transfer, and the centrifugal field allows for sufficiently high velocities of the adsorbent particles in counter-current motion. In CAT, substantial reduction in equipment volume is achieved, by using

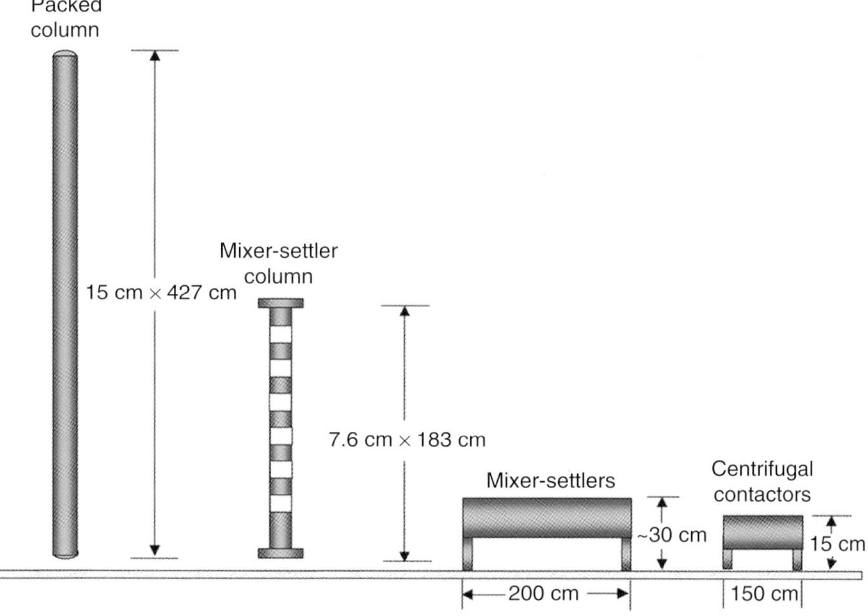

Figure 4.75 Comparison of sizes of various separators used in actinide processing. Source: Yarbro and Schriber 2003 [441]. Reproduced with permission of John Wiley and Sons.

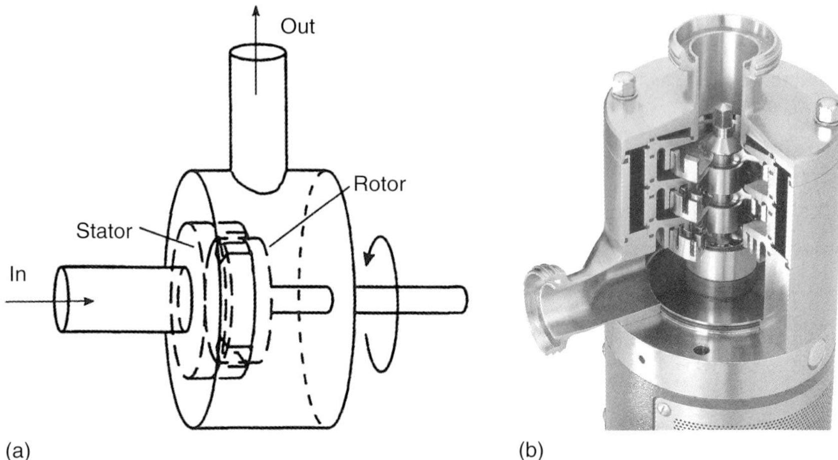

Figure 4.76 Rotor–stator mixer. Source: (a) Sparks et al. 1995 [447]. Presented and published at BHR Group's conference. (b) Used with permission of IKA Works, Inc., USA, www.ika.net.

adsorbent particles with diameters typically between 1 and 100 μm (microsorbents). A schematic view of the concept of CAT is shown in Figure 4.79. Fresh adsorbent particles enter the column close to the axis of rotation and are dragged outward toward the rim by the centrifugal force. The liquid feed enters the column at the outside and is pumped toward the axis of rotation. The liquid flow is chosen such that the adsorbent material is fluidized in the contact zone. The saturated adsorbent is collected at the rim and is transported back by means of a separate transport channel toward the axis of rotation where it is discharged. As liquid and adsorbent enter and leave the rotor via the axis of rotation, the pressure drop over the contactor is kept very low.

According to the inventors, the combined effects of counter-current flows and small particle diameters lead to very high mass transfer rates and extremely compact adsorption equipment with very short contact times and high capacities (typically 10–50 m^3/h). It is anticipated that a CAT apparatus can be 2 to 3 orders of magnitude smaller than the conventional equipment equivalent – see Figure 4.80.

4.7.5 High Gravity Fields in Microprocessing Systems

In microchannel systems, centrifugal and Coriolis forces generated by the rotation are primarily used to achieve intensified mixing at high throughputs and to realize specific contacting patterns. In Chapter 3, when discussing the microstructures, two examples of the use of high-gravity fields in microchannel systems, rotating spiral microchannel distillation column (Figure 3.30) and co-mix Coriolis microreactor (Figure 3.48), are presented. Next to the above,

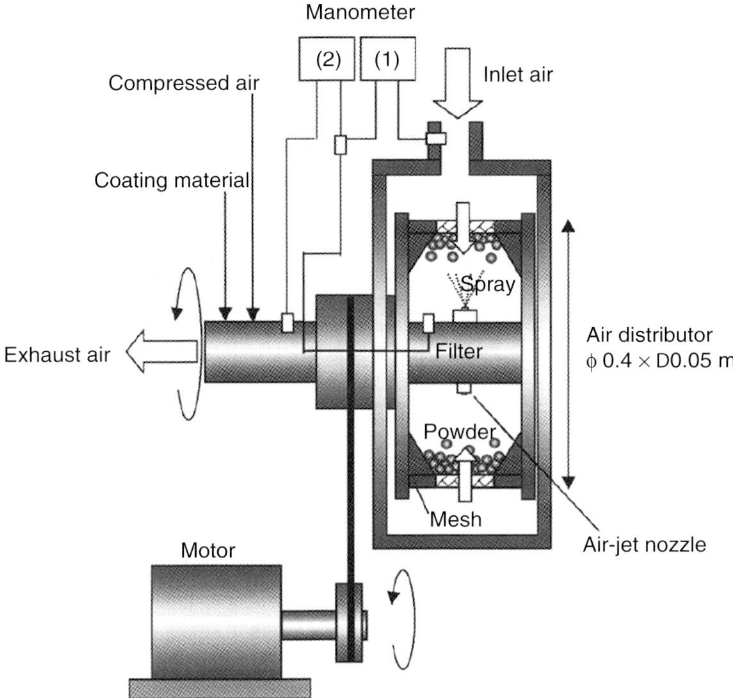

Figure 4.77 Rotating fluidized bed. Source: Watano et al. 2004 [452]. Reproduced with permission of Elsevier.

potential applications of the rotating microchannel systems can be primarily sought in the field of microseparations, such as chromatography [461] or extraction [462]. More information on phenomena occurring in rotating microchannels and potential applications can be found in the review paper by Roy et al. [463].

References

1 Grave, K., Hazrat, M., Boeve, S. et al. (2015). *Electricity Costs of Energy Intensive Industries: An International Comparison.* Fraunhoffer ISI, http://www.ecofys.com/files/files/ecofys-fraunhoferisi-2015-electricity-costs-of-energy-intensive-industries.pdf.

2 Mai, T., Wiser, R., Sandor, D. et al. (2012). Exploration of High-Penetration Renewable Electricity Futures. Renewable Electricity Futures Study. *NREL/TP-6A20-52409-1*, Vol. 1. Golden, CO: National Renewable Energy Laboratory

3 Gill, H. (1983). The electrothermic system for enhancing oil recovery. *J. Microwave Power* 18: 107–110.

4 Rehman, M.M. and Meribout, M. (2012). Conventional versus electrical enhanced oil recovery: a review. *J. Petrol. Explor. Prod. Technol.* 2: 169–179.

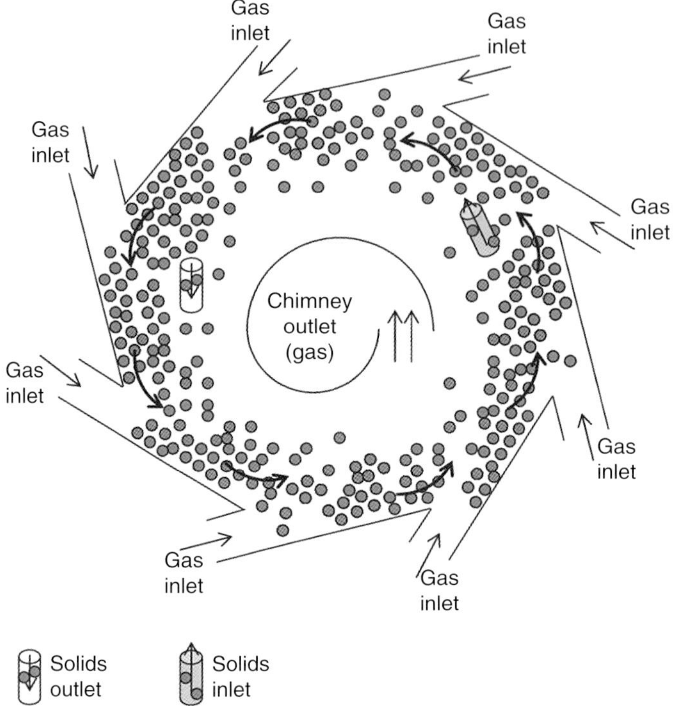

Figure 4.78 Scheme of the rotating fluidized bed in a static geometry. Source: De Broqueville and De Wilde 2009 [456]. Reproduced with permission of Elsevier.

5 Franssen, I.S., Irimia, D., Stefanidis, G.D., and Stankiewicz, A.I. (2014). Practical challenges in energy-based control of molecular transformations in chemical reactors. *AIChE J.* 60: 3392–3405.

6 Thornton, J.D. (1968). The applications of electrical energy to chemical and physical rate processes. *Rev. Pure Appl. Chem.* 18: 197–218.

7 Taylor, G.I. (1964). Disintegration of water drops in electric field. *Proc. R. Soc. London, Ser. A* 280: 383–397.

8 Ptasinski, K.J. and Kerkhof, P.J.A.M. (1992). Electric field driven separations: phenomena and applications. *Sep. Sci. Technol.* 27: 995–1021.

9 Scott, T.C. (1989). Use of electric fields in solvent extraction: a review and prospectus. *Sep. Purif. Methods* 18: 65–109.

10 Weatherley, L.R. (1991). Electrically enhanced mass transfer. *Heat Recovery Syst. CHP* 13: 515–537.

11 He, W., Baird, H.I., and Chang, J.S. (1993). The effect of electric field on mass transfer from drops dispersed in a viscous liquid. *Can. J. Chem. Eng.* 71: 366–376.

12 Glitzenstein, A., Tamir, A., and Oren, Y. (1995). Mass transfer enhancement of acetic acid across a plane kerosene/water by an electric field. *Can. J. Chem. Eng.* 73: 95–102.

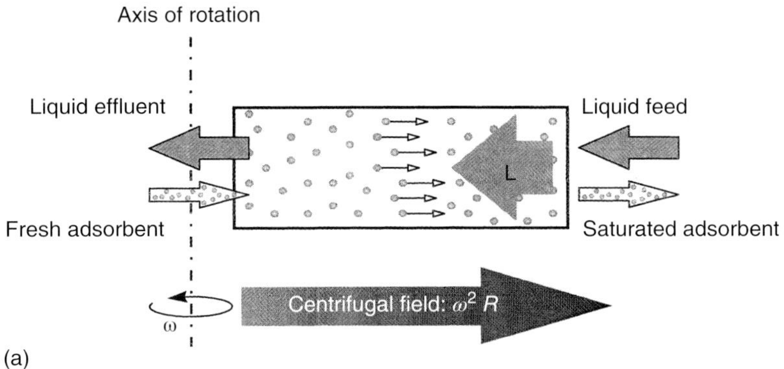

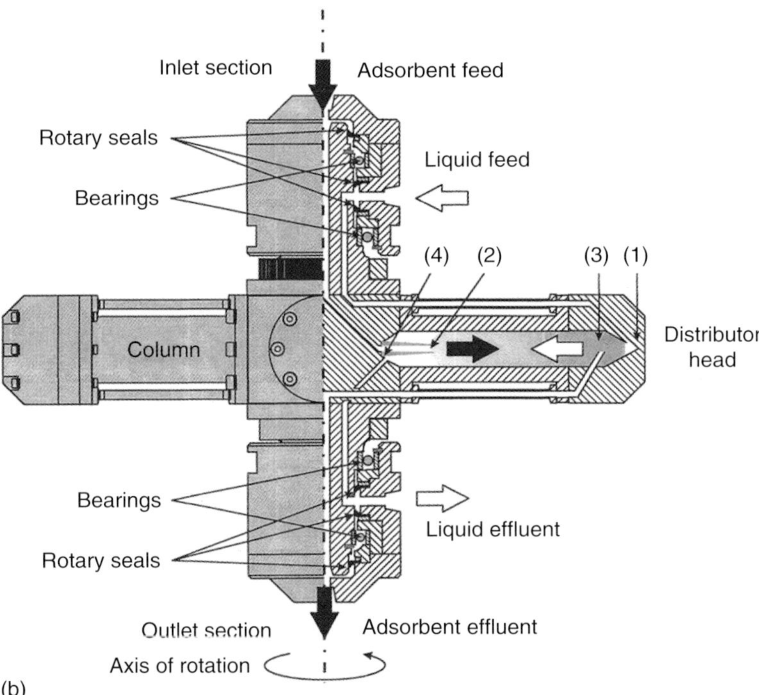

Figure 4.79 Working principle (a) and an apparatus prototype (b) of the centrifugal adsorption technology. Source: Bisschops et al. 2000 [460]. Reproduced with permission of American Chemical Society.

13 Eow, S., Ghadiri, M.J., Sharif, A.O., and Williams, T.J. (2001). Electrostatic enhancement of coalescence of water droplets in oil: a review of the current understanding. *Chem. Eng. J.* 84: 173–192.

14 Eow, S. and Ghadiri, M.J. (2003). Drop-drop coalescence in an electric field: the effects of applied electric field and electrode geometry. *Colloids Surf., A* 219: 253–279.

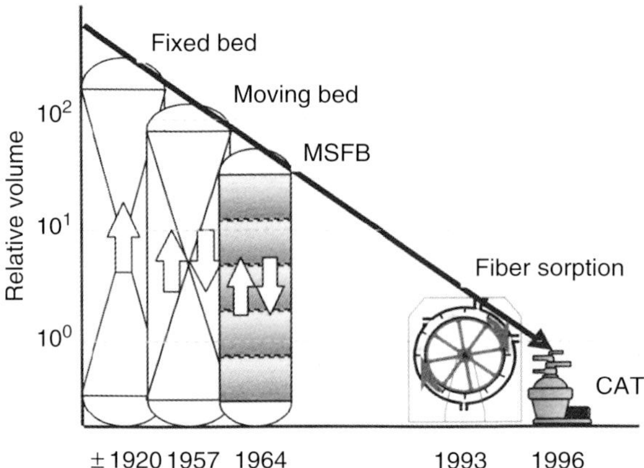

Figure 4.80 Illustration of the compactness of the centrifugal adsorption technology. Source: Bisschops 1999 [459].

15 Savova, M., Bart, H., and Seikova, I. (2005). Enhancement of mass transfer in solid-liquid extraction systems by pulsed electric fields. *J. Univ. Chem. Technol. Metal.* 40: 329–334.

16 Xue, D. and Farid, M.M. (2015). Pulsed electric field extraction of valuable compounds from white button mushrooms (*Agaricus bisporus*). *Innov. Food Sci. Emerg. Technol.* 29: 178–186.

17 Senftleben, H. (1932). Über den Einfluss von magnetischen und elektischen Feldern auf den Wärmestrom in Gasen. *Z. Phys.* 74: 757–769.

18 Senftleben, H. and Braun, W. (1936). Der Einfluss elektrischer Felder auf dem Wäremtestrom in Gasen. *Z. Phys.* 102: 480–506.

19 Senftleben, H. and Bültmann, E. (1953). Die Einwirkung elektrischer Felder auf den Wärmeübergang in Gasen. *Z. Phys.* 136: 389–401.

20 Senftleben, H. and Lange-Hanh, R. (1958). Der Einfluss elektrischer Felder auf dem Wärmeübergang in Flüssigkeiten. *Z. Naturforsch. A* 13: 99–105.

21 Ryde, T., Arajs, S., and Nunge, R.J. (1991). Heat transfer from a wire to hexane in a nonuniform electric field. *J. Appl. Phys.* 69: 606–609.

22 Kaji, N., Mori, Y.H., Tochitani, Y., and Komotori, K. (1980). Augmentation of direct-contact heat transfer with an intermittent electric field. *Trans. ASME* 102: 32–37.

23 Bochirol, L., Bonjour, E., and Weil, L. (1962). Amélioration des échanges thermiques par application d'un champ électrique dans les gas liquefies bouillants. *Pure Appl. Cryog.* 2: 251–256.

24 Bonjour, E., Verdier, J., and Weil, L. (1962). Electroconvection effects on heat transfer. *Chem. Eng. Prog.* 58: 63–66.

25 Markels, M. Jr., and Durfee, R.L. (1965). Studies of boiling heat transfer with electrical fields: Part II. Mechanistic interpretations of voltage effects on boiling heat transfer. *AIChE J.* 11: 720–723.

26 Wolny, A. and Kaniuk, R. (1996). The effect of electric field on heat and mass transfer. *Dry. Technol.* 14: 195–216.

27 Darabi, J., Ohadi, M.M., and Dessiatoun, S.V. (1998). Heat transfer enhancement with falling film evaporation on vertical tubes using electric fields.

In: *Heat Transfer in Flowing Systems*, vol. 1 (ed. ASME -PUBLICATIONS-HTD); 361, 331–338. ASME. ISBN: 0791815978.

28 Tsouris, C., Culbertson, C.T., DePaoli, D.W. et al. (2003). Electrohydrodynamic mixing in microchannels. *AIChE J.* 49: 2181–2186.

29 El Moctar, A.O., Aubry, N., and Batton, J. (2003). Electro-hydrodynamic micro-fluidic mixer. *Lab Chip.* 3: 273–280.

30 Kirko, I.M. and Filippov, M.V. (1960). The specific features of a fluidized bed of ferromagnetic particles in a magnetic field. *J. Techn. Fiziki (Russia)* 30: 1081–1084.

31 Rosensweig, R.E. (1979). Fluidization: hydrodynamics stabilization with a magnetic field. *Science* 204: 57–60.

32 Lucchesi, P.J., Hatch, W.H., Mayer, F.X., and Rosensweig, R.E. (1979). Magnetically stabilized beds – new gas-solids contacting technology. *Proceedings of the 10th World Petroleum Congress*, Bucharest, Heyden, Philadelphia, PA, 419–425.

33 Liu, Y.A., Hamby, R.K., and Colberg, R.D. (1991). Fundamental and practical developments of magnetofluidized beds: a review. *Powder Technol.* 64: 3–41.

34 Siegell, J.H. (1982). Radial dispersion and flow distribution of gas in magnetically stabilized beds. *Ind. Eng. Chem. Process Des. Dev.* 21: 135–141.

35 Geuzens, P. and Thoenes, D. (1988). Magnetically stabilized fluidization, Part I: Gas and solids flow. *Chem. Eng. Commun.* 67: 217–228.

36 Moffat, G., Williams, R.A., Webb, C., and Stirling, R. (1994). Selective separation in environmental and industrial processes using magnetic carrier technology. *Miner. Eng.* 7: 1039–1056.

37 Neff, J. and Rubinsky, B. (1983). The effect of a magnetic field on the heat transfer characteristics of an air fluidized bed of ferromagnetic particles. *Int. J. Heat Mass Transfer* 16: 1885–1889.

38 Qian, R.Z. and Saxena, S.C. (1993). Heat transfer from an immersed surface in a magnetofluidized bed. *Int. Commun. Heat Mass Transfer* 20: 859–869.

39 Al-Quodah, Z. and Al-Busoul, M. (2001). The effect of magnetic field on local heat transfer coefficient in fluidized beds with immersed heating surface. *J. Heat Transfer* 123: 157–161.

40 Li, W., Zong, B., Li, X. et al. (2006). Interphase mass transfer in G-L-S magnetically stabilized bed with amorphous alloy SRNA-4 catalyst. *Chin. J. Chem. Eng.* 14: 734–739.

41 Ivanova, V., Hristov, J.Y., Dobreva, E. et al. (1996). Performance of a magnetically stabilized bed reactor utilizing immobilized yeast cells. *J. Appl. Biochem. Biotechnol.* 59: 187–199.

42 Meng, X., Mu, X., Zong, B. et al. (2003). Purification of caprolactam in magnetically stabilized bed reactor. *Catal. Today* 79–80: 21–27.

43 Sikavitsas, V.I., Yang, R.T., Burns, M.A., and Langenmayr, E.J. (1995). Magnetically stabilized fluidized bed for gas separations: Olefin – Paraffin separations by π-complexation. *Ind. Eng. Chem. Res.* 34: 2873–2880.

44 De Brito, J.F., De Oliveira Ferreira, L., Ragozoni Pereira, M.C. et al. (2012). Adsorption of aromatic compounds under magnetic field influence. *Water Air Soil Pollut.* 223: 3545–3551.

45 Yavuz, C.T., Prakash, A., Mayo, J.T., and Colvin, V.L. (2009). Magnetic separations: from steel plants to biotechnology. *Chem. Eng. Sci.* 64: 2510–2521.

46 Broomberg, J., Gélinas, S., Finch, J.A., and Xu, Z. (1999). Review of magnetic carrier technologies for metal ion removal. *Magn. Electr. Sep.* 9: 169–188.

47 Azizian, R., Doroodchi, E., McKrell, T. et al. (2014). Effect of magnetic field on laminar convective heat transfer of magnetite nanofluids. *Int. J. Heat Mass Transfer* 68: 94–109.

48 Samadi, Z., Haghshenasfard, M., and Moheb, A. (2014). CO_2 absorption using nanofluids in a Wetted-Wall column with external magnetic field. *Chem. Eng. Technol.* 37: 462–470.

49 Pamme, N. (2005). Magnetism and microfluidics. *Lab Chip* 6: 24–38.

50 Suh, Y.K. and Kang, S. (2010). A review on mixing in microfluidics. *Micromachines* 1: 82–111.

51 Ward, K. and Fan, Z.H. (2015). Mixing in microfluidic devices and enhancement methods. *J. Micromech. Microeng.* 25: 1–17.

52 Kim, T., Lee, J., and Lee, K.H. (2014). Microwave heating of carbon-based solid materials. *Carbon Lett.* 15 (1): 15–24.

53 Milestone SRL (2009). MicroSYNTH—microwave synthesis labstation. http://www.milestonesrl.com/analytical/products-microwave-synthesis-microsynth.html (Retrieved 1 February 2013).

54 Sineo Corp. (2009). Ultra high throughput closed microwave digestion/extraction workstation. from http://www.sineomicrowave.com/Products_px_en.asp?pid=60 (Retrieved 1 February 2013).

55 CEM Corp. (2011). Discover legacy systems – microwave synthesizers. http://www.cem.com/discover-legacy-systems.html (Retrieved 20 December 2011).

56 CEM Corp. (2011). MARS 6—one touch makes all the difference. http://cem.com/mars6.html (Retrieved 1 February 2013).

57 Radoiu, M., Depagneux, B., Trescarte, E. et al. (2009). The scale-up of microwave-assisted processes. The 12th International Conference on Microwave and High Frequency Heating AMPERE.

58 SAIREM S (2010). Labotron extraction & synthesis 2450 MHz. http://www.sairem.com/fiches_techniques/equipements/Labotron%20X%20EN.pdf (Retrieved accessed 1 June 2012).

59 Brown, G.H., Hoyler, C.N., and Bierwirth, R.A. (1947). *Theory and Application of Radio-Frequency Heating*, 328–344. New York: van Nostrand.

60 Osepchuk, J.M. (2002). Microwave power applications. *IEEE Trans. Microw. Theory Tech.* 50 (3): 975–985.

61 Regier, M., Knoerzer, K., and Schubert, H. (eds.) (2016). *The Microwave Processing of Foods*. Woodhead Publishing.

62 Vadivambal, R. and Jayas, D.S. (2010). Non-uniform temperature distribution during microwave heating of food materials—a review. *Food Bioproc. Technol.* 3 (2): 161–171.

63 Li, Z.Y., Wang, R.F., and Kudra, T. (2011). Uniformity issue in microwave drying. *Dry. Technol.* 29 (6): 652–660.

64 Datta, A.K. and Anantheswaran, R.C. (2001). *Handbook of Microwave Technology for Food Application*. New York: Marcel Dekker.

65 Ahmed, J. and Ramaswamy, H.S. (2004). Microwave pasteurization and sterilization of foods. In: *Food Science and Technology*, vol. 167, 691. New York: Marcel Dekker.

66 Haghi, A.K. (2005). Application of microwave techniques in textile chemistry. *Asian J. Chem.* 17 (2): 639.

67 Kudra, T. and Mujumdar, A.S. (2009). *Advanced Drying Technologies*. CRC Press.

68 Duan, X., Zhang, M., Mujumdar, A.S., and Wang, R. (2010). Trends in microwave-assisted freeze drying of foods. *Dry. Technol.* 28 (4): 444–453.

69 Zhang, M., Tang, J., Mujumdar, A.S., and Wang, S. (2006). Trends in microwave-related drying of fruits and vegetables. *Trends Food Sci. Technol.* 17 (10): 524–534.

70 Stefanidis, G.D., Muñoz, A.N., Sturm, G.S., and Stankiewicz, A. (2014). A helicopter view of microwave application to chemical processes: reactions, separations, and equipment concepts. *Rev. Chem. Eng.* 30 (3): 233–259.

71 Kappe, C.O. (2004). Controlled microwave heating in modern organic synthesis. *Angew. Chem. Int. Ed.* 43 (46): 6250–6284.

72 Leadbeater, N.E., Torenius, H.M., and Tye, H. (2004). Microwave-promoted organic synthesis using ionic liquids: a mini review. *Comb. Chem. High Throughput Screen.* 7 (5): 511–528.

73 Nüchter, M., Ondruschka, B., Bonrath, W., and Gum, A. (2004). Microwave assisted synthesis–a critical technology overview. *Green Chem.* 6 (3): 128–141.

74 Kappe, C.O. and Van der Eycken, E. (2010). Click chemistry under non-classical reaction conditions. *Chem. Soc. Rev.* 39 (4): 1280–1290.

75 Kumar Gupta, A., Singh, N., and Nand Singh, K. (2013). Microwave assisted organic synthesis: cross coupling and multicomponent reactions. *Curr. Org. Chem.* 17 (5): 474–490.

76 Kranjc, K. and Kocevar, M. (2013). From conventional reaction conditions to microwave-assisted catalytic transformations of various substrates. State of the Art in 2012 (Part A: general). *Curr. Org. Chem.* 17 (5): 448–456.

77 Kranjc, K. and Kocevar, M. (2013). From conventional reaction conditions to microwave-assisted catalytic transformations of various substrates. State of the art in 2012 (Part B: catalysis). *Curr. Org. Chem.* 17 (5): 457–473.

78 Gedye, R.N., Smith, F.E., and Westaway, K.C. (1988). The rapid synthesis of organic compounds in microwave ovens. *Can. J. Chem.* 66 (1): 17–26.

79 Larhed, M., Moberg, C., and Hallberg, A. (2002). Microwave-accelerated homogeneous catalysis in organic chemistry. *Acc. Chem. Res.* 35 (9): 717–727.

80 Johnstone, M.D., Lowe, A.J., Henderson, L.C., and Pfeffer, F.M. (2010). Rapid synthesis of cyclobutene diesters using a microwave-accelerated ruthenium-catalysed [2+ 2] cycloaddition. *Tetrahedron Lett.* 51 (45): 5889–5891.

81 He, P., Haswell, S.J., and Fletcher, P.D. (2004). Microwave heating of heterogeneously catalysed Suzuki reactions in a micro reactor. *Lab Chip* 4 (1): 38–41.

82 Comer, E. and Organ, M.G. (2005). A microreactor for microwave-assisted capillary (continuous flow) organic synthesis. *J. Am. Chem. Soc.* 127 (22): 8160–8167.

83 Jacob, J., Chia, L.H.L., and Boey, F.Y.C. (1995). Thermal and non-thermal interaction of microwave radiation with materials. *J. Mater. Sci.* 30 (21): 5321–5327.

84 Perreux, L. and Loupy, A. (2008). Nonthermal effects of microwaves in organic synthesis. In: *Microwaves in Organic Synthesis*, 2e (ed. A. Loupy), 134–218. Wiley-VCH.

85 Howarth, P. and Lockwood, M. (2004). Come of age. *Chem. Eng.* 756: 29–31.

86 De la Hoz, A., Diaz-Ortiz, A., and Moreno, A. (2005). Microwaves in organic synthesis. Thermal and non-thermal microwave effects. *Chem. Soc. Rev.* 34 (2): 164–178.

87 Kuhnert, N. (2002). Microwave-assisted reactions in organic synthesis—are there any nonthermal microwave effects? *Angew. Chem. Int. Ed.* 41 (11): 1863–1866.

88 Stuerga, D.A.C. and Gaillard, P. (1996). Microwave athermal effects in chemistry: a Myth's autopsy: Part I: Historical background and fundamentals of wave-matter interaction. *J. Microw. Power Electromagn. Energy* 31 (2): 87–100.

89 Stuerga, D.A.C. and Gaillard, P. (1996). Microwave athermal effects in chemistry: a Myth's autopsy: Part II: Orienting effects and thermodynamic consequences of electric field. *J. Microw. Power Electromagn. Energy* 31 (2): 101–113.

90 Obermayer, D., Gutmann, B., and Kappe, C.O. (2009). Microwave chemistry in silicon carbide reaction vials: separating thermal from nonthermal effects. *Angew. Chem.* 121 (44): 8471–8474.

91 Kappe, C.O., Pieber, B., and Dallinger, D. (2013). Microwave effects in organic synthesis: myth or reality? *Angew. Chem. Int. Ed.* 52 (4): 1088–1094.

92 Bond, G., Moyes, R.B., and Whan, D.A. (1993). Recent applications of microwave heating in catalysis. *Catal. Today* 17 (3): 427–437.

93 Chen, C., Hong, P., Dai, S., and Kan, J. (1995). Microwave effects on the oxidative coupling of methane over proton conductive catalysts. *J. Chem. Soc., Faraday Trans.* 91 (7): 1179–1180.

94 Zhang, X., Hayward, D.O., Lee, C., and Mingos, D.M.P. (2001). Microwave assisted catalytic reduction of sulfur dioxide with methane over MoS_2 catalysts. *Appl. Catal., B* 33 (2): 137–148.

95 Zhang, X., Hayward, D.O., and Mingos, D.M.P. (2003). Effects of microwave dielectric heating on heterogeneous catalysis. *Catal. Lett.* 88 (1): 33–38.

96 Roussy, G., Thiebaut, J.M., Souiri, M. et al. (1994). Controlled oxidation of methane doped catalysts irradiated by microwaves. *Catal. Today* 21 (2–3): 349–355.

97 Ramírez, A., Hueso, J.L., Mallada, R., and Santamaría, J. (2016). Ethylene epoxidation in microwave heated structured reactors. *Catal. Today* 273: 99–105.

98 Ramirez, A., Hueso, J.L., Mallada, R., and Santamaria, J. (2017). In situ temperature measurements in microwave-heated gas-solid catalytic systems.

Detection of hot spots and solid-fluid temperature gradients in the ethylene epoxidation reaction. *Chem. Eng. J.* 316: 50–60.

99 Fidalgo, B., Domínguez, A., Pis, J.J., and Menéndez, J.A. (2008). Microwave-assisted dry reforming of methane. *Int. J. Hydrogen Energy* 33 (16): 4337–4344.

100 Fidalgo, B., Arenillas, A., and Menéndez, J.A. (2011). Mixtures of carbon and Ni/Al_2O_3 as catalysts for the microwave-assisted CO_2 reforming of CH_4. *Fuel Process. Technol.* 92 (8): 1531–1536.

101 Ioffe, M.S., Pollington, S.D., and Wan, J.K. (1995). High-power pulsed radio-frequency and microwave catalytic processes: selective production of acetylene from the reaction of methane over carbon. *J.f Catal.* 151 (2): 349–355.

102 Zhang, X., Lee, C.S.M., Mingos, D.M.P., and Hayward, D.O. (2003). Carbon dioxide reforming of methane with Pt catalysts using microwave dielectric heating. *Catal. Lett.* 88 (3): 129–139.

103 Fidalgo, B., Fernández, Y., Domínguez, A. et al. (2008). Microwave-assisted pyrolysis of CH_4/N_2 mixtures over activated carbon. *J. Anal. Appl. Pyrolysis* 82 (1): 158–162.

104 Chen, C.L., Hong, P.J., Dai, S.S. et al. (1997). Microwave effects on the oxidative coupling of methane over Bi_2O_3-WO_3 oxygen ion conductive oxides. *React. Kinet. Catal. Lett.* 61 (1): 175–180.

105 Zhang, X., Lee, C.S.M., Mingos, D.M.P., and Hayward, D.O. (2003). Oxidative coupling of methane using microwave dielectric heating. *Appl. Catal., A* 249 (1): 151–164.

106 Koch, T.A., Krause, K.R., and Mehdizadeh, M. (1997). Improved safety through distributed manufacturing of hazardous chemicals. *Process Saf. Prog.* 16 (1): 23–24.

107 Hayward, D. (1999). Apparent equilibrium shifts and hot-spot formation for catalytic reactions induced by microwave dielectric heating. *Chem. Commun.* (11): 975–976.

108 Beckers, J., van der Zande, L.M., and Rothenberg, G. (2006). Clean diesel power via microwave susceptible oxidation catalysts. *ChemPhysChem* 7 (3): 747–755.

109 Sinev, I., Kardash, T., Kramareva, N. et al. (2009). Interaction of vanadium containing catalysts with microwaves and their activation in oxidative dehydrogenation of ethane. *Catal. Today* 141 (3): 300–305.

110 Liu, Y., Lu, Y., Liu, S., and Yin, Y. (1999). The effects of microwaves on the catalyst preparation and the oxidation of o-xylene over a V_2O_5/SiO_2 system. *Catal. Today* 51 (1): 147–151.

111 Seyfried, L., Garin, F., Maire, G. et al. (1994). Microwave electromagnetic-field effects on reforming catalysts.: 1. Higher selectivity in 2-methylpentane isomerization on alumina-supported Pt catalysts. *J. Catal.* 148 (1): 281–287.

112 Roussy, G., Hilaire, S., Thiebaut, J.M. et al. (1997). Permanent change of catalytic properties induced by microwave activation on 0.3% $PtAl_2O_3$ (EuroPt-3) and on 0.3% Pt-0.3% $ReAl_2O_3$ (EuroPt-4). *Appl. Catal., A* 156 (2): 167–180.

113 Perry, W.L., Datye, A.K., Prinja, A.K. et al. (2002). Microwave heating of endothermic catalytic reactions: reforming of methanol. *AIChE J.* 48 (4): 820–831.

114 Chen, W.H. and Lin, B.J. (2010). Effect of microwave double absorption on hydrogen generation from methanol steam reforming. *Int. J. Hydrogen Energy* 35 (5): 1987–1997.

115 Durka, T., Stefanidis, G.D., Van Gerven, T., and Stankiewicz, A.I. (2011). Microwave-activated methanol steam reforming for hydrogen production. *Int. J. Hydrogen Energy* 36 (20): 12843–12852.

116 Lucchesi, M.E., Chemat, F., and Smadja, J. (2004). Solvent-free microwave extraction of essential oil from aromatic herbs: comparison with conventional hydro-distillation. *J. Chromatogr. A* 1043 (2): 323–327.

117 Mato, R., Martin, V., Navarrete, A. et al. (2008). Solvent-free microwave extraction of essential oil from lavandin super. In: *Presentations at the 2008 AIChE Annual Meeting*, 24. Curran Associates, Inc. ISBN: 978-1-61567-226-4.

118 Lucchesi, M.E., Smadja, J., Bradshaw, S. et al. (2007). Solvent free microwave extraction of *Elletaria cardamomum* L.: a multivariate study of a new technique for the extraction of essential oil. *J. Food Eng.* 79 (3): 1079–1086.

119 Bousbia, N., Vian, M.A., Ferhat, M.A. et al. (2009). A new process for extraction of essential oil from citrus peels: microwave hydrodiffusion and gravity. *J. Food Eng.* 90 (3): 409–413.

120 Sahraoui, N., Vian, M.A., Bornard, I. et al. (2008). Improved microwave steam distillation apparatus for isolation of essential oils: comparison with conventional steam distillation. *J. Chromatogr. A* 1210 (2): 229–233.

121 Pan, X., Niu, G., and Liu, H. (2003). Microwave-assisted extraction of tea polyphenols and tea caffeine from green tea leaves. *Chem. Eng. Process.* 42 (2): 129–133.

122 Chemat, S., Aït-Amar, H., Lagha, A., and Esveld, D.C. (2005). Microwave-assisted extraction kinetics of terpenes from caraway seeds. *Chem. Eng. Process.* 44 (12): 1320–1326.

123 Jain, T., Jain, V., Pandey, R. et al. (2009). Microwave assisted extraction for phytoconstituents-an overview. *Asian J. Res. Chem. (AJRC)* 2 (1): 19–25.

124 Pasquet, V., Chérouvrier, J.-R., Farhat, F. et al. (2011). Study on the microalgal pigments extraction process: performance of microwave assisted extraction. *Process Biochem.* 46 (1): 59–67.

125 Terigar, B.G., Balasubramanian, S., Sabliov, C.M. et al. (2011). Soybean and rice bran oil extraction in a continuous microwave system: from laboratory-to pilot-scale. *J. Food Eng.* 104 (2): 208–217.

126 Yao, H., Du, X., Yang, L. et al. (2012). Microwave-assisted method for simultaneous extraction and hydrolysis for determination of flavonol glycosides in Ginkgo foliage using Brönsted acidic ionic-liquid $[HO_3S(CH_2)_4mim]HSO_4$ aqueous solutions. *Int. J. Mol. Sci.* 13 (7): 8775–8788.

127 Veggi, P.C., Martinez, J., and Meireles, M.A.A. (2012). Fundamentals of microwave extraction. In: *Microwave-Assisted Extraction for Bioactive Compounds*, Food Engineering Series (ed. F. Chemat and G. Cravotto), 15–52. Boston, MA: Springer.

128 Mandal, V., Mohan, Y., and Hemalatha, S. (2007). Microwave assisted extraction—an innovative and promising extraction tool for medicinal plant research. *Pharmacogn. Rev.* 1 (1): 7–18.

129 Yuen, F.K. and Hameed, B.H. (2009). Recent developments in the preparation and regeneration of activated carbons by microwaves. *Adv. Colloid Interface Sci.* 149 (1): 19–27.

130 Cherbański, R. and Molga, E. (2009). Intensification of desorption processes by use of microwaves—an overview of possible applications and industrial perspectives. *Chem. Eng. Process.* 48 (1): 48–58.

131 Cherbanski, R., Komorowska-Durka, M., Stefanidis, G.D., and Stankiewicz, A.I. (2011). Microwave swing regeneration vs temperature swing regeneration - comparin of desorption kinetics. *Ind. Eng. Chem. Res.* 50 (14): 8632–8644.

132 Chowdhury, T., Shi, M., Hashisho, Z. et al. (2012). Regeneration of Na-ETS-10 using microwave and conductive heating. *Chem. Eng. Sci.* 75: 282–288.

133 Foo, K.Y. and Hameed, B.H. (2012). Microwave-assisted regeneration of activated carbon. *Bioresour. Technol.* 119: 234–240.

134 Foo, K.Y. and Hameed, B.H. (2012). A cost effective method for regeneration of durian shell and jackfruit peel activated carbons by microwave irradiation. *Chem. Eng. J.* 193: 404–409.

135 Pinard, M.A. and Aslan, K. (2010). Metal-assisted and microwave-accelerated evaporative crystallization. *Cryst. Growth Des.* 10 (11): 4706–4709.

136 Alabanza, A.M. and Aslan, K. (2011). Metal-assisted and microwave-accelerated evaporative crystallization: application to l-alanine. *Cryst. Growth Des.* 11 (10): 4300–4304.

137 Pinard, M.A., Grell, T.A., Pettis, D. et al. (2012). Rapid crystallization of L-arginine acetate on engineered surfaces using metal-assisted and microwave-accelerated evaporative crystallization. *CrystEngComm* 14 (14): 4557–4561.

138 Radacsi, N., Ter Horst, J.H., and Stefanidis, G.D. (2013). Microwave-assisted evaporative crystallization of niflumic acid for particle size reduction. *Cryst. Growth Des.* 13 (10): 4186–4189.

139 Gao, X., Li, X., Zhang, J. et al. (2013). Influence of a microwave irradiation field on vapor–liquid equilibrium. *Chem. Eng. Sci.* 90: 213–220.

140 Tompsett, G.A., Conner, W.C., and Yngvesson, K.S. (2006). Microwave synthesis of nanoporous materials. *ChemPhysChem* 7 (2): 296–319.

141 Li, Y. and Yang, W. (2008). Microwave synthesis of zeolite membranes: a review. *J. Membr. Sci.* 316 (1): 3–17.

142 Patete, J.M., Peng, X., Koenigsmann, C. et al. (2011). Viable methodologies for the synthesis of high-quality nanostructures. *Green Chem.* 13 (3): 482–519.

143 Nakai, Y., Yoshimizu, H., and Tsujita, Y. (2006). Enhancement of gas permeability in HPC, CTA and PMMA under microwave irradiation. *Polym. J.* 38 (4): 376.

144 Ji, Z., Wang, J., Hou, D. et al. (2013). Effect of microwave irradiation on vacuum membrane distillation. *J. Membr. Sci.* 429: 473–479.

145 Scapinello, M., Delikonstantis, E., and Stefanidis, G.D. (2017). Microwave heating of carbon-based solid materials. *Chem. Eng. Process.* 117: 120–140.

146 Gutsol, A. (2010). *Handbook of Combustion. New Technology*, vol. 5, 323. Wiley-VCH.

147 Fridman, A., Nester, S., Kennedy, L.A. et al. (1999). Gliding arc gas discharge. *Prog. Energy Comb. Sci.* 25 (2): 211–231.

148 Goldman, M., Goldman, A., and Sigmond, R.S. (1985). The corona discharge, its properties and specific uses. *Pure Appl. Chem.* 57 (9): 1353–1362.

149 Raizer, Y.P. and Bazelyan, E.M. (1998). *Spark Discharge*. CRC Press.

150 Pai, D.Z., Lacoste, D.A., and Laux, C.O. (2010). Transitions between corona, glow, and spark regimes of nanosecond repetitively pulsed discharges in air at atmospheric pressure. *J. Appl. Phys.* 107 (9): 093303.

151 Lebedev, Y.A. (2010). Microwave discharges: generation and diagnostics. *J. Phys. Conf. Ser.* 257 (1): 012016.

152 Fridman, A. (2008). *Plasma Chemistry*, 1e. Cambridge University Press.

153 Pässler, P., Hefner, W., Buckl, K. et al. (2008). Acetylene. In: *Ullmann's Encyclopedia of Industrial Chemistry*. Weinheim: Wiley-VCH.

154 Kassel, L.S. (1932). The thermal decomposition of methane1. *J. Am. Chem. Soc.* 54 (10): 3949–3961.

155 Gorov, Y.M. (1989). *Kinetics of Reactions in Industrial Plasma Chemistry*, 82. Moscow: Chimia (Chemistry).

156 Polak, L.S. (1966). Chemical processes in low-temperature plasmas. *Pure Appl. Chem.* 13 (3): 345–360.

157 Fincke, J.R., Anderson, R.P., Hyde, T. et al. (2002). Plasma thermal conversion of methane to acetylene. *Plasma Chem. Plasma Process.* 22 (1): 105–136.

158 Agon, N., Hrabovský, M., Chumak, O. et al. (2016). Plasma gasification of refuse derived fuel in a single-stage system using different gasifying agents. *Waste Manage.* 47: 246–255.

159 Materazzi, M., Lettieri, P., Mazzei, L. et al. (2015). Reforming of tars and organic sulphur compounds in a plasma-assisted process for waste gasification. *Fuel Process. Technol.* 137: 259–268.

160 Fabry, F., Rehmet, C., Rohani, V., and Fulcheri, L. (2013). Waste gasification by thermal plasma: a review. *Waste Biomass Valori* 4 (3): 421–439.

161 Uhm, H.S., Na, Y.H., Hong, Y.C. et al. (2014). Production of hydrogen-rich synthetic gas from low-grade coals by microwave steam-plasmas. *Int. J. Hydrogen Energy* 39 (9): 4351–4355.

162 Hrycak, B., Czylkowski, D., Miotk, R. et al. (2015). Hydrogen production from ethanol in nitrogen microwave plasma at atmospheric pressure. *Open Chem.* 13 (1): 317–324.

163 Bongers, W., Bouwmeester, H., Wolf, B. et al. (2017). Plasma-driven dissociation of CO_2 for fuel synthesis. *Plasma Process. Polym.* 14 (6): 1600126.

164 Carp, O., Huisman, C.L., and Reller, A. (2004). Photoinduced reactivity of titanium oxide. *Prog. Solid State Chem.* 32: 33–177.

165 Stankiewicz, A. (2006). Energy matters, alternative sources and forms of energy for intensification of chemical and biochemical processes. *Trans. IChemE, Part A, Chem. Eng. Res. Des.* 84: 511–521.

166 Hoffmann, N. (2008). Photochemical reactions as key steps in organic synthesis. *Chem. Rev.* 108: 1052–1103.

167 Van den Bogaert, B., Havaux, D., Binnemans, K., and Van Gerven, T. (2015). Photochemical recycling of europium from Eu/Y mixtures in red lamp phosphor waste streams. *Green Chem.* 17: 2180–2187.

168 Van den Bogaert, B., Van Meerbeeck, L., Binnemans, K., and Van Gerven, T. (2016). Influence of irradiance on the photochemical reduction of europium(III). *Green Chem.* 18: 4198–4204.

169 Schiavello, M. (1997). *Heterogeneous Photocatalysis*. Chichester: Wiley.

170 Mills, A. and Le Hunte, S. (1997). An overview of semiconductor photocatalysis. *J. Photochem. Photobiol., A* 108: 1–35.

171 Herrmann, J.M. (2005). Heterogeneous photocatalysis: state of the art and present applications. *Top. Catal.* 34: 49–65.

172 Hoffmann, M.R., Martin, S.T., Choi, W.Y., and Bahnemann, D.W. (1995). Environmental applications of semiconductor photocatalysis. *Chem. Rev.* 95: 69–96.

173 de Lasa, H., Serrano, M., and Salaices, M. (2005). *Photocatalytic Reaction Engineering*. New York: Springer.

174 Maldotti, A., Molinari, A., and Amadelli, R. (2002). Photocatalysis with organized systems for the oxofunctionalization of hydrocarbons by O_2. *Chem. Rev.* 102: 3811–3836.

175 Peral, J., Domenech, X., and Ollis, D.F. (1997). Heterogeneous photocatalysis for purification, decontamination and deodorization of air. *J. Chem. Technol. Biotechnol.* 70: 117–140.

176 Van Gerven, T., Mul, G., Moulijn, J.A., and Stankiewicz, A. (2007). A review of intensification of photocatalytic processes. *Chem. Eng. Process. Process Intensif.* 46: 781–789.

177 Wang, W. and Ku, Y. (2003). The light transmission and distribution in an optical fiber coated with TiO_2 particles. *Chemosphere* 50: 999–1006.

178 Marinangeli, R.E. and Ollis, D.F. (1977). Photoassisted heterogeneous catalysis with optical fibers. 1. Isolated single fiber. *AIChE J.* 23: 415–426.

179 Lin, H. and Valsaraj, K.T. (2006). An optical fiber monolith reactor for photocatalytic wastewater treatment. *AIChE J.* 52: 2271–2280.

180 Du, P., Carneiro, J.T., Moulijn, J.A., and Mul, G. (2008). A novel photocatalytic monolith reactor for multiphase heterogeneous photocatalysis. *Appl. Catal., A* 334: 119–128.

181 Schmidt, M., Cubillas, A.M., Taccardi, N. et al. (2013). Chemical and (photo-)catalytical transformations in photonic crystal fibers. *ChemCatChem* 5: 641–650.

182 Coyle, E. and Oelgemöller, M. (2008). Micro-photochemistry: photochemistry in microstructured reactors. The new photochemistry of the future? *Photochem. Photobiol. Sci.* 7: 1313–1323.

183 Cambié, D., Bottecchia, C., Straathof, N.J.W. et al. (2016). Applications of continuous-flow photochemistry in organic synthesis, material science, and water treatment. *Chem. Rev.* 116: 10276–10341.

184 Matsushita, Y., Ohba, N., Kumada, S. et al. (2008). Photocatalytic reactions in microreactors. *Chem. Eng. J.* 135: S303–S308.

185 Lu, H., Schmidt, M.A., and Jensen, K.F. (2001). Photochemical reactions and on-line UV detection in microfabricated reactors. *Lab Chip* 1: 22–28.

186 Gorges, R., Meyer, S., and Kreisel, G. (2004). Photocatalysis in microreactors. *J. Photochem. Photobiol., A* 167: 95–99.

187 Barthe, P.J., Letourneur, D.H., Themont, J.P., and Woehl, P. (2004). Method and microfluidic reactor for photocatalysis. European Patent EP1415707, filed 29 October 2002 and issued 6 May 2004.

188 Li, X., Wang, H., Inoue, K. et al. (2003). Modified micro-space using self-organized nanoparticles for reduction of methylene blue. *Chem. Commun.* 8: 964–965.

189 Takei, G., Kitamori, T., and Kim, H.B. (2005). Photocatalytic redox-combined synthesis of l-pipecolinic acid with a titania-modified microchannel chip. *Catal. Commun.* 6: 357–360.

190 Gole, J.L., Fedorov, A., Hesketh, P., and Burda, C. (2004). From nanostructures to porous silicon: sensors and photocatalytic reactors. *Phys. Status Solidi C* 1: S188–S197.

191 Liedy, W. (2003). Reactor and method for treating fluids by using photocatalysts coupled with phosphorescent solids. International Patent WO 03/086618 A1, filed 10 April 2003 and issued 23 October 2003.

192 Ciambelli, P., Sannino, D., Palma, V. et al. (2011). Intensification of gas-phase photoxidative dehydrogenation of ethanol to acetaldehyde by using phosphors as light carriers. *Photochem. Photobiol. Sci.* 10: 414–418.

193 Zare, R.N. (1998). Laser control of chemical reactions. *Science* 279: 1875–1879.

194 Polanyi, J. (1972). Some concepts in reaction dynamics. *Acc. Chem. Res.* 5: 161–168.

195 Hoffmann, R. (2000). Exquisite control. *Am. Sci.* 88: 14–25.

196 Simpson, W.R., Rakitzis, P., Kandel, A. et al. (1996). Picturing the transition-state region and understanding vibrational enhancement for the $Cl + CH_4 \rightarrow HCl + CH_3$ reaction. *J. Phys. Chem.* 100: 7938–7947.

197 Kandel, S.A. and Zare, R.N. (1998). Reaction dynamics of atomic chlorine with methane: importance of methane bending and torsional excitation in controlling reactivity. *J. Chem. Phys.* 109: 9719–9727.

198 Vander Wal, R.L., Scott, J.L., and Crim, F.F. (1990). Selectively breaking the O—H bond in HOD. *J. Chem. Phys.* 92: 803–805.

199 Metz, R.B., Thoemke, J.D., Pfeiffer, J.M., and Crim, F.F. (1993). Selectively breaking either bond in the bimolecular reaction of HOD with hydrogen atoms. *J. Chem. Phys.* 99: 1744–1751.

200 Levis, R.J., Mankir, G.M., and Rabitz, H. (2001). Selective bond dissociation and rearrangement with optimally tailored, strong-field laser pulses. *Science* 292: 709–713.

201 Lim, T.H. and Kim, S.D. (2004). Trichloroethylene degradation by photocatalysis in annular flow and annulus fluidised bed photoreactors. *Chemosphere* 54: 305–312.

202 Ciambelli, P., Sannino, D., Palma, V. et al. (2009). Improved performances of a fluidized bed photoreactor by a microscale illumination system. *Int. J. Photoenergy* 2009, 1: –7. https://doi.org/10.1155/2009/709365.

203 Vassilikogiannakis, G., Ioannou, G.I., Montagnon, T. et al. (2017). A novel nebulizer-based continuous flow reactor: introducing the use of pneumatically generated aerosols for highly productive photooxidations. *ChemPhotoChem* https://doi.org/10.1002/cptc.201600054.

204 Dionysiou, D.D., Balasubramanian, G., Suidan, M.T. et al. (2000). Rotating disk photocatalytic reactor: development, characterization, and evaluation for the destruction of organic pollutants in water. *Water Res.* 34: 2927–2940.

205 Mozia, S. (2010). Photocatalytic membrane reactors (PMRs) in water and wastewater treatment. A review. *Sep. Purif. Technol.* 73: 71–91.

206 Iglesias, O., Rivero, M.J., Urtiaga, A.M., and Ortiz, I. (2016). Membrane-based photocatalytic systems for process intensification. *Chem. Eng. J.* 305: 136–148.

207 Wang, P. (2016). Membrane photoreactors (MPRs) for photocatalysts separation and pollutants removal: recent overview and new perspectives. *Sep. Sci. Technol.* 51: 147–167.

208 Wang, P. and Lim, T.T. (2012). Membrane vis-LED photoreactor for simultaneous penicillin G degradation and TiO_2 separation. *Water Res.* 46: 1825–1837.

209 Wang, P., Fane, A.G., and Lim, T.T. (2013). Evaluation of a submerged membrane vis-LED photoreactor (sMPR) for carbamazepine degradation and TiO_2 separation. *Chem. Eng. J.* 215–216: 240–251.

210 Leblebici, M.E., Stefanidis, G.D., and Van Gerven, T. (2015). Comparison of photocatalytic space-time yields of 12 reactor designs for wastewater treatment. *Chem. Eng. Process.* 97: 106–111.

211 Borowitzka, M.A. (1999). Commercial production of microalgae: ponds, tanks, tubes and fermenters. *J. Biotechnol.* 70: 313–321.

212 Spolaore, P., Joannis-Cassan, C., Duran, E., and Isambert, A. (2006). Commercial applications of microalgae. *J. Biosci. Bioeng.* 101: 87–96.

213 Li, Y. (2008). Biofuels from microalgae. *Biotechnol. Prog.* 24: 815–820.

214 Xu, L., Weathers, P.J., Xiong, X.-R., and Liu, C.-Z. (2009). Microalgal bioreactors: challenges and opportunities. *Eng. Life Sci.* 9: 178–189.

215 Posten, C. (2009). Design principles of photo-bioreactors for cultivation of microalgae. *Eng. Life Sci.* 9: 165–277.

216 Bahadar, A. and Bilal Khan, M. (2013). Progress in energy from microalgae: a review. *Renew. Sust. Energy Rev.* 27: 128–148.

217 Hadiyanto, H., Elmore, S., Van Gerven, T., and Stankiewicz, A. (2013). Hydrodynamic evaluations of high-rate algae pond (HRAP) design. *Chem. Eng. J.* 217: 231–239.

218 Blanco Gálvez, J. and Malato Rodríguez, S. (2009). Solar photochemistry technology. In: *Solar Energy Conversion and Photoenergy System*, vol. II (ed. J. Blanco Gálvez and S. Malato Rodríguez), 139–162. Oxford: EOLSS Publishers.

219 Miller, J., Coker, E., Ambrosini, A. et al. (2016). Sunshine to petrol: thermochemistry for solar fuels. In: *CO₂ Summit II: Technologies and Opportunities*, ECI Symposium Series (ed. H. Krutka and F. Zhu), http://dc.engconfintl.org/co2_summit2/18.

220 Muhich, C.L., Ehrehart, B.D., Al-Shankiti, I. et al. (2016). A review and perspective of efficient hydrogen generation via solar thermal water splitting. *WIREs Energy Environ.* 5: 261–287.

221 Koepf, E., Villasmil, W., and Meier, A. (2016). Demonstration of a 100-kWth high-temperature solar thermochemical reactor pilot plant for ZnO dissociation. In: *AIP Conference Proceedings*, vol. 1734, 120005. AIP Publishing.

222 Chueh, W.C., Falter, C., Abbott, M. et al. (2010). High-flux solar-driven thermochemical dissociation of CO_2 and H_2O using nonstoichiometric ceria. *Science* 330: 1797–1801.

223 Meier, A., Bonaldi, E., Cella, G.M. et al. (2006). Solar chemical reactor technology for industrial production of lime. *Sol. Energy* 80: 1355–1362.

224 Kodama, T., Kiyama, A., Moriyama, T., and Mizuno, O. (2004). Solar methane reforming using a new type of catalytically-activated metallic foam absorber. *J. Sol. Energy Eng.* 126: 808–811.

225 Berman, A., Karna, R.K., and Epstein, M. (2006). Steam reforming of methane on a Ru/Al_2O_3 catalyst promoted with Mn oxides for solar hydrogen production. *Green Chem.* 9: 626–631.

226 Diver, R.B., Miller, J.E., Allendorf, M.D. et al. (2008). Solar thermochemical water-splitting ferrite-cycle heat engines. *J. Sol. Energy Eng.* 130: 041001.

227 Service, R.E. (2009). Sunlight in your tank. *Science* 326: 1472–1475.

228 Kim, J., Johnson, T.A., Miller, J.E. et al. (2012). Fuel production from CO_2 using solar-thermal energy: system level analysis. *Energy Environ. Sci.* 5: 8417–8429.

229 Kaneko, H., Miura, T., Fuse, A. et al. (2007). Rotary-type solar reactor for solar hydrogen production with two-step water splitting process. *Energy Fuels* 21: 2287–2293.

230 Ermanoski, I. and Siegel, N. (2014). Annual average efficiency of a solar-thermochemical reactor. *Energy Procedia* 49: 1932–1939.

231 Abanades, S., Kimura, H., and Otsuka, H. (2015). A drop-tube particle-entrained flow solar reactor applied to thermal methane splitting for hydrogen production. *Fuel* 153: 56–66.

232 Benkowsky, G. (1990). *Induktionserwärmung: Härten, Glühen, Schmelzen, Löten, Schweißn: Grundlagen und praktische Anleitungen für Induktionserwärmungsverfahren, insbesondere auf dem Gebiet der Hochfrequenzerwärmung*, 5e, 12. Berlin: Verlag Technik.

233 Kirschning, A., Kupracz, L., and Hartwig, J. (2012). New synthetic opportunities in miniaturized flow reactors with inductive heating. *Chem. Lett.* 41: 562–570.

234 Hinzepeter, U. and Wrona, E. (2011). Use of induction heating in plastic injection molding. In: *Advances in Induction and Microwave Heating of Mineral and Organic Materials* (ed. S. Grunds), 339–344. Rijeka: InTech.

235 Bayerl, T., Duhovic, M., Mitschang, P., and Bhattacharyya, D. (2014). The heating of polymer composites by electromagnetic induction. *Composites Part A* 57: 27–40.

236 Archibald, R.C., May, N.C., and Greensfelder, B.S. (1952). Experimental catalytic and thermal cracking. *Ind. Eng. Chem.* 44: 1811–1817.

237 Fireteanu, V., Paya, B., Nuns, J. et al. (2005). Medium frequency induction-heated chemical reactor with cooling metallic envelope of the tank. *COMPEL: Int. J. Comput. Math. Electr. Electron. Eng.* 24: 324–333.

238 Tsai, W.T., Lee, M.K., and Chang, Y.M. (2006). Fast pyrolysis of rice straw, sugarcane bagasse and coconut shell in an induction-heating reactor. *J. Anal. Appl. Pyrolisis* 76: 230–337.

239 Tsai, W.-T., Chang, J.-H., Hsien, K.-J., and Chang, Y.-M. (2009). Production of pyrolytic liquids from industrial sewage sludges in an induction-heating reactor. *Bioresour. Technol.* 100: 406–412.

240 Lee, M.-K., Tsai, W.-T., Tsai, Y.-L., and Lin, S.-H. (2010). Pyrolysis of napier grass in an induction-heating reactor. *J. Anal. Appl. Pyrolysis* 88: 110–116.

241 Muley, P.D., Henkel, C., Abdollahi, K.K., and Boldor, D. (2015). Pyrolysis and catalytic upgrading of pinewood sawdust using an induction heating reactor. *Energy Fuels* 29: 7375–7385.

242 Ceylan, S., Coutable, L., Wegner, J., and Kirschning, A. (2011). Inductive heating with magnetic materials inside flow reactors. *Chem. Eur. J.* 17: 1884–1893.

243 Latifi, M., Berruti, F., and Briens, C. (2014). A novel fluidized bed and induction heated microreactor for catalyst testing. *AIChE J.* 60: 3107–3122.

244 Latifi, M. and Chaouki, J. (2015). A novel induction heating fluidized bed reactor: its design and applications in high temperature screening tests with solid feedstocks and prediction of defluidization state. *AIChE J.* 61: 1507–1523.

245 Keil, F.J. and Swamy, K.M. (1999). Reactors for sonochemical engineering – present status. *Rev. Chem. Eng.* 15: 85–155.

246 Cravotto, G. and Cintas, P. (2006). Power ultrasound in organic synthesis: moving cavitational chemistry from academia to innovative and large-scale applications. *Chem. Soc. Rev.* 35: 180–196.

247 Gogate, P.R. (2008). Cavitational reactors for process intensification of chemical processing applications: a critical review. *Chem. Eng. Process.* 47: 515–527.

248 Thompson, L.H. and Doraiswamy, L.K. (1999). Sonochemistry: science and engineering. *Ind. Eng. Chem. Res.* 38: 1215–1249.

249 Mason, T.J. (1999). *Sonochemistry.* Oxford: University Press.

250 Cravotto, G. and Cintas, P. (2011). *Handbook on Applications of Ultrasound: Sonochemistry for Sustainability: Introduction to Sonochemistry: A Historical and Conceptual Overview.* Taylor and Francis Group LLC Books.

251 Mason, T.J. (1997). Ultrasound in synthetic organic chemistry. *Chem. Soc. Rev.* 26: 443–451.

252 Chen, D., Sharma, S.K., and Mudhoo, A. (2012). *Handbook on Applications of Ultrasound: Sonochemistry for Sustainability.* Boca Raton, FL: CRC Press Taylor & Francis Group.

253 Krishna, R. and Ellenberger, J. (2002). Improving gas-liquid mass transfer in bubble columns by applying low-frequency vibrations. *Chem. Eng. Technol.* 25: 159–162.

254 Riley, N. (1998). Acoustic streaming. *Theor. Comput. Fluid Dyn.* 10: 349–356.

255 Horst, C., Gogate, P.R., and Pandit, A.B. (2007). Ultrasound reactors. In: *Modeling of Process Intensification*, 193–277. Wiley-VCH.

256 Brennen, C.E. (2013). *Cavitation and Bubble Dynamics.* Cambridge: University Press.

257 Wagterveld, R.M., Boels, L., Mayer, M.J., and Witkamp, G.J. (2011). Visualization of acoustic cavitation effects on suspended calcite crystals. *Ultrason. Sonochem.* 18: 216–225.

258 Neppiras, E.A. (1980). Acoustic cavitation. *Phys. Rep.* 61: 159–251.

259 Luther, S., Mettin, R., Koch, P., and Lauterborn, W. (2001). Observation of acoustic cavitation bubbles at 2250 frames per second. *Ultrason. Sonochem.* 8: 159–162.

260 Ashokkumar, M., Lee, J., Iida, Y. et al. (2009). The detection and control of stable and transient acoustic cavitation bubbles. *Phys. Chem. Chem. Phys.* 11: 10118–10121.

261 Ashokkumar, M. (2011). The characterization of acoustic cavitation bubbles - an overview. *Ultrason. Sonochem.* 18: 864–872.

262 Ashokkumar, M. and Pankaj (2011). *Theoretical and Experimental Sonochemistry Involving Inorganic Systems.* New York: Springer.

263 Leighton, T. (1994). *The Acoustic Bubble.* London: Academic Press.

264 Leighton, T. (1995). Bubble population phenomena in acoustic cavitation. *Ultrason. Sonochem.* 2: S123–S136.

265 Mason, T.J. and Lorimer, J.P. (1988). *Sonochemistry: Theory, Applications and Uses of Ultrasound in Chemistry.* Chichester: Ellis Horwood Publishers.

266 Zeiger, B.W. and Suslick, K.S. (2011). Sonofragmentation of molecular crystals. *J. Am. Chem. Soc.* 133: 14530–14533.

267 Sander, J.R.G., Zeiger, B.W., and Suslick, K.S. (2014). Sonocrystallization and sonofragmentation. *Ultrason. Sonochem.* 21: 1908–1915.

268 Fernandez Rivas, D., Cintas, P., and Gardeniers, H.J.G.E. (2012). Merging microfluidics and sonochemistry: towards greener and more efficient micro-sono-reactors. *Chem. Commun. (Camb.)* 48: 10935–10947.

269 Suslick, K.S., Doktycz, S.J., and Flint, E.B. (1990). On the origin of sonoluminescence and sonochemistry. *Ultrasonics* 28: 280–290.

270 Suslick, K.S., McNamara, W.B. III,, and Didenko, Y. (1999). Hot spot conditions during multi-bubble cavitation. In: *Sonochemistry and Sonoluminescence* (ed. L.A. Crum, T.J. Mason, J. Reisse and K.S. Suslick), 191–204. Dordrecht: Kluwer Publishers.

271 Suslick, K.S., Eddingsaas, N.C., Flannigan, D.J. et al. (2011). Extreme conditions during multibubble cavitation: sonoluminescence as a spectroscopic probe. *Ultrason. Sonochem.* 18: 842–846.

272 Adewuyi, Y.G. (2001). Sonochemistry: environmental science and engineering applications. *Ind. Eng. Chem. Res.* 40: 4681–4715.

273 Pflieger, R., Brau, H.P., and Nikitenko, S.I. (2010). Sonoluminescence from $OH(C_2\Sigma+)$ and $OH(A_2\Sigma+)$ radicals in water: evidence for plasma formation during multibubble cavitation. *Chem. Eur. J.* 16: 11801–11803.

274 Ndiaye, A.A., Pflieger, R., Siboulet, B. et al. (2012). Nonequilibrium vibrational excitation of OH radicals generated during multibubble cavitation in water. *J. Phys. Chem. A* 116: 4860–4867.

275 Pflieger, R., Ndiaye, A.A., Chave, T., and Nikitenko, S.I. (2015). Influence of ultrasonic frequency on swan band sonoluminescence and sonochemical activity in aqueous tert-butyl alcohol solutions. *J. Phys. Chem. B* 119: 284–290.

276 Jordens, J., Bamps, B., Gielen, B. et al. (2016). The effects of ultrasound on micromixing. *Ultrason. Sonochem.* 32: 68–78.

277 Neis, U. (2002). Intensification of biological and chemical processes by ultrasound. *TU Hamb.-Harb. Rep. Sanit. Eng.* 35: 79–90.

278 Doktycz, S.J. and Suslick, K.S. (1990). Sonochemistry. *Science* 247: 1067–1069.

279 Prozorov, T., Prozorov, R., and Suslick, K.S. (2004). High velocity inter-particle collisions driven by ultrasound. *J. Am. Chem. Soc.* 126: 13890–13891.

280 Suslick, K.S. and Price, G.J. (1999). Applications of ultrasound to materials chemistry. *Annu. Rev. Mater. Sci.* 29: 295–326.

281 Bang, J.H. and Suslick, K.S. (2010). Applications of ultrasound to the synthesis of nanostructured materials. *Adv. Mater.* 22: 1039–1059.

282 Horst, C., Kunz, U., Rosenplänter, A., and Hoffman, U. (1999). Activated solid-fluid reactions in ultrasound reactors. *Chem. Eng. Sci.* 54: 2849–2858.

283 Gogate, P.R., Mujumdar, S., and Pandit, A.B. (2003). Large-scale sonochemical reactors for process intensification: design and experimental validation. *J. Chem. Technol. Biotechnol.* 78: 685–693.

284 Mujumdar, S., Senthil Kumar, P., and Pandit, A.B. (1997). Emulsification by ultrasound: relation between intensity and emulsion quality. *Indian J. Chem. Technol.* 4: 277–284.

285 Jafari, S.M., He, Y., and Bhandari, B. (2007). Production of sub-micron emulsions by ultrasound and microfluidization techniques. *J. Food Eng.* 82: 478–488.

286 Gaikwad, S.G. and Pandit, A.B. (2008). Ultrasound emulsification: effect of ultrasonic and physicochemical properties on dispersed phase volume and droplet size. *Ultrason. Sonochem.* 15: 554–563.

287 Chisti, Y. (2003). Sonobioreactors: using ultrasound for enhanced microbial productivity. *Trends Biotechnol.* 21: 89–93.

288 Yu, G.-H., He, P.-J., Shao, L.-M., and Lee, D.-J. (2007). Enzyme activities in activated sludge flocs. *Appl. Microbiol. Biotechnol.* 77: 605–612.

289 Muthukumaran, S., Kentish, S.E., Stevens, G.W., and Ashokkumar, M. (2006). Application of ultrasound in membrane separation processes: a review. *Rev. Chem. Eng.* 22: 155–194.

290 Gogate, P.R. and Pandit, A.B. (2004). A review of imperative technologies for wastewater treatment II: hybrid methods. *Adv. Environ. Res.* 8: 553–597.

291 Comninellis, C., Kapalka, A., Malato, S. et al. (2008). Advanced oxidation processes for water treatment: advances and trends for R&D. *J. Chem. Technol. Biotechnol.* 83: 769–776.

292 Swamy, K.M. and Narayana, K.L. (2001). Intensification of leaching process by dual-frequency ultrasound. *Ultrason. Sonochem.* 8: 341–346.

293 Öncel, M.S., Ince, M., and Bayramoglu, M. (2005). Leaching of silver from solid waste using ultrasound assisted thiourea method. *Ultrason. Sonochem.* 12: 237–242.

294 Bese, A.V. (2007). Effect of ultrasound on the dissolution of copper from copper convertor slag by acid leaching. *Ultrason. Sonochem.* 14: 790–796.

295 Xie, F., Li, H., Ma, Y. et al. (2009). The ultrasonically assisted metals recovery treatment of printed circuit board waste sludge by leaching separation. *J. Hazard. Mater.* 170: 430–435.

296 Yao, J. and Li, X. (2011). Study on indium leaching from indium-poor zinc residue enhanced by ultrasonic treatment. *Adv. Mater. Res.* 201–203: 1770–1773.

297 Xie, F., Cai, T., Ma, Y. et al. (2009). Recovery of Cu and Fe from printed circuit board waste sludge by ultrasound: evaluation of industrial application. *J. Clean. Prod.* 17: 1494–1498.

298 Li, C., Xie, F., Ma, Y. et al. (2010). Multiple heavy metals extraction and recovery from hazardous electroplating sludge waste via ultrasonically enhanced two-stage acid leaching. *J. Hazard. Mater.* 178: 823–833.

299 Huang, Z., Xie, F., and Ma, Y. (2011). Ultrasonic recovery of copper and iron through the simultaneous utilization of printed circuit boards (PCB) spent acid etching solution and PCB waste sludge. *J. Hazard. Mater.* 185: 155–161.

300 Wang, L.J. and Weller, C.L. (2006). Recent advances in extraction of nutraceuticals from plants. *Trends Food Sci. Technol.* 17: 300–312.

301 Vilku, K., Mawson, R., Simons, L., and Bates, D. (2008). Applications and opportunities for ultrasound assisted extraction in the food industry – a review. *Innov. Food Sci. Emerg. Technol.* 9: 161–169.

302 Chemat, F., Zill-e-Huma, and Khan, M.K. (2011). Applications of ultrasound in food technology: processing, preservation and extraction. *Ultrason. Sonochem.* 18: 813–835.

303 Cravotto, G., Boffa, L., Mantegna, S. et al. (2008). Improved extraction of vegetable oils under high-intensity ultrasound and/or microwaves. *Ultrason. Sonochem.* 15: 898–902.

304 Narducci, O. (2013). Particle Engineering via Sonocrystallization: the aqueous adipic acid system. Doctoral thesis. University College London.

305 Gielen, B. (2017). Particle engineering by sonocrystallization. Doctoral dissertation. KU Leuven.

306 Richards, W.T. and Loomis, A.L. (1927). The chemical effects of high frequency sound waves – I. A preliminary survey. *J. Am. Chem. Soc.* 49: 3086–3100.

307 Kapustin, A.P. (1963). *The Effects of Ultrasound on the Kinetics of Crystallization*. Moscow: Academy of Sciences Press.

308 de Luque, Castro, M.D. and Priego-Capote, F. (2007). Ultrasound-assisted crystallization (sonocrystallization). *Ultrason. Sonochem.* 14: 717–724.

309 McCausland, L.J. and Cains, P.W. (2004). Power ultrasound--a means to promote and control crystallization in biotechnology. *Biotechnol. Genet. Eng. Rev.* 21: 3–10.

310 Gracin, S., Uusi-Penttilä, M., and Rasmuson, Å.C. (2005). Effect of ultrasonic irradiation on the selective polymorph control in sulfamerazine. *Cryst. Growth Des.* 5: 1787–1794.

311 Ruecroft, G., Hipkiss, D., Ly, T. et al. (2005). Sonocrystallization: the use of ultrasound for improved industrial crystallization. *Org. Process Res. Dev.* 9: 923–932.

312 Li, H., Li, H., Guo, Z., and Liu, Y. (2006). The application of power ultrasound to reaction crystallization. *Ultrason. Sonochem.* 13: 359–363.

313 Miyasaka, E., Ebihara, S., and Hirasawa, I. (2006). Investigation of primary nucleation phenomena of acetylsalicylic acid crystals induced by ultrasonic irradiation-ultrasonic energy needed to activate primary nucleation. *J. Cryst. Growth* 295: 97–101.

314 Guo, Z., Jones, A.G., Li, N., and Germana, S. (2007). High-speed observation of the effects of ultrasound on liquid mixing and agglomerated crystal breakage processes. *Powder Technol.* 171: 146–153.

315 Sivabalan, R., Gore, G.M., Nair, U.R. et al. (2007). Study on ultrasound assisted precipitation of CL-20 and its effect on morphology and sensitivity. *J. Hazard. Mater.* 139: 199–203.

316 Kordylla, A., Koch, S., Tumakaka, F., and Schembecker, G. (2008). Towards an optimized crystallization with ultrasound: effect of solvent properties and ultrasonic process parameters. *J. Cryst. Growth* 310: 4177–4184.

317 Kurotani, M., Miyasaka, E., Ebihara, S., and Hirasawa, I. (2009). Effect of ultrasonic irradiation on the behavior of primary nucleation of amino acids in supersaturated solutions. *J. Cryst. Growth* 311: 2714–2721.

318 Kim, J.M., Chang, S.M., Kim, K.S. et al. (2011). Acoustic influence on aggregation and agglomeration of crystals in reaction crystallization of cerium carbonate. *Colloids Surf., A* 375: 193–199.

319 Narducci, O. and Jones, A.G. (2012). Seeding in situ the cooling crystallization of adipic acid using ultrasound. *Cryst. Growth Des.* 12: 1727–1735.

320 Jordens, J., Gielen, B., Braeken, L., and Van Gerven, T. (2014). Determination of the effect of the ultrasonic frequency on the cooling crystallisation of paracetamol. *Chem. Eng. Process.* 84: 38–44.

321 Zeng, G., Li, H., Luo, S. et al. (2014). Effects of ultrasonic radiation on induction period and nucleation kinetics of sodium sulfate. *Korean J. Chem. Eng.* 31: 807–811.

322 Jordens, J., De Coker, N., Gielen, B. et al. (2015). Ultrasound precipitation of manganese carbonate: the effect of power and frequency on particle properties. *Ultrason. Sonochem.* 26: 64–72.

323 Bhangu, S.K., Ashokkumar, M., and Lee, J. (2016). Ultrasound assisted crystallization of paracetamol: crystal size distribution and polymorph control. *Cryst. Growth Des.* 16: 1934–1941.

324 Bari, A.H., Chawla, A., and Pandit, A.B. (2016). Sono-crystallization kinetics of K_2SO_4: estimation of nucleation, growth, breakage and agglomeration kinetics. *Ultrason. Sonochem.* 35: 196–203.

325 Gielen, B., Thimmesch, Y., Jordens, J. et al. (2016). Ultrasonic precipitation of manganese carbonate: reactor design and scale-up. *Chem. Eng. Res. Des.* 115: 131–144.

326 Jordens, J., Appermont, T., Gielen, B. et al. (2016). Sonofragmentation: effect of ultrasound frequency and power on particle breakage. *Cryst. Growth Des.* 16: 6167–6177.

327 Jordens, J. (2016). Application of acoustic energy in crystallization processes, 229. Doctoral dissertation. KULeuven.

328 Liu, Y. and Hu, J. (2009). Trapping of particles by the leakage of a standing wave ultrasonic field. *J. Appl. Phys.* 106: 034903.

329 Lenshof, A. and Laurell, T. (2010). Continuous separation of cells and particles in microfluidic systems. *Chem. Soc. Rev.* 39: 1203–1217.

330 Lenshof, A., Evander, M., Laurell, T., and Nilsson, J. (2012). Acoustofluidics 5: building microfluidic acoustic resonators. *Lab Chip* 12: 684–695.

331 Lenshof, A., Magnusson, C., and Laurell, T. (2012). Acoustofluidics 8: applications of acoustophoresis in continuous flow microsystems. *Lab Chip* 12: 1210–1223.

332 Guldiken, R., Jo, M.C., Gallant, N.D. et al. (2012). Sheathless size-based acoustic particle separation. *Sensors* 12: 905–922.

333 Laurell, T. Acoustophoresis. http://bme.lth.se/research-pages/ nanobiotechnology-and-lab-on-a-chip/research/acoustophoresis/ (accessed 31 January 2017).

334 Desarabia, E.R.F. and Gallegojuarez, J.A. (1986). Ultrasonic agglomeration of micron aerosols under standing wave conditions. *J. Sound Vib.* 110: 413–327.

335 Dormann, M., Vanderheyden, B., and Steyls, D. (2008). Advanced technique to reduce the emissions of particulate matter (PM). *Revue Métall.* 105: 586–595.

336 Sinisterra, J.V. (1992). Application of ultrasound to biotechnology: an overview. *Ultrasonics* 30: 180–185.

337 Rokhina, E.V., Lens, P., and Virkutye, J. (2009). Low-frequency ultrasound in biotechnology: state of the art. *Trends Biotechnol.* 27: 209–306.

338 Kwiatkowska, B., Bennett, J., Akunna, J. et al. (2011). Stimulation of bioprocesses by ultrasound. *Biotechnol. Adv.* 29: 768–780.

339 Tarasova, I.I., Khavski, N.N., and Dudeney, A.W.L. (1993). The effects of ultrasonics on the bioleaching of laterites. In: *Biohydrometallurgical Technologies* (ed. A.E. Torma, J.E. Wey and V.L. Lakshmanan), 357–361. The Minerals, Metals & Materials Society.

340 Sukla, L.B., Swamy, K.M., Narayana, K.L. et al. (1995). Bioleaching of Sukinda laterite using ultrasonics. *Hydrometallurgy* 37: 387–391.

341 Swamy, K.M., Narayana, K.L., and Misra, V.N. (2005). Bioleaching with ultrasound. *Ultrason. Sonochem.* 12: 301–306.

342 Anjum, F., Shahid, M., and Akcil, A. (2012). Biohydrometallugy techniques of low grade ores: a review on black shale. *Hydrometallurgy* 117–118: 1–12.

343 Zhang, Q.Q. and Jin, R.C. (2015). The application of low-intensity ultrasound irradiation in biological wastewater treatment: a review. *Crit. Rev. Environ. Sci. Technol.* 45: 2728–2761.

344 Gogate, P.R. and Patil, P.N. (2016). Sonochemical reactors. *Top. Curr. Chem.* 374: 61.

345 Wade, E.H.R. and Preece, C.M. (1978). Cavitation erosion of iron and steel. *Metall. Trans. A* 9: 1299–1310.

346 Gogate, P.R. and Pandit, A.B. (2004). Sonochemical reactors: scale up aspects. *Ultrason. Sonochem.* 11: 105–117.

347 Gogate, P.R., Tayal, R.K., and Pandit, A.B. (2006). Cavitation: a technology on the horizon. *Curr. Sci.* 91: 35–46.

348 Hallez, L., Touyeras, F., Hihn, J.Y. et al. (2010). Characterization of HIFU transducers designed for sonochemistry application: cavitation distribution and quantification. *Ultrasonics* 50: 310–317.

349 Hallez, L., Touyeras, F., Hihn, J.Y., and Bailly, Y. (2016). Characterization of HIFU transducers designed for sonochemistry application: acoustic streaming. *Ultrason. Sonochem.* 29: 420–427.

350 Bengtsson, M. and Laurell, T. (2004). Ultrasonic agitation in microchannels. *Anal. Bioanal. Chem.* 378: 1716–1721.

351 Kuhn, S., Noel, T., Gu, L. et al. (2011). A Teflon microreactor with integrated piezoelectric actuator to handle solid forming reactions. *Lab Chip* 11: 2488–2492.

352 Dong, Z.Y., Yao, C.Q., Zhang, X.L. et al. (2015). A high-power ultrasonic microreactor and its application in gas-liquid mass transfer intensification. *Lab Chip* 15: 1145–1152.

353 John, J.J., Kuhn, S., Braeken, L., and Van Gerven, T. (2016). Ultrasound assisted liquid-liquid extraction in microchannels – a direct contact method. *Chem. Eng. Process.* 102: 37–46.

354 Jordens, J., Honings, A., Degreve, J. et al. (2013). Investigation of design parameters in ultrasound reactors with confined channels. *Ultrason. Sonochem.* 20: 1345–1352.

355 Gielen, B., Jordens, J., Janssen, J. et al. (2015). Characterization of stable and transient cavitation bubbles in a milliflow reactor using a multibubble sonoluminescence quenching technique. *Ultrason. Sonochem.* 25: 31–39.

356 Gogate, P.R. and Pandit, A.B. (2001). Hydrodynamic cavitation reactors: a state of the art review. *Rev. Chem. Eng.* 17: 1–84.

357 Gogate, P.R. and Pandit, A.B. (2005). A review and assessment of hydrodynamic cavitation as a technology for the future. *Ultrason. Sonochem.* 12: 21–27.

358 Tang, S.Y. and Sivakumar, M. (2013). A novel and facile liquid whistle hydrodynamic cavitation reactor to produce submicron multiple emulsions. *AIChE J.* 59: 155–167.

359 Ramisetty, K.A., Pandit, A.B., and Gogate, P.R. (2014). Novel approach of producing oil in water emulsion using hydrodynamic cavitation reactor. *Ind. Eng. Chem. Res.* 53: 16508–16515.

360 Maddikeri, G.L., Gogate, P.R., and Pandit, A.B. (2014). Intensified synthesis of biodiesel using hydrodynamic cavitation reactors based on the interesterification of waste cooking oil. *Fuel* 137: 285–292.

361 Nakashima, K., Ebi, Y., Shibasaki-Kitakawa, N. et al. (2015). Hydrodynamic cavitation reactor for efficient pretreatment of lignocellulosic biomass. *Ind. Eng. Chem. Res.* 55: 1866–1871.

362 Sayyaadi, H. (2015). Enhanced cavitation-oxidation process of non-VOC aqueous solution using hydrodynamic cavitation reactor. *Chem. Eng. J.* 272: 79–91.

363 Mancuso, G., Langone, M., Laezza, M., and Andreottola, G. (2016). Decolourization of rhodamine B: a swirling jet-induced cavitation combined with NaCl. *Ultrason. Sonochem.* 32: 18–30.

364 Patil, P.N., Gogate, P.R., Csoka, L. et al. (2016). Intensification of biogas production using pretreatment based on hydrodynamic cavitation. *Ultrason. Sonochem.* 30: 79–86.

365 Badve, M.P., Gogate, P.R., Pandit, A.B., and Csoka, L. (2014). Hydrodynamic cavitation as a novel approach for delignification of wheat straw for paper manufacturing. *Ultrason. Sonochem.* 21: 162–168.

366 Badve, M., Gogate, P., Pandit, A., and Csoka, L. (2013). Hydrodynamic cavitation as a novel approach for wastewater treatment in wood finishing industry. *Sep. Purif. Technol.* 106: 15–21.

367 Kelkar, M.A., Gogate, P.R., and Pandit, A.B. (2008). Intensification of esterification of acids for synthesis of biodiesel using acoustic and hydrodynamic cavitation. *Ultrason. Sonochem.* 15: 188–194.

368 Witte, J.H. (1969). Mixing shocks in two-phase flow. *J. Fluid Mech.* 36: 639–655.

369 Cunningham, R.G. and Dopkin, R.J. (1974). Jet breakup and mixing throat lengths for the liquid jet gas pump. *J. Fluids Eng.* 96: 216–226.

370 Tinge, J.T. and Rodriguez Casado, A.J. (2002). Influence of pressure on the gas hold-up of aqueous activated carbon slurries in a down flow jet loop reactor. *Chem. Eng. Sci.* 57: 3575–3580.

371 Pangarkar, V.G. (2015). *Design of Multiphase Reactors*. Hoboken, NJ: Wiley.

372 Blenke, H., Bohner, K., and Schuster, S. (1965). Beitrag zur optimalen Gestaltung chemischer Reaktoren. *Chem. Ing. Tech.* 37: 289–294.

373 Nagel, O., Kürten, H., and Sinn, R. (1970). Strahldüsenreaktoren Teil I: Die Anwendung des Ejektorprincips zur Verbesserung der Gasbasorption in Blasensäulen. *Chem. Ing. Tech.* 42: 474–479.

374 Zahradník, J., Kaštánek, F., Kratochvíl, J., and Rylek, M. (1982). Hydrodynamic characteristics of gas-liquid beds in contactors with ejector-type gas distributors. *Collect. Czech. Chem. Commun.* 47: 1939–1949.

375 Wachsmann, U., Räbiger, N., and Vogelpohl, A. (1985). Einfluss der Geometrie auf die Hydrodynamik und den Stoffaustausch im Kompaktreaktor. *Chem. Ing. Tech.* 57: 346–347.

376 Dutta, N.N. and Raghavan, K.V. (1987). Mass transfer and hydrodynamic characteristics of loop reactors with downflow liquid jet ejector. *Chem. Eng. J.* 36: 111–121.

377 Warnecke, H.-J., Geisendörfer, M., and Hempel, D.C. (1988). Mass transfer behaviour of gas-liquid jet loop reactors. *Chem. Eng. Technol.* 11: 306–311.

378 Cramers, P.H.M.R., Beenackers, A.A.C.M., and Van Dierendonck, L.L. (1992). Hydrodynamics and mass transfer characteristics of a loop-venturi reactor with a downflow liquid jet ejector. *Chem. Eng. Sci.* 47: 3557–3564.

379 Van Dierendonck, L.L., Zahradnik, J., and Linek, V. (1998). Loop venturi reactors – a feasible alternative to stirred tank reactors? *Ind. Eng. Chem. Res.* 37: 734–738.

380 Cramers, P.H.M.R. and Beenackers, A.A.C.M. (2001). Influence of the ejector configuration, scale and gas density on the mass transfer characteristics of gas-liquid ejectors. *Chem. Eng. J.* 82: 131–141.

381 Zehner, P. and Kraume, M. (2000). Bubble columns. In: *Uhlmann's Encyclopedia of Industrial Chemistry* (ed. G. Bellussi, M. Bohnet, J. Bus, et al.). Wiley-VCH, https://doi.org/10.1002/14356007.b04_275.

382 Gourich, B., Vial, C., Soulami, M.B. et al. (2008). Comparison of hydrodynamic and mass transfer performances of an emulsion loop-venturi reactor in cocurrent downflow and upflow configurations. *Chem. Eng. J.* 140 (1): 439–447.

383 Mattick, A.T., Russell, D.A., Hertzberg, A., and Knowlen, C. (1995). Shock-controlled chemical processing. In: *Shockwaves @ Marseille II*, Proceedings of International Symposium (ed. R. Brun and L.Z. Dumitrescu), 209–214. Berlin: Springer-Verlag.

384 Cheng, A.T.Y. (1997). A high-intensity gas-liquid tubular reactor under supersonic two phase flow conditions. In: *Process Intensification in Practice – Applications and Opportunities* (ed. J. Semel), 205–219. Bury St Edmunds: Mechanical Engineering Publications Limited.

385 Gross, G. (1998). Verfahren und Vorrichtung zur Umwandlung von Schwefelwasserstoff in elementaren Schwefel. DE 197 18 261.

386 Gross, G. (1998). Vorrichtung und Verfahren zur Durchfürung von oxidierenden Reaktionen in fluidisierten Partikelschichten. DE 197 22 382.

387 Gross, G. (2000). Supersonic oxygen injection doubles the capacity of fluidized-bed reactors. In: *ACHEMA 2000, International Meeting on Chemical Engineering, Environmental Protection and Biotechnology, Abstracts of the Lecture Groups Chemical Engineering and Reaction Technology*, 161–162. Frankfurt am Main: DECHEMA.

388 Gross, G. and Ludwig, P. (2003). Transversal oxygen supply. Supersonic injection increases performance of sludge combustion plants. *Chem. Anlagen Verfahren* 36 (3): 84–86.

389 Groen, D.J., Noorman, H.J., and Stankiewicz, A. (2005). Improved method for aerobic fermentation intensification. In: *Proceedings of the International Conference Sustainable (Bio)Chemical Process Technology*, Delft (27–29 September 2005) (ed. P.J. Jansens, A. Green and A. Stankiewicz), 105–112. Cranfield: BHR Group Ltd.

390 Salazar, J.A., Joshi, M., Ard, C.D., and Zelnik, D.J. (2013). Nozzle reactor and method of use. WO 2013/162667.

391 Majumder, D. and Leonard, L.E. (2015). Methane conversion apparatus and process using a supersonic flow reactor. US 2015/0165412.

392 Bowes, H. (1889). Process of facilitating chemical reactions. US 410067.

393 Tamir, A. (1994). *Impinging Stream Reactors: Fundamentals and Applications*. Elsevier.

394 Wu, Y. (2007). *Impinging Streams: Fundamentals, Properties and Applications*. Elsevier.

395 Sprehe, M., Gaddis, E., and Vogelpohl, A. (1998). On the mass transfer in an impinging-stream reactor. *Chem. Eng. Technol.* 21: 19–21.

396 Gaddis, E.S. and Vogelpohl, A. (1992). The impinging-stream reactor: a high performance loop reactor for mass transfer controlled chemical reactions. *Chem. Eng. Sci.* 47: 2877–2882.

397 Mudimu, A.O., Gaddis, E.S., and Vogelpohl, A. (2000). Gas holdup in an impinging-stream reactor. *Chem. Eng. Technol.* 23: 661–663.

398 Dehkordi, A.M. and Savari, C. (2011). Determination of interfacial area and overall volumetric mass-transfer coefficient in a novel type of two impinging streams reactor by chemical method. *Ind. Eng. Chem. Res.* 50: 6426–6435.

399 Wu, Y., Zhou, Y., Guo, J., and Yuan, J. (2010). Features of impinging streams intensifying processes and their applications. *Int. J. Chem. Eng.* 2010: 1–16. https://doi.org/10.1155/2010/681501.

400 Cai, J., Zhang, J., and Song, W. (2012). Characteristics of coal liquefaction in an impinging stream reactor. *Energy Fuels* 26: 3510–3513.

401 Wu, Y., Li, Q., and Li, F. (2007). Desulfurization in the gas-continuous impinging stream gas-liquid reactor. *Chem. Eng. Sci.* 62: 1814–1824.

402 Sohrabi, M. and Marvast, M.A. (2000). Application of a continuous two impinging streams reactor in solid-liquid enzyme reactions. *Ind. Eng. Chem. Res.* 39: 1903–1910.

403 Lishun, H., Xinjun, W., Xiaoming, L. et al. (2007). The characteristics of liquid-phase methanol synthesis in an impinging stream reactor. *Chem. Eng. Process. Process Intensif.* 46: 905–909.

404 Royaee, S.J., Sohrabi, M., and Fallah, N. (2012). A comprehensive study on wastewater treatment using photo-impinging streams reactor: continuous treatment. *Korean J. Chem. Eng.* 29: 1577–1584.

405 Dehkordi, A.M. (2002). Liquid-liquid extraction with an interphase chemical reaction in an air-driven two-impinging-streams reactor: effective interfacial area and overall mass transfer coefficient. *Ind. Eng. Chem. Res.* 41: 4085–4093.

406 Tamir, A., Elperin, I., and Luzzatto, K. (1984). Drying in a new two impinging streams reactor. *Chem. Eng. Sci.* 39: 139–146.

407 Ravi Kumar, D.V., Prasad, B.L.V., and Kulkarni, A.A. (2013). Impinging jet micromixer for flow synthesis of nanocrystalline MgO: role of mixing/impingement zone. *Ind. Eng. Chem. Res.* 52: 17376–17382.

408 Liu, Z.-W., Zhang, Q.-C., Wen, L.-X., and Chen, J.-F. (2016). Preparation of ultrafine manganese dioxide by micro-impinging stream reactors and its electrochemical properties. *Can. J. Chem. Eng.* 94: 461–468.

409 Podbielniak, W.J., Kaiser, H.R., and Ziegenhorn, G.J. (1970). Centrifugal solvent extraction. *Chem. Eng. Prog. Symp. Ser.* 66: 43–50.

410 Ramshaw, C. and Mallison, R.H. (1981). Mass transfer process. US Patent 4,282,255.

411 Ramshaw, C. (1983). HiGee distillation—an example of process intensification. *Chem. Eng.* 389: 13–14.

412 Trent, D. (2004). Chemical processing in high-gravity fields. In: *Re-Engineering the Chemical Processing Plant: Process Intensification* (ed. A. Stankiewicz and J.A. Moulijn), 33–67. New York: Marcel Dekker Inc.

413 Rao, D.P., Bhowal, A., and Goswami, P.S. (2004). Process intensification in rotating packed beds (HIGEE): an appraisal. *Ind. Eng. Chem. Res.* 43: 1150–1162.

414 Zhao, H., Shao, L., and Chen, J.-F. (2010). High-gravity process intensification technology and application. *Chem. Eng. J.* 156: 588–593.

415 Górak, A. and Stankiewicz, A. (2011). Intensified reaction and separation systems. *Annu. Rev. Chem. Biomol. Eng.* 2: 431–451.

416 Guo, F., Zheng, C., Guo, K. et al. (1997). Hydrodynamics and mass transfer in cross-flow rotating packed bed. *Chem. Eng. Sci.* 52: 3853–3859.

417 Lin, C.-C., Ho, T.-J., and Liu, W.-T. (2002). Distillation in a rotating packed bed. *J. Chem. Eng. Jpn.* 35: 1298–1304.

418 Zheng, C., Guo, K., Song, Y. et al. (1997). Industrial practice of HIGRAVITEC, in water deaeration. In: *Process Intensification in Practice: Applications and Opportunities* (ed. J. Semel), 273–287. Bury St. Edmunds: Mechanical Engineering Publications Limited.

419 Chen, J.-F. and Shao, L. (2007). Recent advances in nanoparticles production by high gravity technology – from fundamentals to commercialization. *J. Chem. Eng. Jpn.* 40: 896–904.

420 Hu, T.-T., Wang, J.-X., Shen, Z.-G., and Chen, J.-F. (2008). Engineering of drug nanoparticles by HGCP for pharmaceutical applications. *Particuology* 6: 239–251.

421 Wang, G.Q., Xu, Z.C., and Ji, J.B. (2011). Progress on higee distillation – introduction to a new device and its industrial applications. *Chem. Eng. Res. Des.* 89: 1434–1442.

422 Gebauer, K., Steiner, L., and Hartland, S. (1982). Zentrifugalextraktion – Eine Literaturübersicht. *Chem. Ing. Tech.* 54: 476–496.

423 Ramshaw, C. (2004). The spinning disc reactor. In: *Re-Engineering the Chemical Processing Plant: Process Intensification* (ed. A. Stankiewicz and J.A. Moulijn), 69–119. New York: Marcel Dekker Inc.

424 Aoune, A. and Ramshaw, C. (1999). Process intensification: heat and mass transfer characteristics of liquid films on rotating discs. *Int. J. Heat Mass Transfer* 42: 2543–2556.

425 Brauner, N. and Maron, D.M. (1982). Characteristics of inclined thin films, waviness and the associated mass transfer. *Int. J. Heat Mass Transfer* 25: 99–110.

426 Oxley, P., Brechtelsbauer, C., Ricard, F. et al. (2000). Evaluation of spinning disc reactor technology for the manufacture of pharmaceuticals. *Ind. Eng. Chem. Res.* 39: 2175–2182.

427 Vicevic, M., Jachuck, R.J., and Scott, K. (2001). Process intensification for green chemistry: rearrangement of α-pinene oxide using a catalyzed spinning disc reactors. In: *Better Processes for Better Products*, 4th International Conference on Process Intensification for the Chemical Industry (ed. M. Gough), 201–206. Cranfield: BHR Group.

428 De Caprariis, B., Di Rita, M., Stoller, M. et al. (2012). Reaction-precipitation by a spinning disc reactor: influence of hydrodynamics on nanoparticles production. *Chem. Eng. Sci.* 76: 73–80.

429 Khan, W.H. and Rathod, V.K. (2014). Process intensification approach for preparation of curcumin nanoparticles via solvent-nonsolvent nanoprecipitation using spinning disc reactor. *Chem. Eng. Process. Process Intensif.* 80: 1–10.

430 Mohammadi, S., Harvey, A., and Boodhoo, K.V.K. (2014). Synthesis of TiO_2 nanoparticles in a spinning disc reactor. *Chem. Eng. J.* 258: 171–184.

431 Boiarkina, I., Norris, S., and Patterson, D.A. (2013). The case for the photocatalytic spinning disc reactor as a process intensification technology: comparison to an annular reactor for the degradation of methylene blue. *Chem. Eng. J.* 225: 752–765.

432 Meeuwse, M., Van der Schaaf, J., and Schouten, J.C. (2012). Multistage rotor-stator spinning disc reactor. *AIChE J.* 58: 247–255.

433 De Beer, M.M., Pezzi Martins Loane, L., Keurentjes, J.T.F. et al. (2014). Single phase fluid-stator heat transfer in a rotor–stator spinning disc reactor. *Chem. Eng. Sci.* 119: 88–98.

434 De Beer, M.M., Keurentjes, J.T.F., Schouten, J.C., and Van der Schaaf, J. (2015). Intensification of convective heat transfer in a stator-rotor-stator spinning disc reactor. *AIChE J.* 61: 2307–2317.

435 Meeuwse, M., Van der Schaaf, J., Kuster, B.F.M., and Schouten, J.C. (2010). Gas-liquid mass transfer in a rotor-stator spinning disc reactor. *Chem. Eng. Sci.* 65: 466–471.

436 Meeuwse, M., Lempers, S., Van der Schaaf, J., and Schouten, J.C. (2010). Liquid-solid mass transfer and reaction in a rotor-stator spinning disc reactor. *Ind. Eng. Chem. Res.* 49: 10751–10757.

437 Visscher, F., Van der Schaaf, J., De Croon, M.H.J.M., and Schouten, J.C. (2012). Liquid-liquid mass transfer in a rotor-stator spinning disc reactor. *Chem. Eng. J.* 185–186: 267–273.

438 Stemmet, C.P., Meeuwse, M., Van der Schaaf, J. et al. (2007). Gas–liquid mass transfer and axial dispersion in solid foam packings. *Chem. Eng. Sci.* 62 (18): 5444–5450.

439 Trambouze, P. and Euzen, J.-P. (1988). *Chemical Reactors.* Editions OPHRYS.

440 Schuur, B., Kraai, G.N., Winkelman, J.G.M., and Heeres, H.J. (2012). Hydrodynamic features of centrifugal contactor separators: experimental studies on liquid hold-up, residence time distribution, phase behavior and drop size distributions. *Chem. Eng. Process. Process Intensif.* 55: 8–19.

441 Yarbro, S.L. and Schreiber, S.B. (2003). Using process intensification in the actinide processing industry. *J. Chem. Technol. Biotechnol.* 78: 254–259.

442 Hallett, A.J., Kwant, G.J., and De Vries, J.G. (2009). Continuous separation of racemic 3,5-dinitrobenzoyl-amino acids in a centrifugal separator with the aid of cinchona-based chiral host compounds. *Chem. Eur. J.* 15: 2111–2120.

443 Schuur, B., Winkelman, J.G.M., De Vries, J.G., and Heeres, H.J. (2010). Experimental and modeling studies on the enantio-separation of 3,5-dinitrobenzoyl-(R),(S)-leucine by continuous liquid-liquid extraction in a cascade of centrifugal contactor separators. *Chem. Eng. Sci.* 65: 4682–4690.

444 Abduh, M.Y., Van Ulden, W., Kalpoe, V. et al. (2013). Biodiesel synthesis from *Jatropha curcas* L. oil and ethanol in a continuous centrifugal contactor separator. *Eur. J. Lipid Sci. Technol.* 115: 123–131.

445 Atiemo-Obeng, V.A. and Calabrese, R.V. (2003). Rotor-stator mixing devices. In: *Handbook of Industrial Mixing: Science and Practice* (ed. E.L. Paul, V.A. Atiemo-Obeng and S.M. Kresta), 479–505. Hoboken, NJ: Wiley.

446 Rodgers, T.L., Cooke, M., Hall, S. et al. (2011). Rotor-stator mixers. *Chem. Eng. Trans.* 24: 1411–1416.

447 Sparks, T.G., Brown, D.E., and Green, A.J. (1995). Assessing rotor/stator mixers for rapid chemical reactions using overall power characteristics. In: *1st International Conference on Process Intensification for the Chemical Industry*, BHR Group Conference Series, vol. 18 (ed. C. Ramshaw), 97–113. Cranfield: BHR Group Ltd.

448 Bourne, J.R. and Studer, M. (1992). Fast reactions in rotor-stator mixers of different size. *Chem. Eng. Proc.* 31: 285–296.

449 Qian, G.-H., Bágyi, I., Pfeffer, R. et al. (1998). A parametric study of a horizontal rotating fluidized bed using slotted and sintered metal cylindrical gas distributors. *Powder Technol.* 100: 190–199.

450 Qian, G.-H., Burdick, I.W., Pfeffer, R. et al. (2004). Soot removal from diesel engine exhaust using a rotating fluidized bed filter. *Adv. Environ. Res.* 8: 387–395.

451 Quevedo, J., Pfeffer, R., Shen, Y. et al. (2006). Fluidization of nanoagglomerates in a rotating fluidized bed. *AIChE J.* 52: 2401–2412.

452 Watano, S., Nakamura, H., Hamada, K. et al. (2004). Fine particle coating by a novel fluidized bed coater. *Powder Technol.* 141: 172–176.

453 De Wilde, J. and De Broqueville, A. (2007). Rotating fluidized beds in a static geometry: experimental proof of concept. *AIChE J.* 53: 793–810.

454 Dutta, A., Ekatpure, R.P., Heynderickx, G.J. et al. (2010). Rotating fluidized beds with a static geometry: guidelines for design and operating conditions. *Chem. Eng. Sci.* 65: 1678–1693.

455 De Wilde, J. (2014). Gas-solid fluidized beds in vortex chambers. *Chem. Eng. Process. Process Intensif.* 85: 256–290.

456 De Broqueville, A. and De Wilde, J. (2009). Numerical investigation of gas-solid heat transfer in rotating fluidized beds in a static geometry. *Chem. Eng. Sci.* 64: 1232–1248.

457 Rosales Trujillo, W. and De Wilde, J. (2012). Fluid catalytic cracking in a rotating fluidized bed in a static geometry: a CFD analysis accounting for the distribution of the catalyst coke content. *Powder Technol.* 221: 36–46.

458 Ashcraft, R.W., Heynderickx, G.J., and Marin, G.B. (2012). Modeling fast biomass pyrolysis in a gas-solid vortex reactor. *Chem. Eng. J.* 207–208: 195–208.

459 Bisschops, M.A.T. (1999). Centrifugal adsorption technology. PhD thesis. TU Delft.

460 Bisschops, M.A.T., Van Hateren, S.H., Luyben, K.C.A.M., and Van der Wielen, L.A.M. (2000). Mass transfer performance of centrifugal adsorption technology. *Ind. Eng. Chem. Res.* 39: 4376–4382.

461 Penrose, A., Myers, P., Bartle, K., and McCrosen, S. (2004). Development and assessment of a miniaturized centrifugal chromatograph for reversed-phase separations in micro-channels. *Analyst* 129: 704–709.

462 Haeberle, S., Brenner, T., Zengerle, R., and Ducrée, J. (2006). Centrifugal extraction of plasma from whole blood on a rotating disk. *Lab Chip* 6: 776–781.

463 Roy, P., Anand, N.K., and Banerjee, D. (2013). A review of flow and heat transfer in rotating microchannels. *Procedia Eng.* 56: 7–17.

5

SYNERGY – PI Approaches in Functional Domain

5.1 Combining Functions

Synergy is defined as creation of a whole that is greater than the simple sum of its parts. The term synergy comes from the Attic Greek word "synergia" (συνεργία) meaning "working together." Process intensification in the functional domain aims at achieving synergy by combining one or more functions (e.g. heat transfer and mixing) in a single device or process step. Such combination results primarily in the enhancement of process efficiency (e.g. separation efficiency). It must be noted that there are numerous examples of combined functions, in which the "real" synergy (a whole greater than the simple sum of its parts) cannot be proven, yet in which significant process intensification effects (in terms of efficiency) can clearly be seen.

When integrating functions or process steps, one will always be confronted with two very important issues:

- the *degrees of freedom* of the integrated system
- the *feasible operation window* of the integrated system.

The first issue concerns the flexibility versus stiffness of the integrated system, its adaptability, controllability, etc. Figure 5.1 presents an example of the decreasing number of degrees of freedom (and consequently decreasing design/operational flexibility) in a system that integrates a catalytic equilibrium reaction with a membrane separation [1]. One can clearly see that the highest degree of integration (making the membrane a catalytic reactor – on the right) actually couples all process variables (e.g. temperatures and pressures) and design parameters (e.g. membrane area) that were not coupled when the membrane separator was an autonomous unit operating in the reactor loop (Figure 5.1a).

The feasible operation window concerns the range of process parameter values, within which the integrated system can operate in a feasible way. Let us, for example, consider an integrated reactive separation system shown in Figure 5.2. The feasible pressure and temperature range for the reaction is different than the range for the separation operation and different than the range resulting from constraints in equipment design. The resulting feasible operational window of the integrated system is limited to the overlap area of the above-mentioned partial windows. The size of that resulting operational window

The Fundamentals of Process Intensification, First Edition.
Andrzej Stankiewicz, Tom Van Gerven, and Georgios Stefanidis.
© 2019 Wiley-VCH Verlag GmbH & Co. KGaA. Published 2019 by Wiley-VCH Verlag GmbH & Co. KGaA.

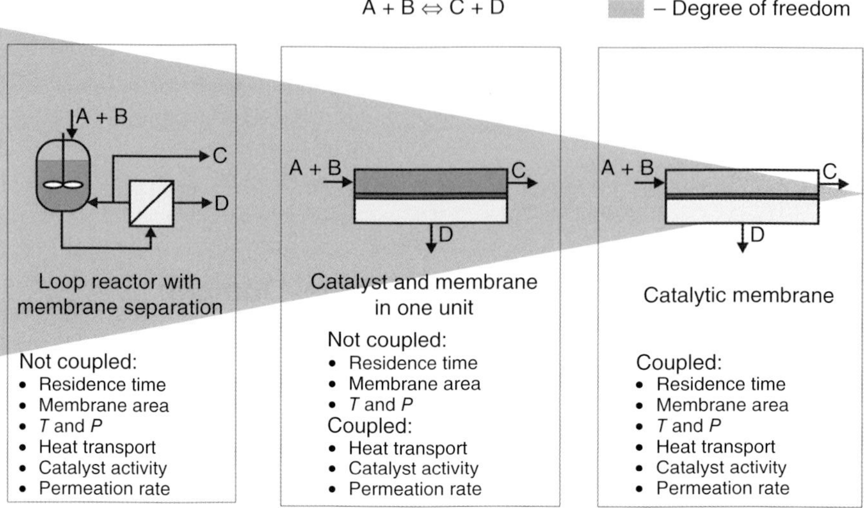

Figure 5.1 The higher degree of integration, the less degrees of freedom: example of a membrane reactor. Source: Schembecker and Tlatlik 2003 [1]. Reproduced with permission of Elsevier.

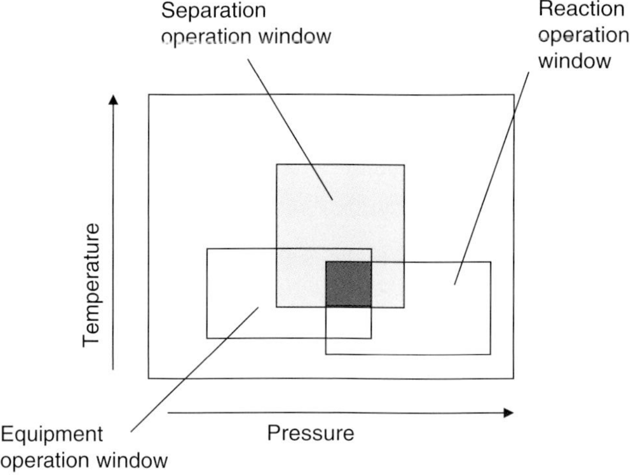

Figure 5.2 Feasible operation window of a reactive separation process. Source: Schembecker and Tlatlik 2003 [1]. Reproduced with permission of Elsevier.

may present a serious limitation of the system integration; often, that window may even not exist. One can try to enlarge the feasible operation window by shifting the operational parameter ranges of the reaction (e.g. by changing the catalyst type), separation (e.g. by adding an extra compound, creating local superheating, or switching to nonequilibrium), and equipment (e.g. by changing the design concept or characteristic dimensions).

5.2 Synergies at Molecular Scale

5.2.1 Multifunctional Catalysts

The concept of multifunctional catalysts is usually understood as a combination of two or more catalytic functions that, acting synergistically, facilitate a sequence of catalytic reactions [2]. The concept is pretty old as it dates from the early 1960s, and the pioneering papers by Weisz [3], who, among other things, showed that the presence of hydrogenation–dehydrogenation function and an acidic function in the catalyst improved the conversion of methylcyclopentadiene. The works by Weisz were followed by the classical paper by Gunn and Thomas [4], in which a model of *bifunctional catalyst* was proposed to show the effect of catalyst formulation on the reaction rates and product yields. Later, bifunctional catalysts have been split further into two subcategories [5]:

− catalysts in which two distinct catalytic sites act simultaneously on *the same* reaction step in a cooperative and synergistic way and
− catalysts in which distinct catalytic sites act on two *different* process steps, one after each other, performing *tandem catalysis* where two or more transformations take place sequentially, without any separation or isolation of the intermediates (*one-pot reactions*).

Bifunctional catalysts offer unprecedented possibilities for intensifying catalytic reactions in terms of reactivity and selectivity [6–8]. Grotjahn [7] states that the combination of metal electrons and a pendant base or acid in the vicinity of an organometallic active site can lead to rate accelerations of 1000- to 10 000-fold in the case of alkyne hydration and alkene isomerization. Corma and coworkers [8] demonstrated that using multifunctional solid catalyst, it was possible to decrease the *E*-factor (defined as the mass ratio of waste to desired product) by one order of magnitude or more, with respect to any conventional routes. In another paper form the same group, a bifunctional metal–organic framework (MOF) compound has been investigated in a one-pot synthesis of methanol from citronellal [9]. Full citronellal conversion with 86% methanol selectivity was achieved in the MOF catalyst after 18 hours, whereas up to 30 hours were necessary for attaining similar values on the reference catalyst.

A classic example of the industrial use of multifunctional catalysts is the manufacturing of gasoline of high antiknock quality in Chevron's *rheniforming* process. Here, the catalyst contains an optimized dehydrogenation/hydrogenation function through a combination of Pt and Re, plus an acidic function on the surface of alumina support, to facilitate the complicated dehydrocyclization reactions of hydrocarbons [3]. Bifunctional acid–base catalysts have found numerous large-scale applications in the industry, as reviewed in [10, 11].

Adsorptive catalysts present another concept of combining different functionalities within a catalyst particle [12, 13]. In this case, specific adsorbent sites are incorporated into the catalyst, as shown in Figure 5.3.

Four most important advantages of the multifunctional adsorptive catalyst with regard to a conventional catalyst-adsorbent system, shown on the left-hand side in Figure 5.3 in the case of an equilibrium-limited reaction, are

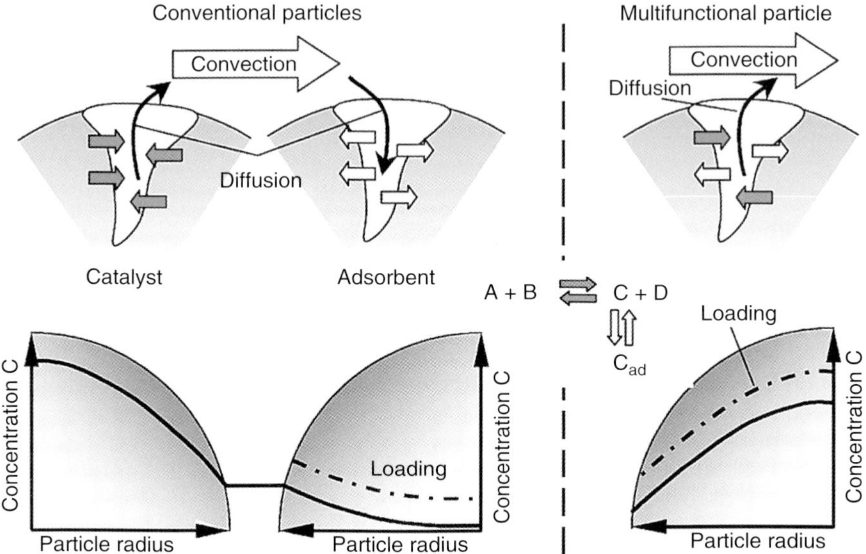

Figure 5.3 Potential advantages of a multifunctional adsorptive catalyst over a conventional one. Source: Dietrich et al. 2005 [13]. Reproduced with permission of Elsevier.

- reduction of the diffusion paths, i.e. lower mass transfer resistance;
- better removal of C because of higher concentrations;
- higher reaction rate because of lower surface concentration of C; and
- better adsorbent utilization (loading).

One of the main questions addressed in the modeling studies of adsorptive catalyst systems concerns the optimum spatial distribution of both functionalities inside the particle. Options range from a shell adsorbent with a catalytic core, through a uniform mixture of both functionalities in the entire particle volume, to a shell catalyst with an adsorbent core (Figure 5.4).

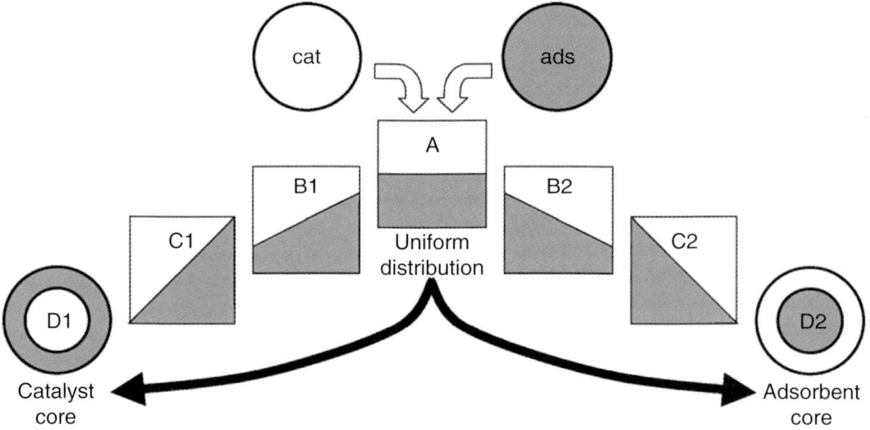

Figure 5.4 Various options for catalyst-adsorbent distribution inside an adsorptive catalyst particle. Source: Dietrich et al. 2005 [13]. Reproduced with permission of Elsevier.

Table 5.1 Yield of hydrogen in an adsorptive catalyst for different adsorbent–catalyst arrangements.

Arrangement	Yield	Fraction of adsorbent from shell to core				
Catalyst core	8.732	0.8006	0.6137	0.4511	0.3229	0.2374
Adsorbent core	401.46	0.2994	0.4863	0.6489	0.7771	0.8626
Uniform mixture	150.76	0.5000	0.5000	0.5000	0.5000	0.5000

Source: Kapil et al. 2008 [14]. Reproduced with permission of John Wiley and Sons.

Kapil et al. [14] modeled, among other things, the natural gas reforming reaction with sorption, to generate pure hydrogen for fuel cells. The amount of hydrogen produced varied strongly with the adsorbent–catalyst distribution, as shown in Table 5.1.

Finally, the multifunctionality in a catalyst particle can enhance the shape-selective control of the reaction taking place. Foley et al. [15] studied the shape-selective methylamine synthesis in a $SiO_2-Al_2O_3$ catalyst integrated with carbon molecular sieves (CMS). The reaction investigated was the ammonia reaction with methanol yielding monomethylamine (MMA), dimethylamine (DMA), and trimethylamine (TMA). The latter presents the unwanted product and its synthesis should be suppressed. The CMS layer provides a "filtration" function: it is permeable to ammonia, methanol, and smaller amines (MMA and DMA). However, the larger molecules of TMA are retained within the catalyst particle where they undergo the backward equilibrium reactions to the desired products (Figure 5.5). Various distributions of oxides and CMS (oxide particle encapsulated by the CMS layer or oxide microcrystals dispersed in a high carbon matrix of CMS) were investigated experimentally. It has been

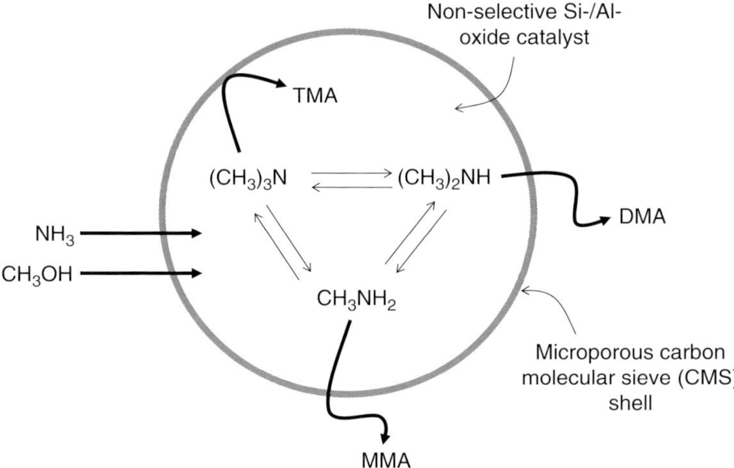

Figure 5.5 Scheme of methylamine synthesis in a bifunctional oxide-CMS catalyst. Source: Adapted from Agar 1999 [143].

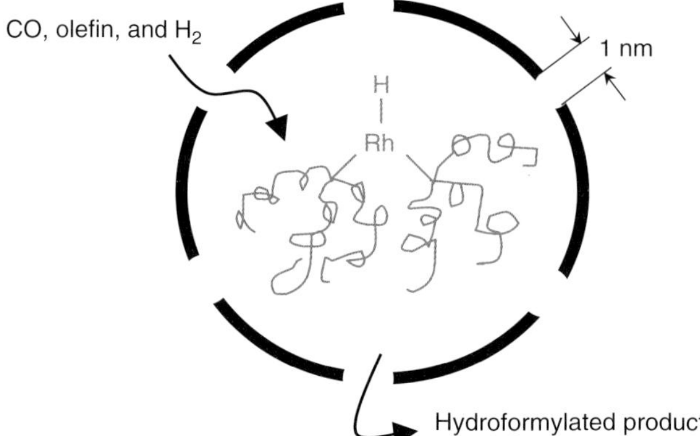

Figure 5.6 Encapsulation of homogeneous Rh complex catalyst inside a porous silica microsphere. Source: Adapted from Dautzenberg and Mukherjee 2001 [16].

shown that the coverage of acid sites with CMS provided an effective shielding against the production of TMA. One of the integrated catalysts delivered a 2.7-fold increase in the (MMA + DMA) to TMA ratio, compared to a standard, non-shape-selective catalyst.

A concept similar to the above was also reported in the case of the hydroformylation process [16] where homogeneous Rh complex-based catalyst was encapsulated inside a porous silica microsphere, which additionally introduced advantageous diffusional limitations for carbon monoxide (Figure 5.6).

5.2.2 Synergistic Use of Alternative Energy Forms

A combination of alternative energy forms can be considered "an intensified form of process intensification." Already fundamental research in the field of chemical physics provides examples of the synergistic use of two different energy forms. One of such examples is shown in Figure 5.7 where carbonyl sulfide molecules are first aligned and oriented in a hexapole *electric field* (left) and then dissociated by a *laser beam* [17].

On a slightly larger scale, interesting effects have been reported by combining acoustic (ultrasound) and electromagnetic (microwave or UV) energies. These effects are mostly due to the fact that various energy forms address various elementary limitations in a chemical process, as shown in Table 5.2. For example, the functional addition of ultrasonic field to a photocatalytic system can result in the following advantageous effects:

- increased local T and P
- cleaning and sweeping of photocatalysts (e.g. TiO_2)
- improved mass transport toward the catalyst
- creation of new solid surfaces (because of fragmentation/deagglomeration of catalyst)
- cavitation-induced radical formation

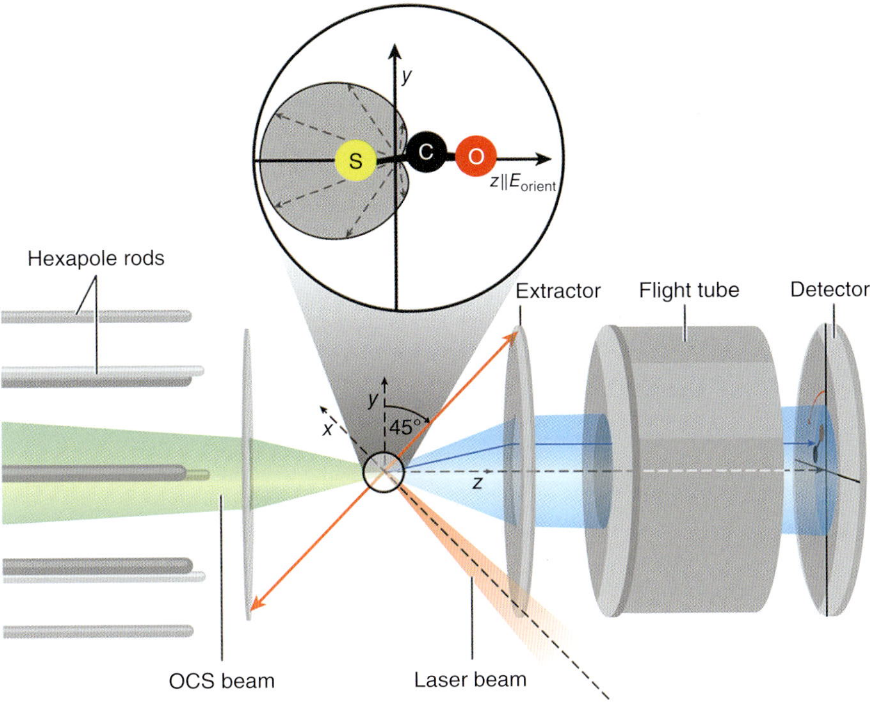

Figure 5.7 Alignment and dissociation of carbonyl sulfide (OCS) molecules by hexagonal electric field and laser beam. Source: Górak and Stankiewicz 2011 [144]. Reproduced with permission of Annual Reviews.

Table 5.2 Elementary limitations in a process addressed by different energy forms.

Energy form	Elementary limitations addressed				
	Mixing	Mass transfer	Heat transfer	Reaction mechanism	Reaction kinetics
Ultrasound (US)	×	×	(×)	×	×
Electric field (EF)	×	×	(×)	×	×
Microwave (MW)			×		×
Light (UV)				×	×

Some examples of the reported effects resulting from a combination of different energy forms are presented in Table 5.3. As one can see, a vast majority of those literature examples concern environmental engineering processes.

The practical application of combined alternative energy forms is obviously far more complex than the use of one or another form of energy alone and has not been seen on the industrial scale yet. As discussed in Chapter 4, modeling and scale-up of operations using single alternative energy forms is often still in early stage. Combining them increases the difficulties more than linearly. Besides, issues such as integrated equipment design or process instrumentation and control must be adequately addressed.

Table 5.3 Examples of synergistic use of alternative energy forms.

Energy forms involved	Reaction/process studied	Reported effect	References
US + UV	Degradation of formaldehyde in water	Reaction rate constant US: <0.1/h; UV: 4.6/h; US + UV: 5.8/h	[153]
	Degradation of benzaldehyde in water	Reaction rate constant US: 2.9/h; UV: 4.0/h; US + UV: 12/h	[153]
	Mineralization of bisphenol A	Rate of DOC (dissolved organic carbon) removal with combined US/UV up to c. 3.5 times higher than the sum of corresponding efficiencies using US and UV separately.	[154]
	Degradation of 2,4,6-trichlorophenol	Percentage of degradation using combined US/UV two to three times higher than using any of those techniques separately	[155]
US + MW	Suzuki reactions	US yield: 54% MW yield: 64% US + MW yield: 88%	[156]
	Hydrazinolysis of methyl salicylate	US: 79% yield after 1.5 h; MW: 80% yield after 18 min; US + MW: 84% yield after 40 s	[157]
	Heterogeneous Williamson synthesis of benzyl phenyl ether	US: 67% yield after 2 h; MW: 46% yield after 50 min; US + MW: 83% yield after 60 s	[158]
	Degradation of phenol	Degradation rate under three reaction conditions, respectively: US: 17%, 28%, and 15% MW: 0%, 7%, and 59% US + MW: 27%, 65%, and 76%	[159]
	Digestion of sunflower oil and sesame oil	Times to achieve 90% yield: Sunflower oil – 40 min MW, 25 min MW + US; Sesame oil – 50 min MW, 35 min MW + US	[160]
	Pyrolysis of urea	Yield after one hour: MW: 46% MW + US: 57%	[161]
	Esterification of propanol	Yield after one hour: MW: 91% MW + US: 99%	[161]

5.3 Synergies in Processing Units – Multifunctional Equipment and Integrated Operations

5.3.1 Integrating Catalysis and Mixing – The Monolithic Stirrer Concept

Stirred tank reactors for carrying out heterogeneous catalytic reactions present a standard technology used in the fine chemical and pharmaceutical industries, thanks to their application flexibility in a multiproduct environment. However, these reactors exhibit several important drawbacks, such as the necessity of catalyst separation from the reaction mixture as well as attrition and agglomeration of catalyst particles. A concept of the *monolithic stirrer reactor* (MSR), introduced by Edvinsson-Albers [18], addresses those shortcomings. The concept combines the catalytic and the mixing functions by employing pieces of a monolithic catalyst as the impeller paddles. Both vertical (Figure 5.8a) and horizontal (Figure 5.8b) shaft configurations can be realized. The latter allows carrying gas–liquid reactions on a solid catalyst in a staged co-current or counter-current system, without the need of gas sparging.

Several types of chemical reactions were investigated in the MSR, including hydrogenation of 3-methyl-1-pentyn-3-ol and etherification of 1-octanol [19], enzyme-catalyzed acylation of butanol with vinyl acetate [20], hydrogenation of sunflower oil [21], and alcoholysis of urea with 1,2-propylene glycol to propylene carbonate [22].

5.3.2 Integrating Mixing and Heat Exchange – Static Mixer Reactors and Heat Exchangers

In the Sulzer SMR™ (*Static Mixer Reactor*) class of static mixers, the mixing function has been advantageously combined with the heat exchange function [23]. The elements (Figure 5.9) form a complex bundle of bended tubes with heating/cooling medium flowing through the tubes. The tube bundles are positioned perpendicularly to each other, in order to enhance the radial mixing across the volume highly packed with heat transfer surface area. Obviously, as it is often the case in multifunctional devices, there is a trade-off aspect and mixing efficiency in an SMR unit is slightly lower than in classical static mixers (Chapter 3). Nevertheless, the SMR is an effective reactor–heat exchanger for highly exothermic reactions, even involving viscous liquids in laminar flow regime. As the heat

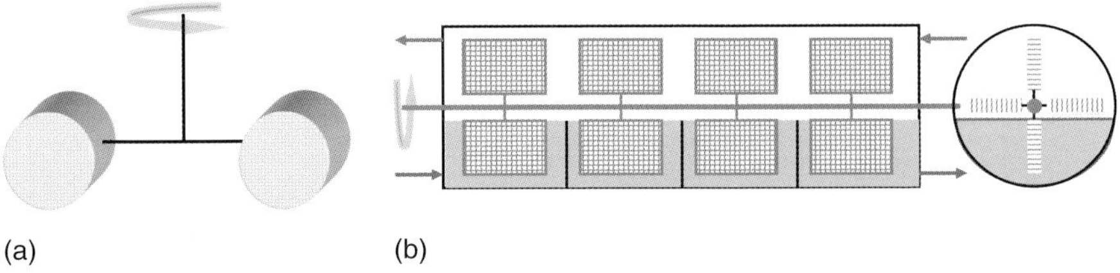

(a)   (b)

Figure 5.8 Monolithic stirrer reactor; (a) vertical shaft configuration, (b) horizontal shaft configuration (counter-current case).

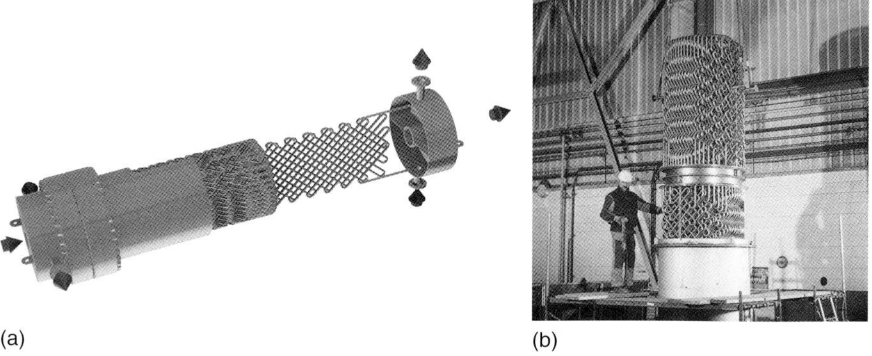

(a)   (b)

Figure 5.9 Integrating mixing and heat exchange – Sulzer SMR™ static mixer reactor. Source: Courtesy of (a) AGENS Stratmann GmbH, Germany, www.agens-stratmann.com; (b) Sulzer Chemtech Ltd., www.sulzer.com.

transfer fluid water, steam, thermal oil, or other special cooling media can be used. The diameter of SMR heat exchangers is ranging from 80 to 1600 mm and the length is ranging from 1 m (for a compact heat exchanger) to 20 m (for a reactor tower).

Another Swiss company Fluitec Georg AG has developed a static mixer heat exchanger with U-shaped heat transfer tubes. Mixing elements form a kind of "baffles" in the intertubular space, see Figure 5.10.

In tube-in-tube and multitube *static mixer heat exchangers* (Figure 5.11), the mixing elements are placed inside the tubes and are used not so much for the mixing of different components but rather to create a radial mixing effect, which results in a considerably increased heat transfer and a narrow residence time distribution. The continuous renewal of the thermal boundary layer on the pipe walls prevents thermal damages to heat-sensitive compounds.

Another interesting concept of a mixer–heat exchanger is the *coiled flow inverter* (CFI), introduced by Nigam [24]. The key components of this concept

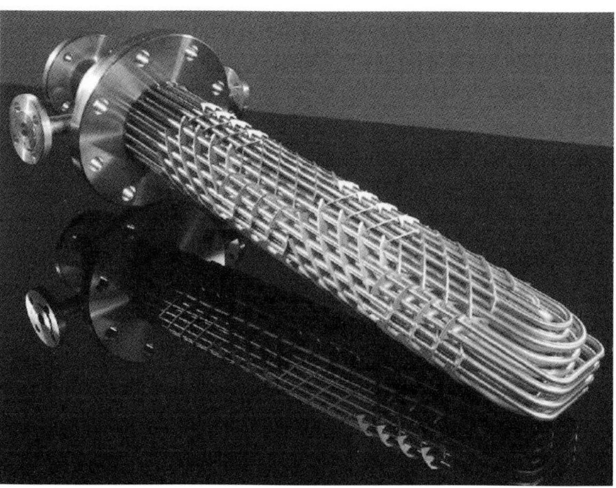

Figure 5.10 CSE-XR mixer heat exchanger from Fluitec Georg AG. Source: Courtesy of Fluitec Solutions AG, Switzerland, www.fluitec.ch.

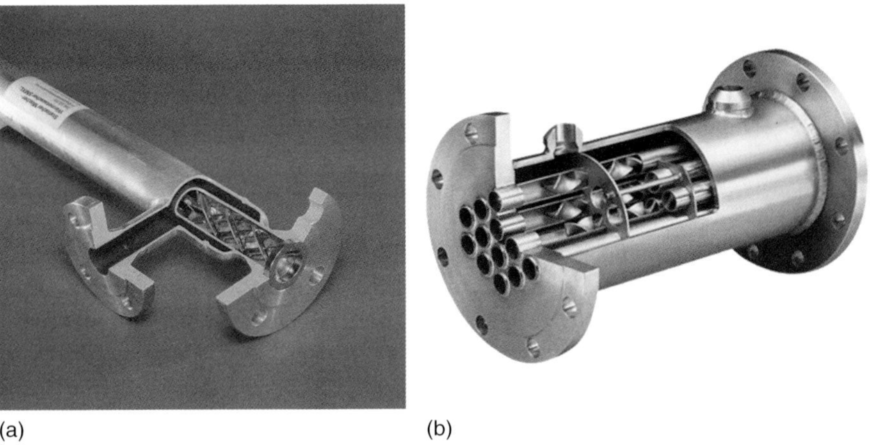

(a) (b)

Figure 5.11 Sulzer SMXL mixer – (a) tube-in-tube heat exchanger. Source: Courtesy of Sulzer Chemtech Ltd., Switzerland, www.sulzer.com; and (b) Kenics multitube mixer heat exchanger. Source: Courtesy of National Oilwell Varco, L. P., USA, www.nov.com.

are helical coils and 90° bends introduced between the coils. The introduction of 90° bends creates random mixing and complete flow inversion. The flow generated due to curvature of a stationary surface bounding the flow changes direction continuously. By shifting the plane of curvature from one bend to the next, one can induce a class of trajectories in one bend and then deform it to another type in the next bend, and so on. The number of 90° bends in the heat exchanger configuration varies from 1 to 57. A standard stainless tube is limited to 6 m in length, whereas the length of the tube for the CFI varies from 12 to 45 m, and therefore joints are provided between two units. The can be seen in Figure 5.12.

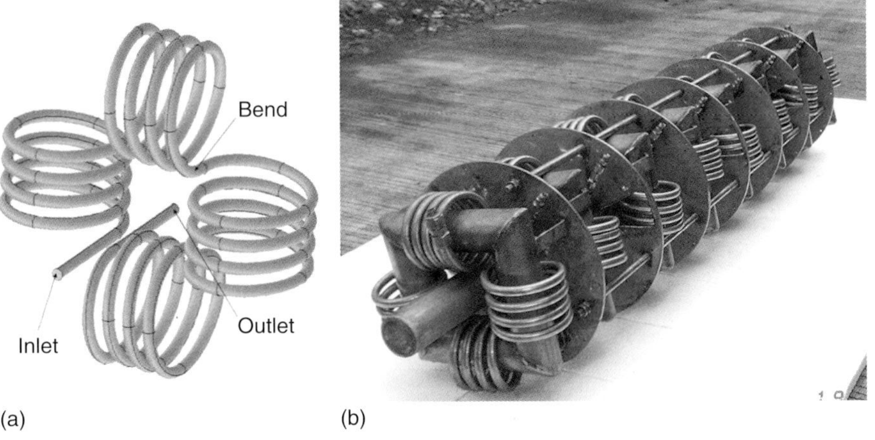

(a) (b)

Figure 5.12 Coiled flow inverter. Source: (a) Kumar et al. 2007 [24]. Reproduced with permission of Elsevier; (b) Photograph courtesy of K. D. P. Nigam, http://web.iitd.ac.in/~kdpnigam/).

5.3.3 Heat Exchangers as Chemical Reactors

The concept of a plate heat exchanger transformed into a reactor *(heat exchanger (HEX) reactor)* is not new and utilizes high thermal transfer performances of plate heat exchangers because of their large heat transfer surface areas per unit volume [25]. Although the direct use of a conventional corrugated plate heat exchanger (Figure 3.22) as a chemical reactor is possible [26, 27], most developers focus on special plate designs. Two of these designs are presented in Figure 5.13.

The reported intensification effects resulting from the employment of the HEX reactors are impressive. For example, the Marbond/ShimTec HEX reactor (Figure 5.14) developed by Chart Industries, Inc., was able to decrease

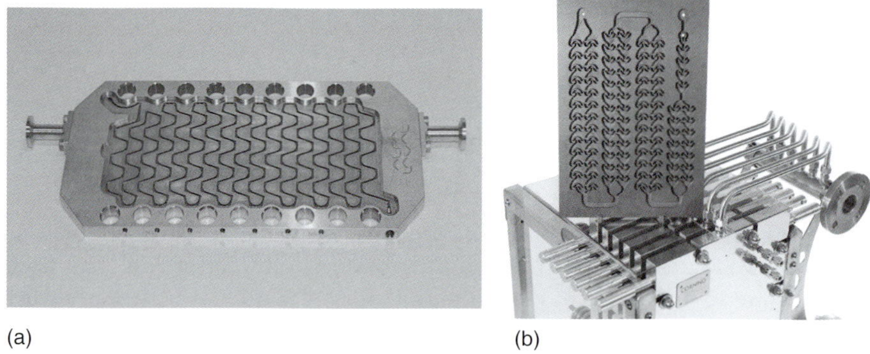

(a)   (b)

Figure 5.13 Plate designs in heat exchanger reactors: (a) Advanced Reactor Technology (ART™) developed by Alfa-Laval. Source: Courtesy of Ehrfeld Mikrotechnik BTS GmbH, www.ehrfeld.com; (b) Mersen/Boostec SiC plate made for Advanced-Flow Reactors™ of Corning Inc. Source: Courtesy of Corning Proprietary: Advanced Flow Reactor with Boostec® Silicon Carbide modules, www.corning.com/.

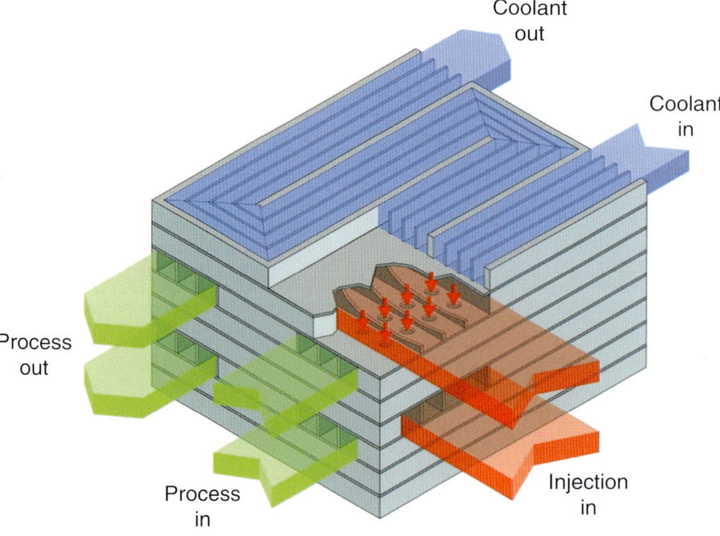

Figure 5.14 Scheme of the ShimTec heat exchanger reactor. Source: Courtesy of Chart Energy & Chemicals Inc., USA, www.chartindustries.com.

Table 5.4 Comparison of the heat transfer performance of various reactor types.

Devices	Stirred tank	Tubular (40 mm)	ART (Alfa Laval)	Advanced Flow (Corning)	ShimTec (Chart)	Boostec SiC (Mersen/Boostec)
Construction material	Stainless steel	Stainless steel	Stainless steel	Glass	Stainless steel	SiC
Overall heat transfer coefficient U (W/m^2 K)	400	500	2 500	660	1 500	10 000
Specific heat transfer area A (m^2/m^3)	2.5	100	400	2 500	2 000	2 000
Performance indicator $U^*A/1\,000$	1	50	1 000	1 650	3 000	20 000

Source: Adapted from Despènes et al. 2012 [30].

the by-product formation in one of the ICI Acrylics processes by 75% [28] and shorten the processing time in a Hickson & Welch two-stage-catalyzed oxidation of a thioether to a sulfone, from 18 hours to 15 minutes, thus saving 98.6% of the batch time [29]. An interesting feature of the ShimTec and other HEX reactors is that they enable a distributed feed (injection) of one of the reactants along the reaction channel. This has an advantageous effect on the heat evolution in the reactor and the reaction selectivity to the required product.

A comparison of the heat transfer performance of various HEX reactor designs with that of the traditional, stirred tank or tubular reactors is presented in Table 5.4. The importance of the plate constructional material in terms of thermal conductivity is clearly visible here, as the SiC-made HEX reactor performs an order of magnitude better than the similar reactors made of glass or stainless steel [30].

5.3.4 Heat Pumping in Distillation Systems

Distillation is definitely the most energy-demanding separation process. A conventional distillation column requires high-quality energy input in the reboiler and discharges the lower temperature heat in the condenser. In order to upgrade that discharged energy and to reuse the heat from the reboiler, integration of various heat pump concepts in distillation systems has been proposed. *Heat pumps* are designed to move thermal energy opposite to the direction of spontaneous heat flow by absorbing heat from a cold space and releasing it to a warmer one. There is a lot of research literature available on this subject, and the paper by Kiss [31] brings an excellent review of the available options and reported effects. These options are schematically presented in Figure 5.15 and include the following:

- *Vapor compression* (VC): already proven at industrial scale; uses a compressor and a specific fluid as a heat transfer medium.

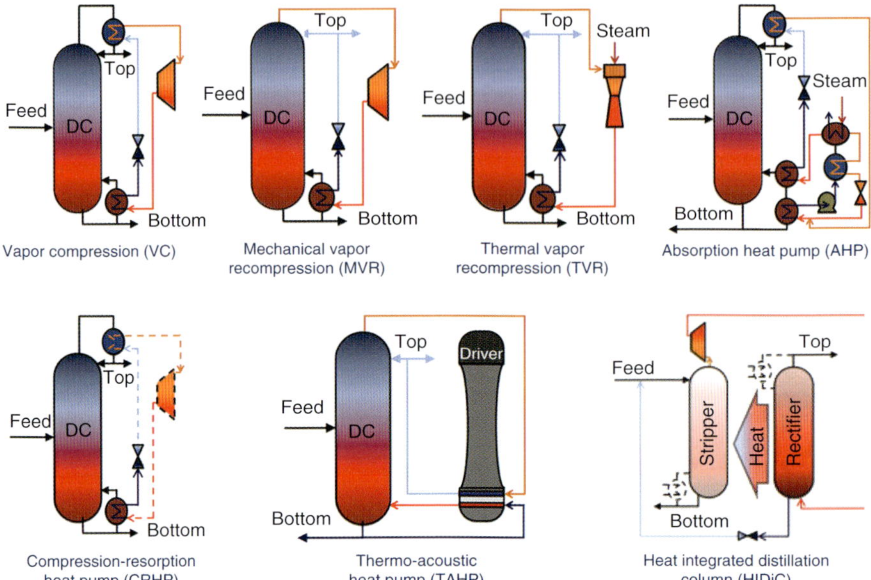

Figure 5.15 Different concepts of heat pump-assisted distillation systems. Source: Kiss 2014 [31]. Reproduced with permission of John Wiley and Sons.

- *Mechanical vapor recompression* (MVR): state of the art in binary distillation of closely boiling components; similar to the previous one but using vapor as the heat transfer medium, fed directly to the compressor.
- *Thermal vapor recompression* (TVR): similar to the previous one; however, the compressor is replaced by a Venturi steam ejector;
- *Absorption heat pump* (AHP): similar to the refrigeration cycle, uses absorption pairs (e.g. ammonia and water) as heat transfer fluids; the key component is the steam-driven desorber that separates the absorption pair;
- *Compression–resorption heat pump* (CRHP): uses the VC scheme, in which the working fluid is replaced by the absorption pair; in the reboiler condensation and absorption run simultaneously enhancing the heat transfer; the heat is taken from the condenser by evaporating both components of the absorption pair.
- *Thermoacoustic heat pump* (TAHP): uses high-amplitude sound waves to pump heat from one place to another; consists of heat exchangers, a resonator (e.g. based on electric driver/linear motor) and a regenerator (on traveling wave devices) or a stack (on standing-wave devices); air or noble gases are used as the medium.
- *Heat-integrated distillation column* (HIDiC): the most radical approach making use of the heat integration; instead of using a point heat source/sink, here the entire rectifying section acts as the heat source while the entire stripping section acts as the heat sink; different configurations have been developed to tackle the problem of different sizes for the rectifying and stripping sections.

Heat pump-assisted distillation systems allow for large primary energy savings; figures of up to 80% have been reported [31].

5.3.5 Integrating Reactions and Separation – Reactive Separations

Reactive separations present an industrially important type of multifunctionality in chemical processes [32]. They can bring a number of advantages to the process under consideration, which are not only limited to a decrease in the size of equipment. In fact, the most important advantage of reactive separations is their ability to increase the yield of the equilibrium-limited reactions by the in-situ product removal. Other potential incentives for considering an application of a reactive separation include lower investment costs (due to compact, integral design), improved heat management/energy utilization, easier separation (e.g. in case of close boiling points or azeotrope problems), extension of the catalyst lifetime (due to milder reaction conditions), etc. The choice of the type of reactive separation depends primarily on the type and number of phases involved, as illustrated in Figure 5.16.

The interaction between the reaction and separation determines to a large extent the selection and the design of the corresponding equipment. Reaction performance can be influenced by such factors as the amount of the catalyst, the residence time, or the temperature, while separation performance can be influenced by the reflux, number of stages, etc. [33]. Eventually, the choice and the design of equipment will depend on the *reaction rate* on the one hand and the *separation factor* (e.g. relative volatility or solubility) on the other hand, as illustrated for the case of reactive distillation in Figure 5.17 [34].

5.3.5.1 Reactive Distillation

Reactive distillation is nowadays widely applied in chemical processes [35]. Harmsen [36] reported in 2007 more than 150 industrial-scale reactive

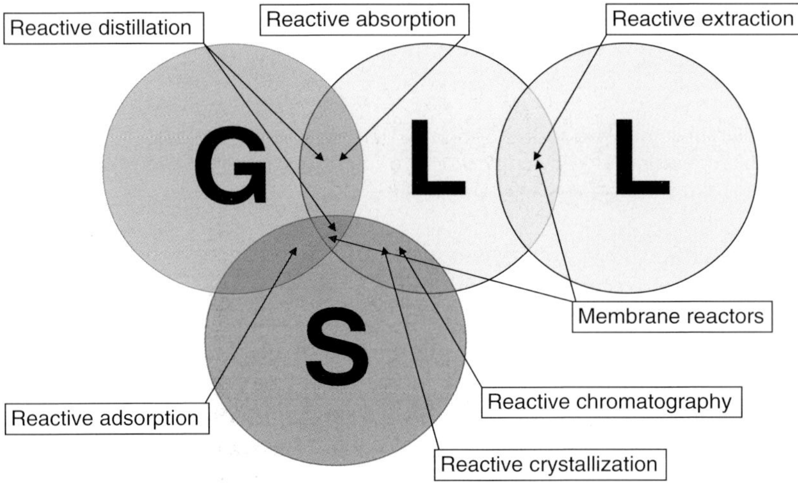

Figure 5.16 Applicability of different types of reactive separations depending on the phases present in the system.

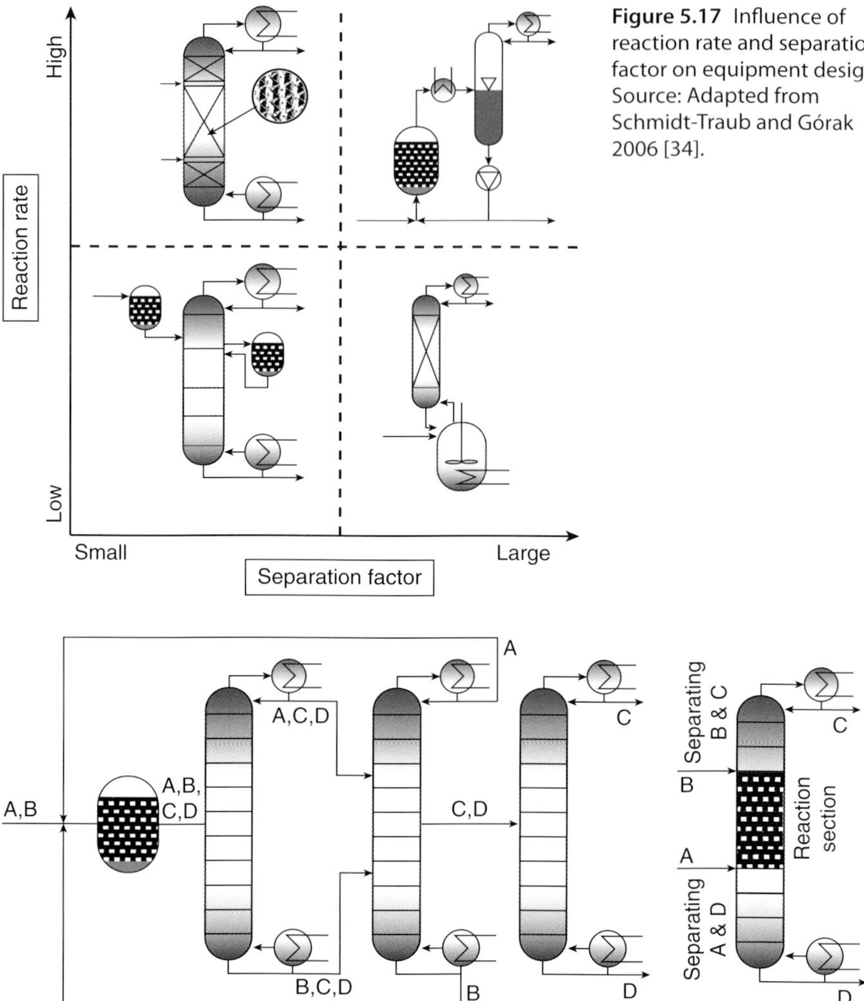

Figure 5.17 Influence of reaction rate and separation factor on equipment design. Source: Adapted from Schmidt-Traub and Górak 2006 [34].

Figure 5.18 Basic concept of reactive distillation for the case of $A + B \leftrightarrow C + D$ equilibrium-limited reaction: (a) conventional system and (b) reactive distillation system. Source: Adapted from Stichlmair and Frey 1999 [145].

distillation systems (both homogeneously and heterogeneously catalyzed) operated worldwide at capacities of 100-3000 kilotons/year. The basic concept of the reactive distillation is presented in Figure 5.18. Placing the reactive section inside the distillation column allows in situ removal of the products of an equilibrium-limited reaction, that in the ideal case, should lead to 100% conversion and full product separation within a single process stage.

Figure 5.18 presents an ideal case. Usually, such a reduction from a four-unit to a single-unit system is not possible. Yet there are examples of industrial processes, in which spectacular reductions in the number of process steps have been achieved. The most prominent of those examples is the methyl acetate process

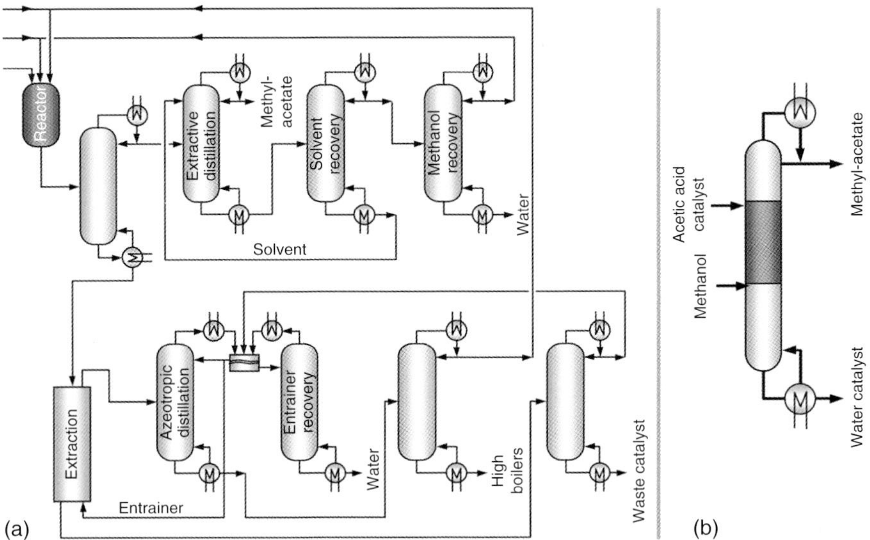

Figure 5.19 (a) Old and (b) new methyl acetate process at Eastman Chemical. Source: Schoenmakers and Bessling 2003 [33]. Reproduced with permission of Elsevier.

developed by Eastman Chemical, where the introduction of reactive distillation and its integration with extractive distillation in a single unit totally solved the azeotrope problem [37]. As the result, the number of apparatuses in the plant was reduced from 28 to 3 (a single column with a reboiler and condenser), as shown in Figure 5.19. Next to the plant-size reduction, savings on investment and energy in the range of 80% were reported.

Reactive distillation can also be used as a powerful separation method in the case of mixtures containing reactive and inert components with close boiling points. The method is schematically depicted in Figure 5.20. Here, a reactive entrainer is introduced to the first reactive distillation column to form an intermediate product having the boiling point much more distant from the boiling point of the inert component. In the first column, the inert component is therefore easily separated, while the intermediate product is fed to the second reactive distillation step, where the reverse reaction takes place and the original reactive component is recovered and separated from the entrainer. Stein et al. [38] investigated the application of the above principle to the separation of the close boiling *i*-butene and *n*-butene, using methanol as a reactive entrainer.

The research on reactive distillation is still expanding. New application areas such as enzymatic processes [39, 40] are addressed. Also, new concepts of reactive distillation equipment, such as Dow's horizontal reactive distillation system for multicomponent chiral resolution (Figure 5.21), are investigated.

5.3.5.2 Membranes in Chemical Reactors

A membrane can play diverse functions in a reactor system, as described in the excellent review paper by Sirkar et al. [41] (Figure 5.22, Table 5.5).

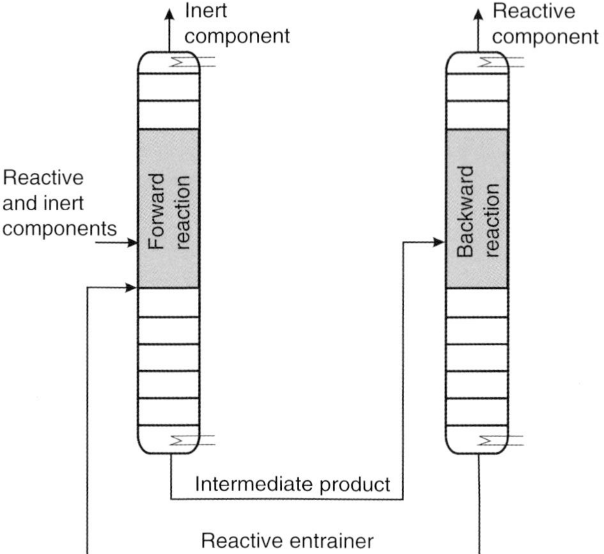

Figure 5.20 Separation of close-boiling components facilitated by reactive distillation. Source: Stankiewicz 2003, [56]. Reproduced with permission of Taylor and Francis Group.

Membrane reactors can be divided in three broad categories depending on the type of the function played by the membrane, as presented in Table 5.6 [42]. Typical (potential) applications of the *extractor membrane reactors* include catalytic dehydrogenation of light alkanes or reactions for hydrogen generation, such as steam reforming or water–gas shift processes using H_2-permselective membranes. Also, water removal in, e.g., esterification reactions or in Fischer–Tropsch synthesis can be realized in *extractor membrane reactors*. The *distributor membrane reactors*, on the other hand, present an interesting option for partial oxidation processes, in which a controlled addition of oxygen along the reactor can prevent complete oxidation. Finally, the *contactor membrane reactors* can be applied in multiphase processes as interfacial contactors in gas–liquid systems, providing bubble-free transport of the gaseous component into the liquid phase with reduced mass transfer resistances. Examples of processes that can potentially benefit from the application of contactor membrane reactors include selective oxidation of light alkanes or liquid-phase hydrogenations.

Hydrogen generation presents an important application area of membrane reactors. In those processes, dense metal membranes are used to selectively separate pure hydrogen from the reaction environment. Drioli et al. [43] analyze various membrane reactor-based processes for hydrogen generation including the water–gas shift process. A scheme of a dense palladium membrane-based membrane reactor and a traditional process for water–gas shift process is shown in Figure 5.23. According to Drioli and coworkers, the CO conversion in the membrane reactor is expected to be two to three times higher, whereas the reaction volume can be as low as 9% of the traditional reactor. Another large

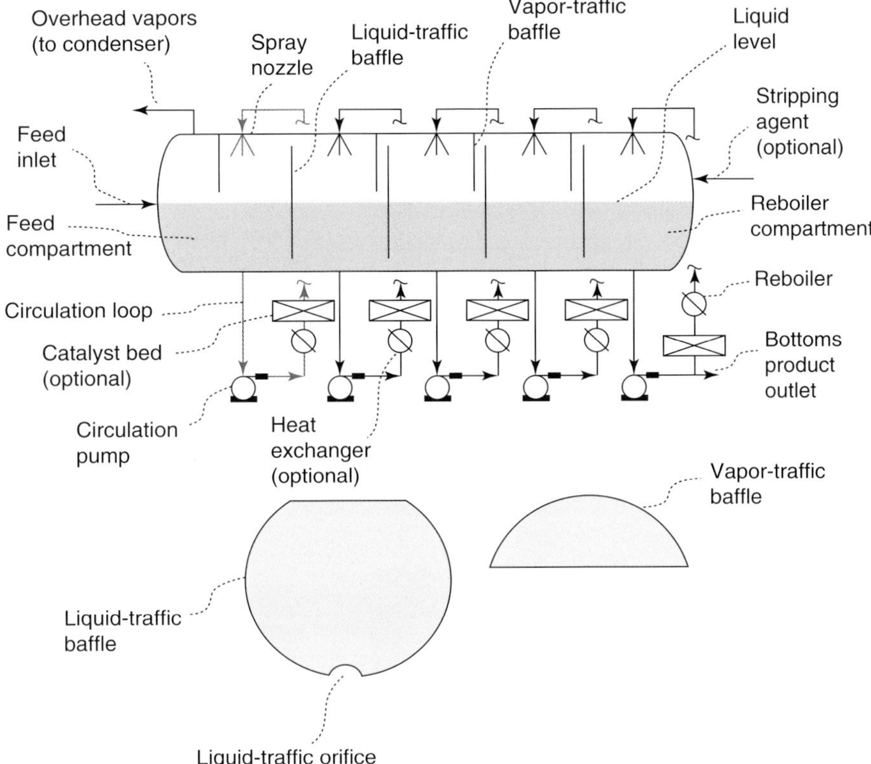

Figure 5.21 Horizontal reactive distillation system for multicomponent chiral resolution developed at the Dow Chemical Company. Source: Au-Yeung et al. 2013 [146]. Reproduced with permission of John Wiley and Sons.

application area for membrane reactor technologies is biocatalysis, both in the manufacturing and in the wastewater treatment [44].

Regardless of the targeted flow characteristics (CSTR or plug-flow), a membrane separator can be integrated with a reactor as an external separation unit (ESU) or as an in situ separation unit (ISU, Figure 5.24). The choice between both types of configurations may have important implications because it affects the degrees of freedom in system design, operation, and control, as discussed at the beginning of this chapter.

5.3.5.3 Reactive Adsorption

The combination of catalytic and adsorptive functionalities within a catalyst pellet has already been discussed in Section 5.2.1. On the equipment scale, such a combination may assume different forms depending, among other things, on whether the given process takes place in the gas or in the liquid phase.

For the gas-phase reaction/adsorption processes, the so-called *gas–solid–solid trickle flow reactor* has been proposed, in which fine adsorbent trickles through the fixed bed of the catalyst (Figure 5.25), removing selectively and in situ one or more products from the reaction zone. In the case of methanol synthesis,

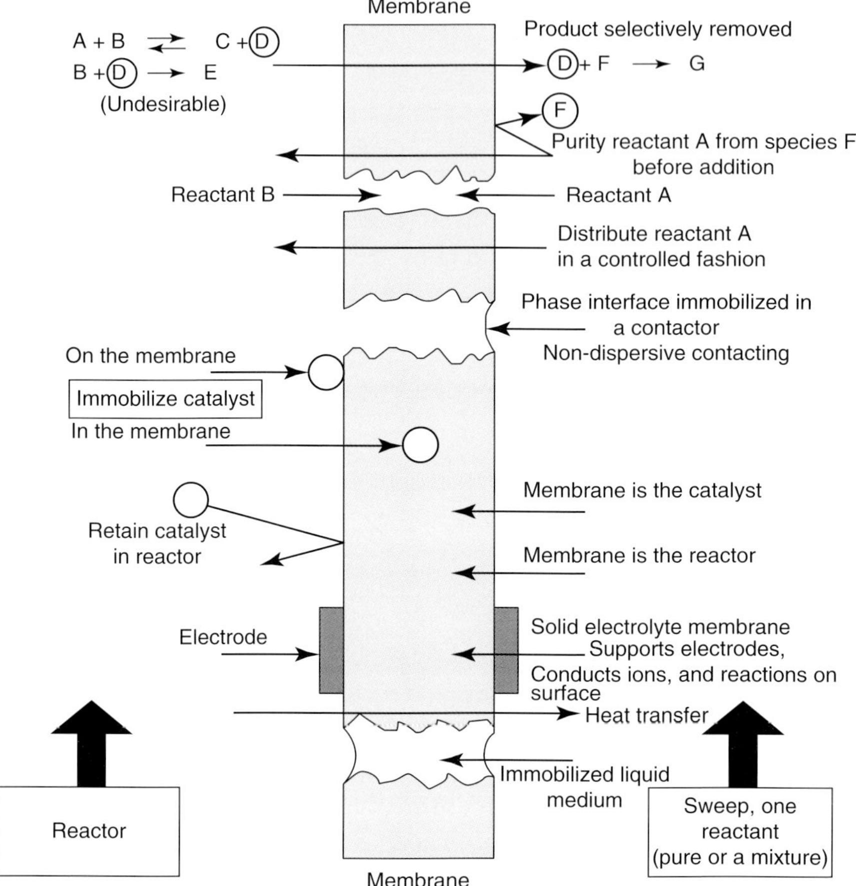

Figure 5.22 Diverse membrane functions in a chemical reactor. Source: Sirkar et al. 1999 [41]. Reproduced with permission of American Chemical Society.

this led to conversions significantly exceeding the equilibrium conversions under given conditions [45]. The economics of the methanol process based on the gas–solid–solid trickle flow reactor was evaluated and compared with the conventional low-pressure Lurgi process [46]. For the production scale of 1000 ton/day, the new technology offered considerable reductions of cooling water consumption (50%), recirculation energy (70%), raw materials (12%), and catalyst amount (70%).

Chromatographic reactors can be used in both gas-phase and, more often, liquid-phase processes. In those reactors, fluid (mobile phase) moves through the stationary phase being adsorbent or a mixture of an adsorbent and a catalyst (in the case of heterogeneous catalytic reactions). Three most important types of chromatographic reactors include the batch chromatographic reactor (BCR), the true moving bed chromatographic reactor (TMBCR), and the simulated moving bed chromatographic reactor (SMBCR) [47].

Table 5.5 Membrane functions in chemical reactor systems.

Function	Examples
1. Separation of products from the reaction mixture	In situ product removal from enzymatic reactor via a nanofiltration or ultrafiltration membrane
	Removal of selected enantiomer via a liquid membrane
	Removal of water in esterification reactions via a pervaporation membrane
	Hydrogen removal in catalytic dehydrogenation reactions
2. Separation of a reactant from a mixed stream for introduction into the reactor	Separation of oxygen from air for oxidizing methane to syngas
	Separation of hydrogen from dehydrogenation reaction to oxidize it with oxygen on permeate side
	Separation of organic priority pollutants from wastewater for biological purification
3. Controlled addition of one reactant or two reactants	Controlled oxygen addition in partial oxidation reactions (to increase selectivity)
	Controlled air introduction in oxidative dehydrogenations
4. Nondispersive phase contacting, with reaction at the phase interface or in the bulk phases	Emulsion-free enzymatic splitting of fats
	Bubble-free oxygen/ozone supply in wastewater treatment via hollow-fiber membranes
5. Segregation of a catalyst (and cofactor) in a reactor	Segregation of enzymes with respect to molecular weight on ultrafiltration membranes
6. Immobilization of a catalyst in (or on) a membrane	Immobilization of enzymes or cells on polymeric membranes
	Immobilization of metals (Pd, Pt) on ceramic membranes
7. Membrane is the catalyst	Cation exchange membranes for esterification reactions
	Palladium membranes for hydrogenation/dehydrogenation reactions
8. Membrane is the reactor	Reactions in flow-through membrane systems ("pore flow through reactors")
9. Solid–electrolyte membrane supports the electrodes, conducts ions, and achieves the reactions on its surface	Solid electrolyte membranes as H^+ and O^{2-} conductors in fuel cells
10. Transfer of heat	Membranes coupling endo- and exothermic reaction zones (e.g. hydrogenation–dehydrogenation)
11. Immobilizing the liquid reaction medium	Supported liquid membranes (SLM) for homogeneous catalytic processes

Source: Adapted from Sirkar et al. 1999 [41].

Table 5.6 Basic categories of membrane reactors.

Reactor category	Functions		Examples of applications
Extractor membrane reactors	Selective product removal 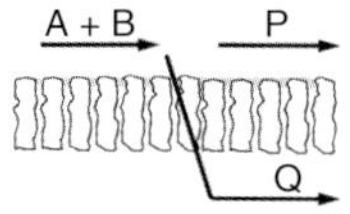 A + B → P Q 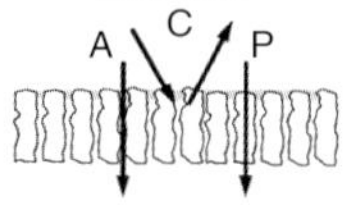 $A + B \leftrightarrow P + Q$	Catalyst retention A, C, P $A \xrightarrow{C} P$	• Dehydrogenations (hydrogen removal) • Reforming/water–gas shift (hydrogen removal) • Esterifications (water removal) • Fischer–Tropsch (water removal)
Distributor membrane reactors	Controlled reactant addition 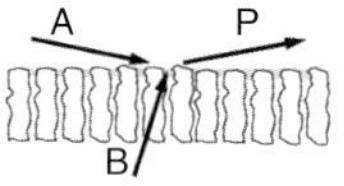 A → P B $A + B \rightarrow P$ $P + B \rightarrow Q$	Selective addition from mixture 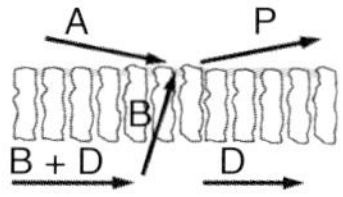 A → P B B + D → D $A + B \rightarrow P$	• Partial oxidations (oxygen feed distribution)
Contactor membrane reactors	Interfacial contactor 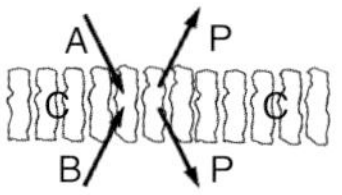 A, C, P B, C, P $A + B \xrightarrow{C} P$	Forced flow-through 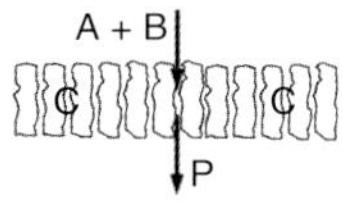 A + B, C C P $A + B \xrightarrow{C} P$	• Oxidations or hydrogenations in gas–liquid systems (bubble-free contacting)

Source: Westermann and Melin 2009 [42]. Reproduced with permission of Elsevier. Text is adapted also from [42].

Figure 5.26 illustrates the operation of a BCR for the case of a reversible dissociation reaction. Here, the reactant A is injected in the form of a sharp pulse into the desorbent stream flowing through the bed of an adsorbent. The products B and C adsorb in the bed at different rates so that their propagation velocities in the bed are different and the separation occurs. Although very straightforward, the BCR technique has a number of disadvantages, such as low yields, high product dilution, and low overall capacities.

One way to increase the efficiency of the use of the stationary phase is to move it in counter-current with respect to the fluid flow. This way, a TMBCR can be realized. The counter-current flow maximizes the driving force. However, physical moving of solid particles is usually difficult to realize. The back-mixing of the solid phase may reduce the efficiency of the process. Also, attrition/abrasion of particles may occur [47]. These shortcomings can be avoided in a SMBCR. The simulated moving bed adsorption technology (without reaction) was originally patented in 1961 by UOP [48] and since then has been applied on industrial scale in petrochemical industries as well as in food and life sciences sectors. The basic idea of the SMBCR consists in dividing the entire length of the adsorbent bed in a number of shorter segments. All inlet/outlet streams are fed/removed

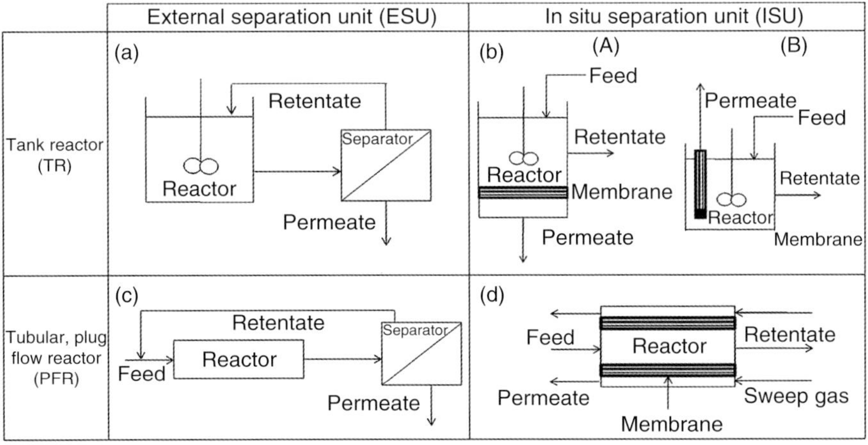

Figure 5.23 showing (a) Traditional process and (b) Pd-based MR:

Figure 5.23 Water–gas shift process in a traditional and in membrane reactor. Source: Adapted from Barbieri et al. 2011 [147].

Figure 5.24 Various membrane reactor – separator configurations. Source: Diban et al. 2013 [148]. Reproduced with permission of American Chemical Society.

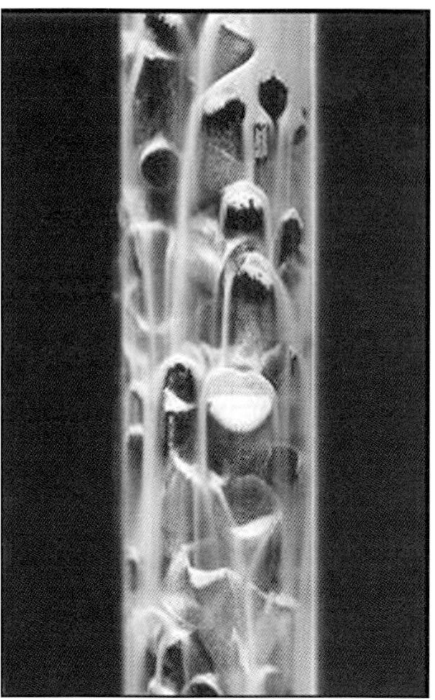

Figure 5.25 Trickle flow of silica/alumina powder through a fixed bed consisting of 5×5 mm catalyst pellets and $7 \times 7 \times 1$ mm Raschig rings. Source: Kuczynski 1987 [45]. Reproduced with permission of Taylor and Francis.

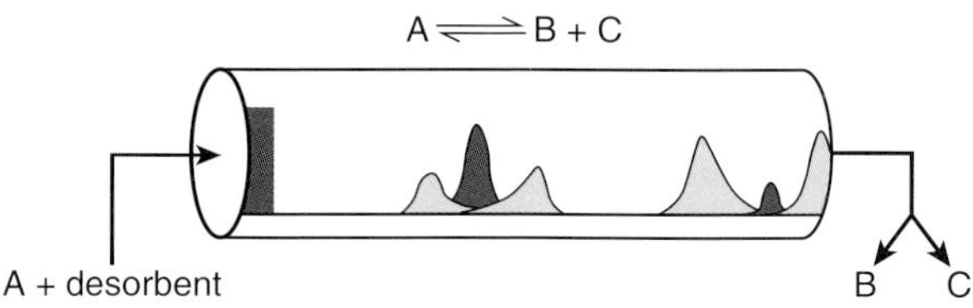

Figure 5.26 Schematic operation of a batch chromatographic reactor. Source: Adapted from Fricke et al. 2008 [47].

from the system via the ports positioned between those segments, and in regular time intervals (switching time), they are shifted from one port to another in the direction of the fluid flow, as illustrated in Figure 5.27. The figure presents the basic SMBCR configuration for a reversible $A = B + C$ reaction, consisting of four sections:

- *Section I*: between the desorbent and extract nodes, in which the more strongly adsorbed product B is removed from the adsorbent;
- *Section II*: between the extract and the feed nodes, in which components B and C are formed and the less strongly adsorbed product C is desorbed and carried downstream, while B is held on the adsorbent; this is why the liquid concentration of B in the extract stream is high;
- *Section III*: between the feed and raffinate nodes, in which A is still converted and B is adsorbed; the liquid concentration of C (less strongly adsorbed and carried from Section II) in the raffinate is high;
- *Section IV*: between the raffinate and desorbent nodes, in which desorbent is regenerated by adsorption of the less adsorbed product and recycled to Section I.

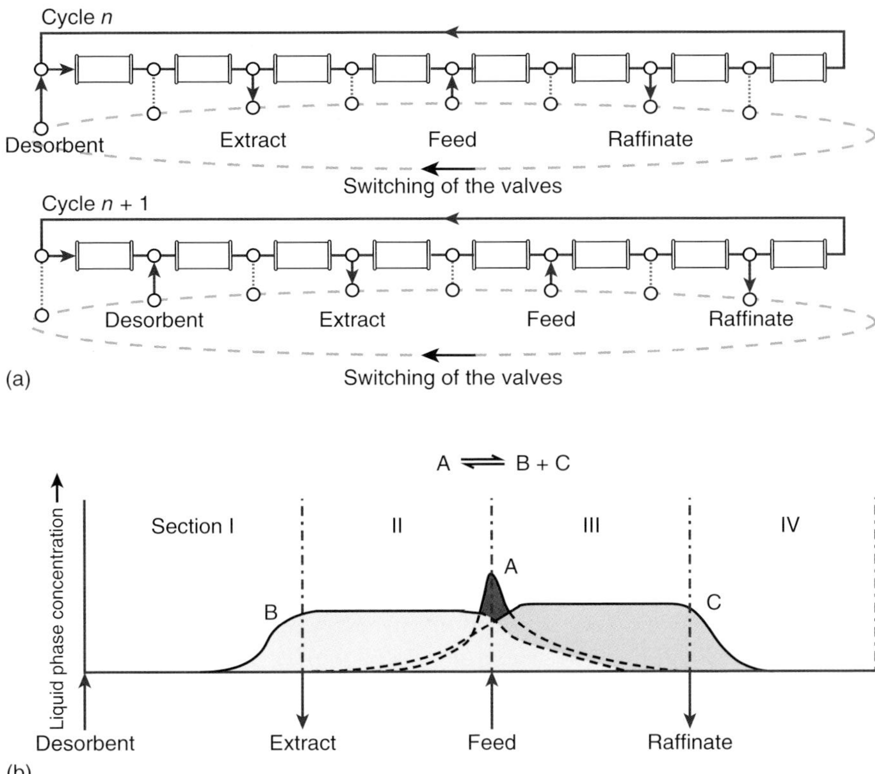

Figure 5.27 Operational scheme of a simulated moving bed chromatographic reactor: (a) configuration of apparatus in two consecutive cycles; (b) axial concentration profiles at the end of a cycle. Source: Adapted from Fricke et al. 2008 [47].

SMBCRs are mostly used in the laboratory scale for both analytical and preparative applications.

Rotating cylindrical annulus chromatographic reactor (RCACR, Figure 5.28) presents another interesting concept of a reactive adsorption system [49, 50]. Here, the rotating annulus is packed with the adsorbent bed. The reactants are fed at a fixed point on the annulus perimeter. The movement of the reactants and products in the bed is two-dimensional, which is caused by the gravity on one the hand and by the annulus rotation on the other hand. Taking different trajectories through the bed and having different residence times in the stationary phase, the components can be collected at the fixed points at the bottom of the annulus. The production rate of the rotating cylindrical annulus is comparable to that of the BCR [47].

Despite the very rich research literature, industrial-scale applications of *reactive* adsorption are scarce. Inversion of sucrose to glucose and fructose, developed by Boehringer Mannheim, is probably the best-known example of a commercial-scale application of reactive adsorption. This technology utilizes the batch reactive chromatography concept [47]. Other potential applications investigated in the literature include in gas-phase processes: dehydrogenation

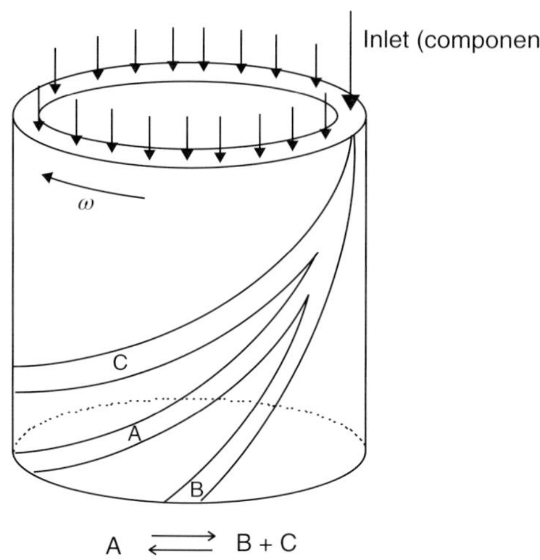

Figure 5.28 Scheme of a rotating cylindrical annulus chromatographic reactor.

of cyclohexane, oxidative coupling of methane, ammonia synthesis, various reduction and oxidation reactions, and in liquid-phase processes: isomerizations, esterifications, chlorinations, and racemization reactions [47]. Reactive adsorption can also find its applications in environmental engineering [51] and in green fuels production, where *simulated moving bed membrane reactor* technology (PermSMBR) has been proposed [52, 53]. In some applications, the productivity of chromatographic reactors could be improved by operating them in the reverse-flow mode [54].

5.3.5.4 Reactive Extraction

A *reactive extraction* system consists of two immiscible phases: the solvent (transport) and the reaction (stationary) phase ([55], Figure 5.29). The product is extracted from the stationary phase, which may contain catalyst, to the transport phase.

Similar to reactive adsorption, reactive extraction can be applied in multireaction systems, for improvement in yields and selectivities to desired products. The combination of a reaction with liquid–liquid extraction can also be used for separation of waste by-products that are hard to separate using conventional

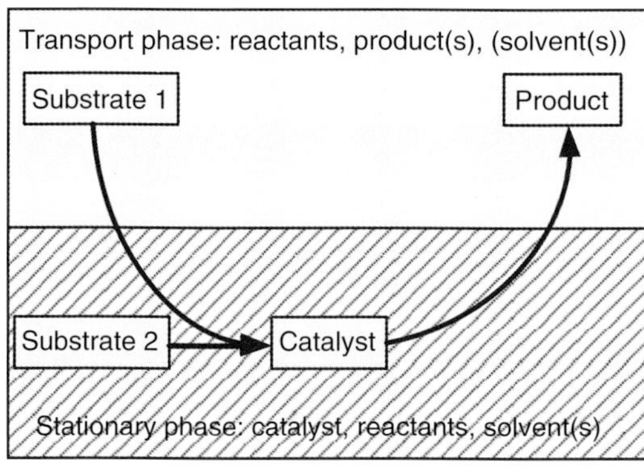

Figure 5.29 The simplest scheme of a reactive extraction system. Source: Adapted from Behr and Seuster 2006 [55].

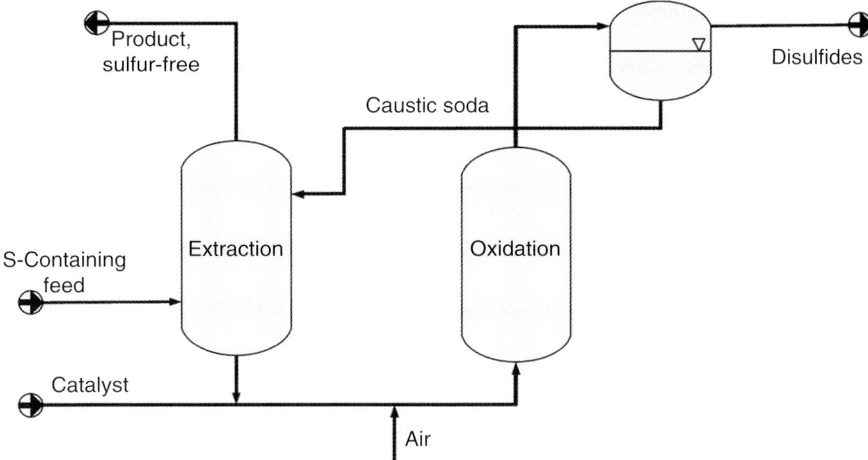

Figure 5.30 UOP's reactive extraction Merox process. Source: Adapted from Behr and Seuster 2006 [55].

techniques. An overview of processes studied in reactive extraction systems can be found in ref. [56]. An example of a reactive extraction process applied on industrial scale is the UOP Merox process to remove mercaptans from different hydrocarbon fractions by converting them to liquid hydrocarbon disulfides (Figure 5.30). In that process, the feedstock is contacted in a trayed extractor vessel with the aqueous caustic solution containing UOP's proprietary liquid catalyst. The caustic solution reacts with mercaptans and extracts them. The aqueous solution is then regenerated by oxidizing mercaptans to water-insoluble disulfides that can be easily separated from the aqueous phase.

Other interesting examples of industrial-scale reactive extraction systems include Shell–Higher–Olefin–Process (SHOP) and Ruhrchemie/Rhône-Poulenc technology for hydroformylation of propylene to butyraldehyde [55]. In biochemical processing, reactive extraction has been proposed, among other things, as an intensification method in enzymatic hydrolysis of penicillin G (Gaidhani et al. [57] – Figure 5.31) and in selective separation of amino acids (Cascaval and Galaction [58] – Figure 5.32).

5.3.5.5 Reactive Crystallization

Reactive crystallization/precipitation plays a role in a number of industrially relevant processes, such as the liquid-phase oxidation of *para*-xylene to produce technical grade terephthalic acid, the acidic hydrolysis of sodium salicylate to salicylic acid, and the absorption of ammonia in aqueous sulfuric acid to form ammonium sulfate [59, 60]. Figure 5.33 presents a scheme of the Meissner process for production of hexamine [61].

Reactive crystallization/precipitation is also commercially applied in the resolution of enantiomers (so-called diastereomeric crystallization). Here, the racemate is reacted with a specific optically active material (resolving agent) to produce two diastereomeric derivatives (usually salts) that are easily separated

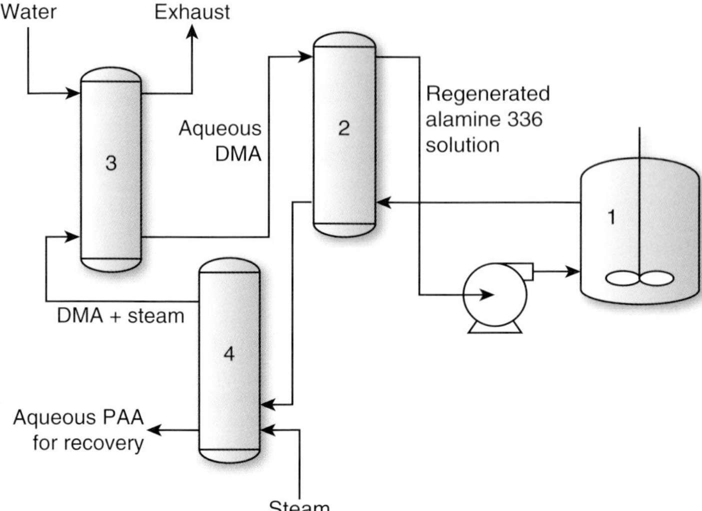

Figure 5.31 Process flow sheet for the enzymatic production of 6-aminopencillanic acid (6-APA) accompanied by reactive extraction and recovery of phenylacetic acid (PAA). 1 - Fermenter; 2 - regenerator for alamine 336; 3 - absorber for DMA; 4 - stripper for DMA; DMA – dimethyl amine. Source: Adapted from Gaidhani et al. 2002 [57].

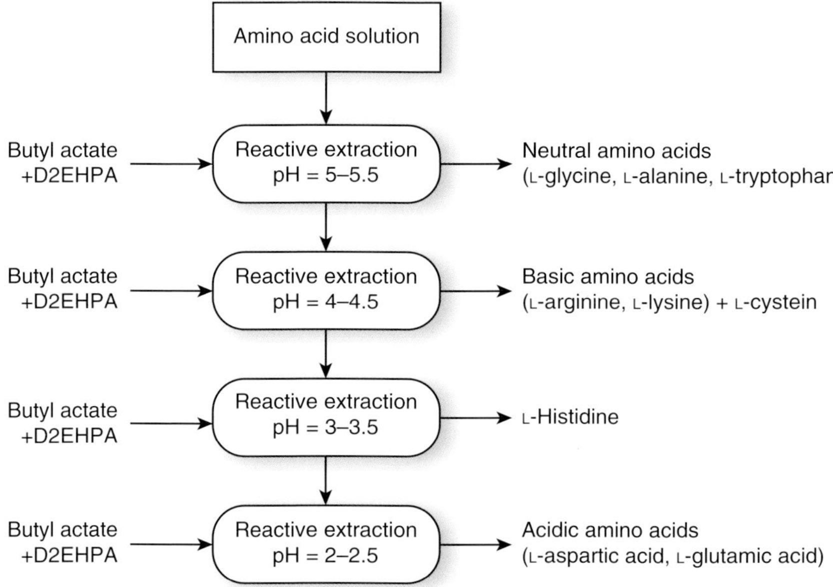

Figure 5.32 Selective separation of amino acids by reactive extraction with D2EHPA dissolved in butyl acetate. Source: Cascaval and Galaction 2004 [58]. Open access journal by Association of Chemical Engineers.

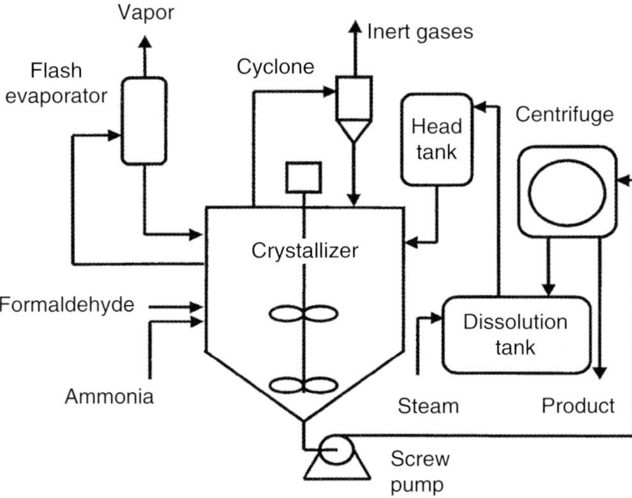

Figure 5.33 Flow diagram of the industrial process for hexamine production. Source: Alamdari and Tabkhi 2004 [61]. Reproduced with permission of Elsevier.

by crystallization:

$$(\text{DL})\text{-A} \;+\; (\text{L})\text{-B} \;\rightarrow\; (\text{D})\text{-A}\,(\text{L})\text{-B} \;+\; (\text{L})\text{-A}\,(\text{L})\text{-B}$$

racemate   resolving   *n*-salt   *p*-salt

agent

$$(5.1)$$

Slightly similar are the so-called *adductive crystallization* processes, often (wrongly) called *extractive crystallization*, where reactions of complex/adduct formation are used to separate compounds that are otherwise difficult to separate [56]. Recently, Encarnación-Gómez et al. developed a new route for resolution of amino acids that form racemic compounds using chemoenzymatic stereoinversion reactions in saturated or supersaturated solutions [62]. The reaction sequence first oxidizes the undesired D-enantiomer with D-amino acid oxidase, and the resulting intermediate imino acid is then reduced to a racemic mixture of the enantiomers by ammonia borane. The result of this reaction network is fully enantioselective because the first step is enantioselective. In the work, L-phenylalanine and L-methionine crystals were recovered with chemical and enantiomeric purities greater than 99%.

Obviously, the introduction of a chemical reaction in the already complex crystallization system increases its complexity further. Next to usual crystallization elements, such as supersaturation creation, crystal nucleation, growth, polymorphism, attrition, or agglomeration, the reaction kinetics and (in particular) mezzo- and micro-mixing play an important role. This results in clear challenges with respect to both modeling and control of the reactive crystallization systems [60, 63].

5.3.5.6 Reactive Absorption

Reactive absorption is the oldest reactive separation technology and has been applied in a number of classical bulk chemical processes, such as nitric acid, sulfuric acid, or hydroxylamine [56, 64, 65]. Figure 5.34 presents a scheme of a sulfuric acid plant, in which SO_3 formed in a multistage fixed bed reactor with intermediate cooling is reactively absorbed: once – after the fourth bed of the catalyst (Abs1), to shift the equilibrium conversion, and then in the end-stage absorber (Abs2), which also purifies the outlet gas.

In the Raschig hydroxylamine process, water, ammonia, and carbon dioxide react together in an absorption column to yield a solution of ammonium carbonate, which subsequently forms an alkaline solution of ammonium nitrite by the reactive absorption of nitrous oxide at low temperature. In a further step, the ammonium nitrite is converted to hydroxylamine disulfonate with sulfur dioxide. The hydroxylamine disulfonate solution is drawn off and the salt is hydrolyzed and neutralized to give hydroxylamine sulfate and ammonium sulfate as the coproduct.

Waste gas treatment presents another important application area of reactive absorption. Processes here include removal of carbon dioxide by reactive absorption in amine solutions or desulfurization of flue gas by reactive absorption of SO_2 in ammonia. In Figure 5.35, a process scheme is shown for the removal of NO_x from the waste gas in the adipic acid plant. The reactive absorption system consists of four units containing eight absorption columns. Each column, packed with INTALOX ceramic saddles, is 2.2 m in diameter and 7 m high [65].

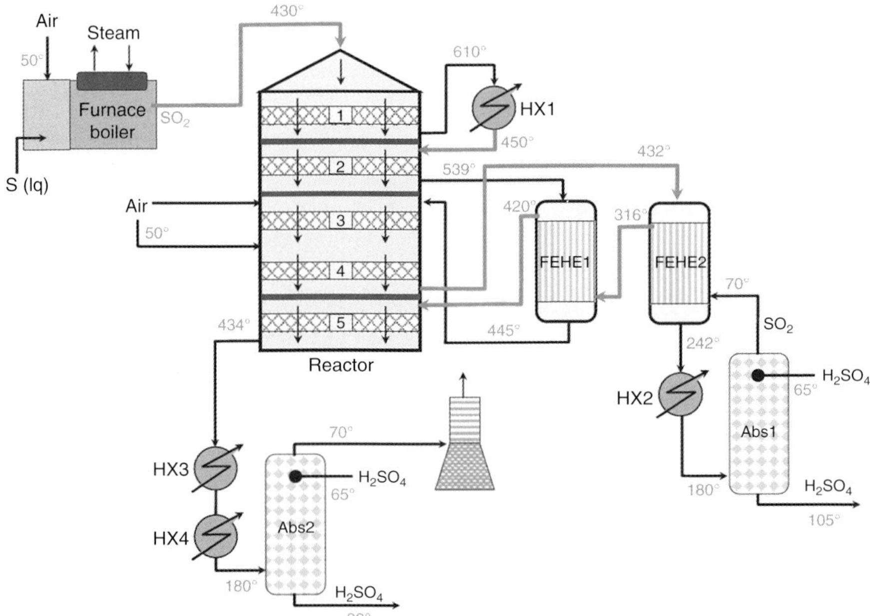

Figure 5.34 Scheme of a sulfuric acid plant with two reactive absorption stages. Source: Kiss et al. 2010 [149]. Reproduced with permission of Elsevier.

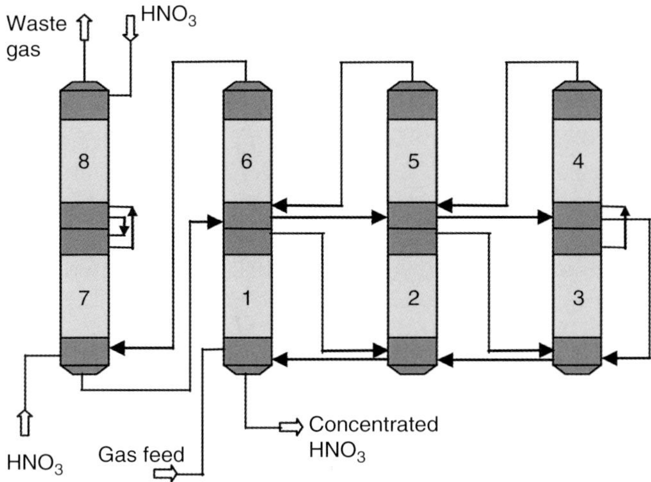

Figure 5.35 Reactive absorption process for NO$_x$ removal from the waste gas in an adipic acid plant. Source: Adapted from Kenig 2003 [65].

5.3.6 Reactive Comminution

Reactive comminution, also known under the more common name of mechanochemistry, is a branch of chemistry that is concerned with (physico-) chemical transformations of substances in all states of aggregation produced by the effect of mechanical energy [66, 67]. These (physico-)chemical transformations include the accumulation of defects (amorphization process), the formation of polymorphs (change of crystal structure), and activated chemical reactions [68]. There are also purely physical changes in the material accompanying the (physico-)chemical changes, such as particle size reduction and surface generation [69].

The first known example of mechanochemistry dates from ancient Greece. Theophrastus of Ersus (371-286 BC), a student from Aristotle, explains in his book "On Stones" how he rubs native cinnabar (mercury sulfide) with vinegar in a copper mortar with a copper pestle yielding elemental mercury. In the nineteenth century, Faraday performed a mechanochemical process in a mortar for the extraction of silver from silver chloride with the aid of elemental zinc. Later that century, several papers were written by the American scientist Lea (1823–1897), who is considered the founder of mechanochemistry [70].

Although originally mortar and pestle were used for mechanochemistry, the first tumbling mill was designed in 1870 and used small balls for the first time [71]. In the 1920s, the stirred ball attritor mill (1922) and the first vibration mills (1930) were invented in Germany. The planetary ball mill was first developed in 1961 by the Fritsch company and is still the most applied device in mechanochemical activation research nowadays. Other devices have been designed since. The Uni-Ball-Mill uses magnets to control the motion of the balls in a tumbler mill [72]. The Zoz company assembles ball mills for scale-up and the Japanese company Nissin Giken developed the "Super Misuni," a device able to control the temperature while milling using liquid nitrogen and electrical heating [67]. Recently, a lot of research goes into in situ and real-time monitoring of mechanical milling reactions with X-ray diffraction [73–76].

Ball milling is a type of high-energy milling where the energy comes from the impact of balls inside a moving vial. There are several types of ball mills available with each their own characteristics. The most commonly used ball mills are normal ball mill, planetary ball mill, vibration mill, and attritor or stirring ball mill. An attritor produces more shear force, giving more new surface area and less crystal disorders, in comparison to a planetary ball mill, which induces more impact forces generating more disordering and less new surfaces [77]. An overview is given in Figure 5.36.

A normal ball mill usually operates at a speed of 4–20 rpm. The planetary ball mill is a more advanced type of ball mill, where grinding balls are accelerated because of the rotation of the vessels, which are mounted on a disc rotating in the opposite direction. The random trajectories of the balls result in frictional forces and impact forces between the balls themselves and between balls and vessel wall. The high energy of the grinding balls now resides in the rotation speed, going up to 1100 rpm. The impact energy of the milling balls can reach up to 40 times higher than the impact energy because of gravitational acceleration caused by normal ball milling. Energy density in planetary mills is 100–1000 times greater than the energy density in normal ball mills [78]. At first, the powder undergoes comminution. This causes a reduction of particle size, an increase in specific surface area, and it promotes reactivity. After some time, molecular defects start to emerge, which opens up the possibility of changing the molecular structure. This change in structure can be purely physical and this

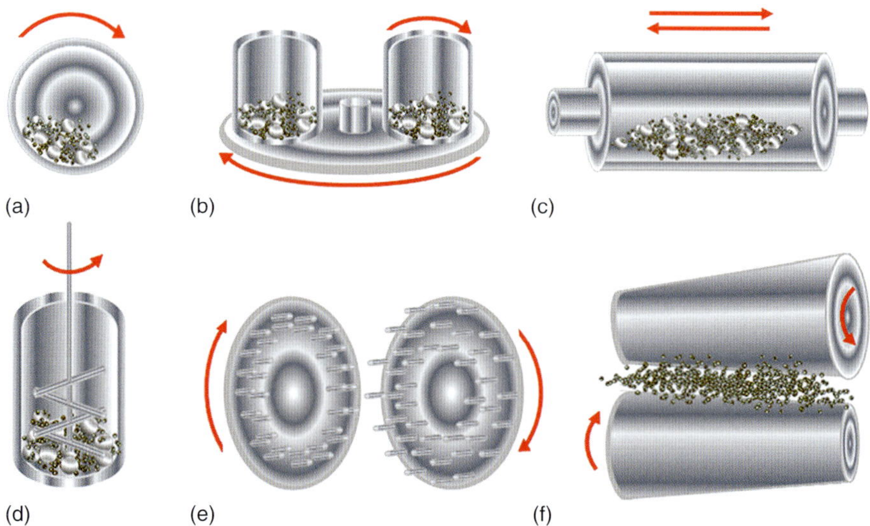

Figure 5.36 Different types of ball mills: (a) normal ball mill, (b) planetary ball mill, (c) vibration mill, (d) attritor/stirring ball mill, (e) pin mill, (f) rolling mill. Source: Baláž et al. 2013 [66]. Reproduced with permission of Royal Society of Chemistry.

is called plastic deformation. It is either random (creation of crystals) or regular (creation of polymorphs). During the entire milling process, there is generation of heat. At high speeds, temperatures over 200 °C can be achieved in a planetary ball mill [67].

The main stress types in a ball mill are compression, shear, shock, cutting, and impact as can be seen in Figure 5.37 [79]. The different crystallographic effects of ball milling, on the other hand, are given in Figure 5.38. First, there are point defects, which are defects on a single lattice point. These can be vacancies, substitutional atoms, and interstitial atoms. Next are dislocations, which are line defects where a crystal row stops and the continuous rows have to move around this stop to bind. Then, there are grain boundaries, which are planar defects where the direction of the lattice changes. Also, amorphization occurs when the crystalline structure, and the ordering of it, is lost. Lastly, there can be two-phase regions when two different phases are formed in the crystal structure.

Mechanical activation via ball milling causes physicochemical changes. Common effects are particle size reduction, surface generation, crystalline structure defects, and phase transformations. However, all the previously mentioned effects have different relaxation times. An overview can be seen in Figure 5.39. The fact that many of these effects are short-lived requires the reactions to take place, and the reagents to be present during mechanical activation. A study by Van Loy et al. on the recovery of rare earth elements from green lamp phosphors shows that the integration of mechanical activation and leaching in one single step allowed a sixfold reduction in treatment time compared to multistage sequential activation and leaching steps [80].

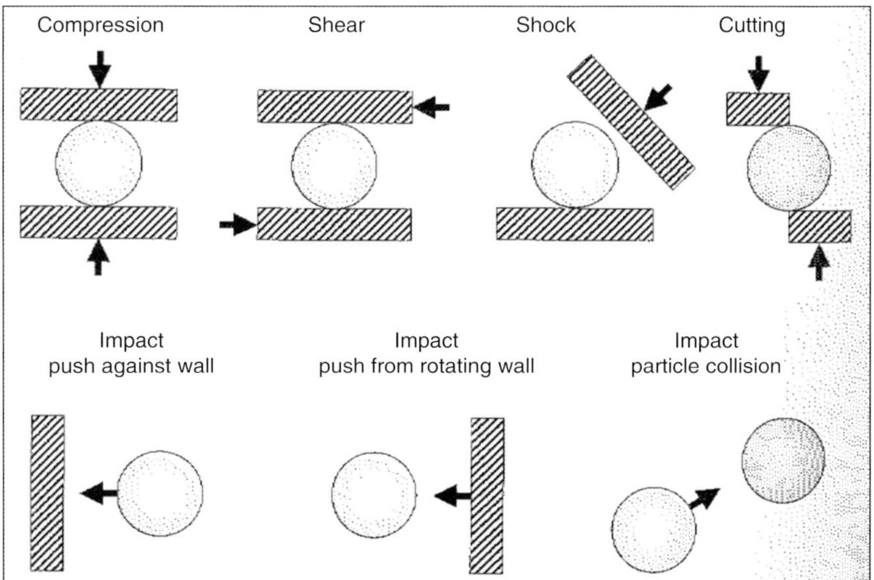

Figure 5.37 Fundamental load modes in milling equipment. Source: Hoffmann et al. 2005 [79]. Reproduced with permission of John Wiley and Sons.

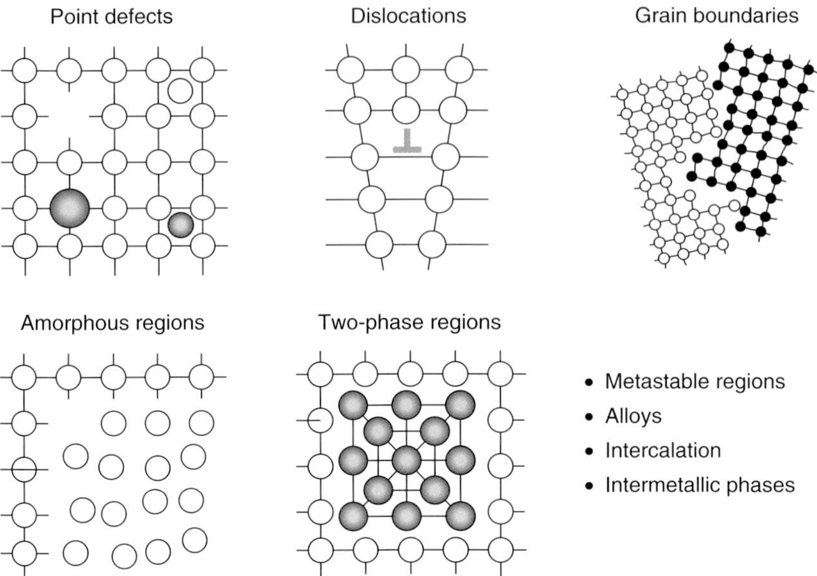

Figure 5.38 Defects created by mechanical activation of solids. Source: Hoffmann et al. 2005 [79]. Reproduced with permission of John Wiley and Sons.

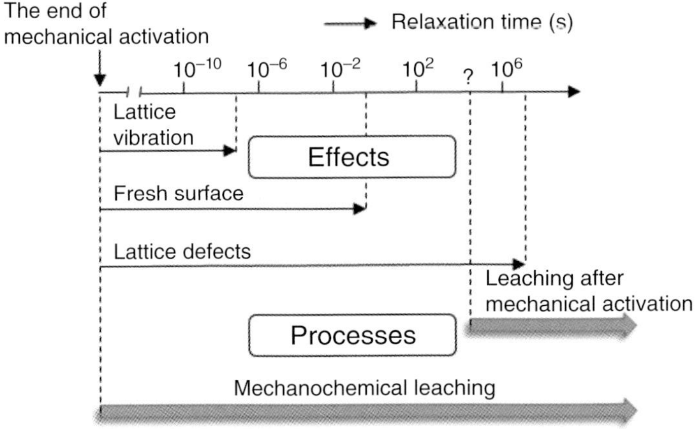

Figure 5.39 Relaxation times of the different effect after ball milling. Source: Adapted from Baláž 2008 [67].

Sasikumar et al. [81] analyzed where the energy goes in a high energy ball milling processes. He evaluated the energy balance for zircon during six hours of planetary ball milling. He concluded that only 13% of the energy input is transferred to the material, of which 90% is used for the breakage of bonds, which is mainly released as heat. The remaining 10% of energy is truly stored in the material in the form of additional surfaces and interfaces, point, line and volume defects, high energy structures, and nonuniform strain.

Different models have been developed over time to describe the mechanisms and effects of mechanochemistry. The first theory, the hotspot model, was first created by Bowden, Tabor, and Yoffe in the 1950s [67, 82]. The model describes the occurrence of very high temperatures (>1000 K), caused by very short friction processes (<1 ms). The magma-plasma model by Thiessen in 1967 suggests that large quantities of energy are set free at the contact spot of colliding particles. This implies direct impact, instead of lateral friction. A plasmatic transient state forms and energetic species including free electrons are ejected. This induces local temperatures of over 10 000 K.

If hotspots and magma-plasmatic sites would be the primary cause for mechanochemical reactions, one would expect extensive decomposition, which in most cases does not happen. More recently, the concept of "hierarchy" of energetic states was proposed for the analysis of different processes. In this concept, a large number of excitation processes occur because of the mechanical activation, which are characterized by different relaxation times. The most highly excited states having the shortest excitation times stand at the beginning. This is called the hierarchic model. The most relevant for ball milling would be processes occurring over areas of c. 1 mm^2, but models for these large areas have not yet been applied [67, 82, 83].

Different operational parameters can be optimized in ball milling [84]: the time, the (rotation or revolution) speed, the speed ratio, the ball to powder ratio, the ball size, and the ball filling ratio. All have an influence on how much energy is transferred to the material. The influence of the parameters is usually studied by using a mechanically induced self-sustaining reaction. The time at which this reaction starts is called the ignition time (t_{ig}). As there is a direct inverse relationship between this ignition time and the power of the planetary ball mill, it can be used as an indicator on how the parameters influence the mill feed.

Time is an important parameter in ball milling. It is independent of the other parameters and can easily be changed. Time has a positive effect on the wanted effects as more energy is dissipated with a higher milling time.

The ignition time strongly declines with increasing rotation speed, which means that the energy transferred strongly increases with increasing rotation speed. The speed ratio (k) is the ratio of the vial angular velocity to the sun wheel angular velocity. Findings of Mio et al. [85, 86] show that there is an optimum value. Their explanation is that the ball motion shifts from cascading to cataracting to rolling, with most ball collisions in the cataracting ball motion (Figure 5.40).

The ball to powder ratio (B:P ratio) is the weight of the balls divided by the weight of the loaded powder. Most of the data between B:P ratios of 5 and 40 follow an inverse relationship with the ignition time [87, 88]. Doubling the B:P ratio allows a rotation speed reduction with 100 rpm. The powder charge gave a linear dependence with the ignition time. The ball type and number have a more complex relationship with ignition time.

The main applications of mechanochemistry can be found in the fields of mechanical alloying, extractive metallurgy, and the synthesis of pharmaceuticals and nanomaterials.

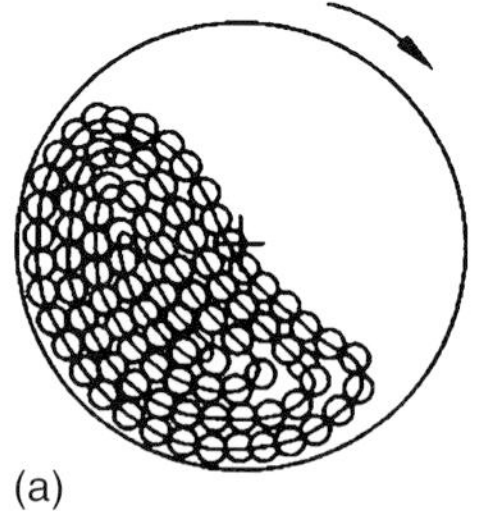 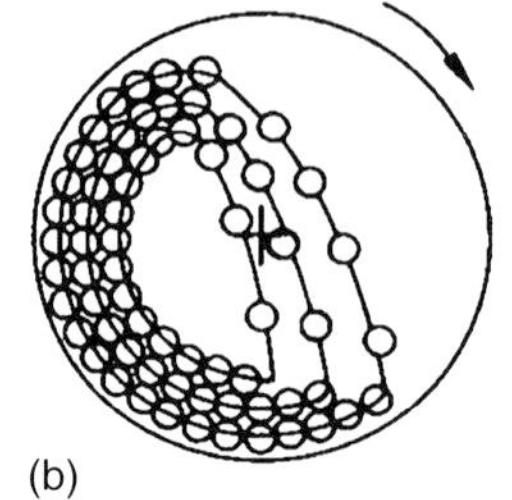 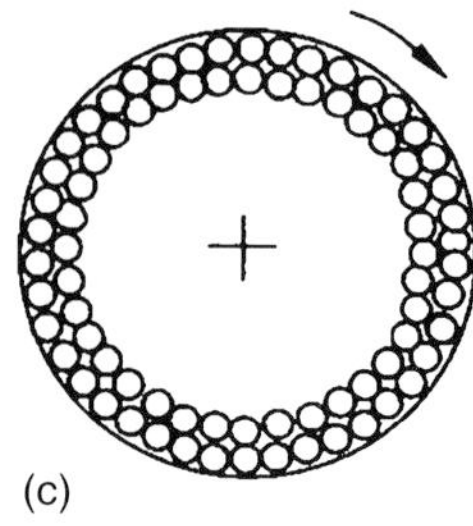

(a) (b) (c)

Figure 5.40 Different ball motions: (a) cascading, (b) cataracting, (c) centrifugal or rolling. Source: Bernotat and Schonert 1998 [150]. Reproduced with permission of John Wiley and Sons.

Mechanical alloying is a powder-processing technology for combining elements or alloys to produce a single homogeneous alloy in high-energy ball mills. It involves repeated cold welding, fracturing, and rewelding of powder particles. Mechanical alloying started in 1966 as a necessity in industry for producing oxide dispersion strengthened nickel and iron-based superalloys. This is done to obtain very high strength at both room and elevated temperatures [89]. Nowadays, the technique is mostly used to synthesize nanocomposites, such as nickel, iron, aluminum, or magnesium-based alloys or for the production of corrosion- and wear-resistant coatings [90]. Mechanical alloying can also be used for glass formation or to make feedstock for powder injection because it can achieve a very well mixed blend of the components [91]. Lastly, it has its function in combustion applications. Mechanical alloying is generally used to combine the best properties of different elements [72, 92–94].

Extractive metallurgy includes removing and refining valuable metals from ores or wastes. This field can be divided into three categories, which are hydrometallurgy (extraction of metals by leaching), pyrometallurgy (chemical reactions caused by high temperatures and smelting), and electrometallurgy (deposition by electrolysis) [92]. An example of mechanochemistry in hydrometallurgy is the Activox process. This process succeeds in leaching metals from a sulfide matrix by milling to a particle size of c. 10 µm in a low-temperature and low-pressure environment. It was first introduced by Corrans et al. in 1991, and since then, various improvements and extensions have been invented to optimize this process [95]. The Melt (MEchanical Leaching of Tetrahedrites) process, developed by Baláž et al. in 1994, is an example of mechanochemistry in pyrometallurgy [66], using attrition grinding in alkaline leaching of tetrahedrite at high temperatures [96]. The Lurgi-Mitterberg process combines aspects of hydrometallurgy and electrometallurgy for the treatment of sulfide concentrates. Leaching of copper was improved by mechanical activation with a vibration mill [92].

In the *pharmaceutical industry*, about 85% of the used chemicals are solvents, which make it very interesting to look for liquid-free solutions. For example, the production of Viagra reduced the use of solvents from 1700 to 7 l/kg with the aid of mechanochemistry [83]. In addition to reducing solvent usage, it is also important that pharmaceuticals exhibit a high bioavailability. Bioavailability depends on three factors: permeability, dose, and solubility. Solubility can be highly altered by mechanochemical treatment. Milling not only increases the surface area and reactivity but also causes amorphization, which raises solubility [92]. Mechanochemistry in pharmaceutical applications is usually correlated with the production of amorphous phases, but lately, a lot of research is focused on its use in cocrystallization. Instead of modifying the molecular structure of the active pharmaceutical ingredient, mechanochemical cocrystallization attempts to change the solid-state arrangements. This change in crystal structure improves several properties relevant to these applications, such as solubility, compressibility, and thermal stability [97]. Finally, it is important to mention the use of liquid-assisted grinding (LAG) to make the mechanochemical approach a viable alternative to solution synthesis [75]. The use of a small amount of liquid while milling can greatly improve results. For example, in cocrystallization reactions, LAG can accelerate the reaction or even lead to higher crystallinity in comparison with regular or neat grinding [98].

Grinding by ball milling is a very effective way for reducing the particle size, and thus, the *production of nanomaterials* as a separate branch of application is evident [99]. Figure 5.41 gives an overview of all the different types of mechanochemically synthesized nanomaterials, mostly catalysts. These include

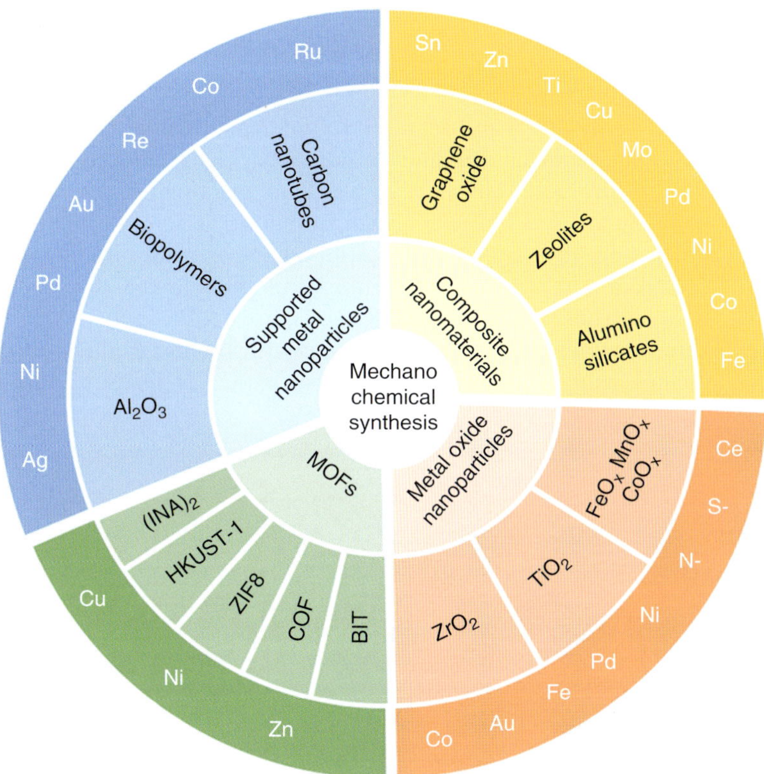

Figure 5.41 The different types of nanomaterials synthesized by mechanochemical methods. Source: Xu et al. 2015 [100]. Reproduced with permission of Royal Society of Chemistry.

supported metal nanoparticles, composite nanomaterials, metal oxide nanoparticles, and MOFs. The first mechanochemically synthesized MOF was recorded in 2006 by Pichon et al. They simply ground copper acetate and isonicotinic acid for 10 minutes in the absence of heating to produce a crystalline $Cu(INA)_2$ [100].

The biggest drawback of mechanochemistry lies in the difficulties in scaling up the device that is used for the milling. A disadvantage of planetary ball mills is that it is a batch operation instead of a, for industry, more convenient continuous operation. However, some continuous planetary ball mills have been studied and even manufactured [79]. There are also challenges in terms of increasing the sizes of ball mills as, for large installations, technical difficulties arise in the mill's drive systems and in cooling [101]. Other challenges include the control of internal heat generation and the process control for uniform processing. For these new instrumentations, exact theoretical models and integrated control models are required.

An example of a mill that is used on an industrial scale is the rod mill. It consists of a hollow cylindrical shell that rotates around an axis that is filled with long rods. When the cylinder is turned, the rods tumble against each other and grind the ore that is between them. The size of the ground material depends on the time in the mill, the amount of rods in the mill, and the initial size of the particles [102].

5.3.7 Handling Chemical Reactions in Highly Viscous Media – Reactive Extrusion

Extrusion technology, used for many years in polymer processing, has advantageously expanded on chemical reactions in complex, highly viscous media. Extruders-reactors (Figure 5.42) present an excellent example of intensified multifunctional equipment that integrates such functions as solids conveying, melting, mixing, reacting, devolatilizing, or pumping. Such multifunctionality implies in turn complex interactions between various phenomena that occur simultaneously in an extruder-reactor, as shown in Figure 5.43.

Extruders-reactors enable continuous processing with intense laminar mixing, which intensifies transport processes and consequently leads to lower

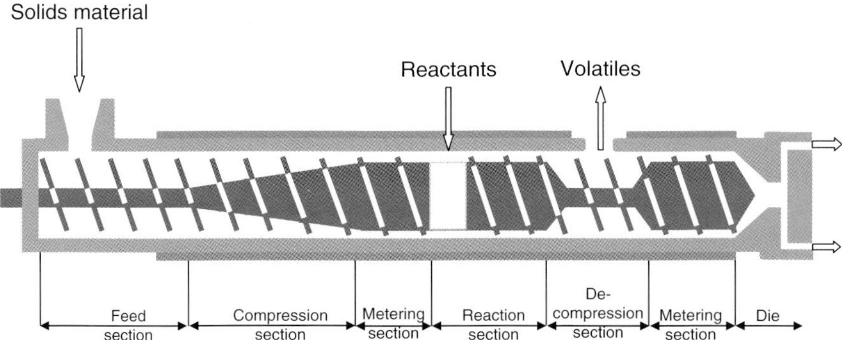

Figure 5.42 Scheme of an extruder-reactor. Source: Bouvier and Campanella 2014 [103]. Reproduced with permission of John Wiley and Sons.

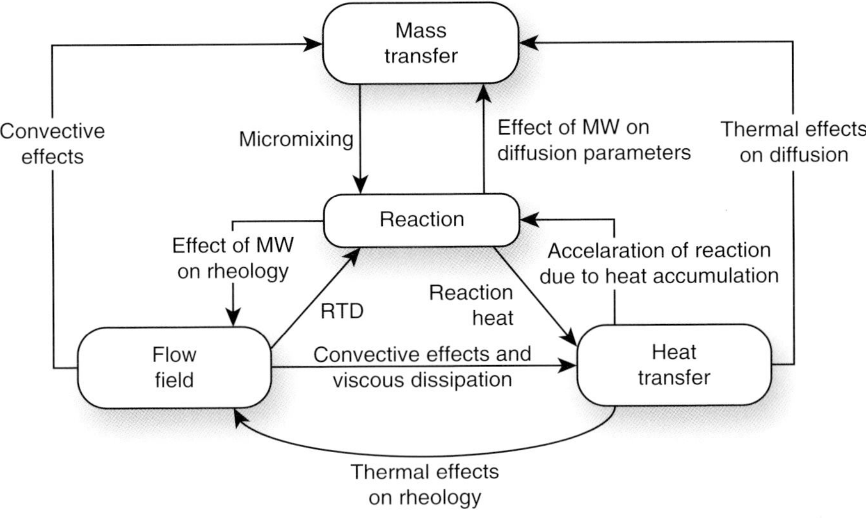

Figure 5.43 Interactions in reactive extrusion; RTD – residence time distribution; MW – molecular weight. Source: Adapted from Tsoganakis 1989 [151].

processing volumes. Thanks to their ability of dealing with highly viscous media, extruders-reactors largely eliminated the necessity of using solvents or diluents, which usually comprised 5–20 times the weight of the desired polymer product and required heavy and costly facilities for recovery, vaporization, condensation, purification, storage, metering, and reuse [103]. Residence times in extruder-reactors vary between a few seconds and several minutes, and, depending on the design/geometry, the flow characteristics are close to either an ideal continuous stirred tank reactor or to an ideal plug-flow reactor.

Extruders-reactors have found many applications in industry, mostly in polymer processing. Types of applications include grafting, interchain copolymer formations, coupling/crosslinking reactions, bulk polymerizations, controlled degradations, functionalizations, and functional group modifications [104]. In addition to conventional polymer processing, reactive extrusion has also been successfully applied in chemical, food, and pulp and paper industries.

5.3.8 Integrating Separation Techniques – Hybrid Separations

Hybrid separations integrate different separation techniques in a single operation, making use of the synergy between them. Possible combinations are shown in Figure 5.44, where all overlaps between the individual separation techniques can yield a hybrid separation.

5.3.8.1 Extractive Distillation

Extractive distillation is one of the most widely applied types of hybrid separations, particularly useful in close boiling point problems or in systems in which components form azeotropes. In the method, an extra component (solvent) is added to the system, which does not form azeotropes with feed components. The

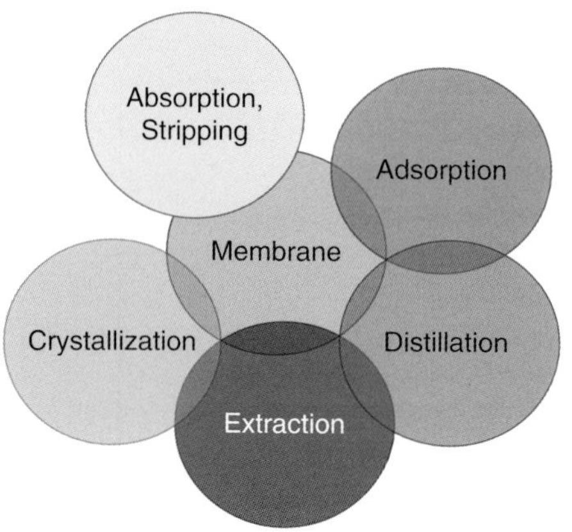

Figure 5.44 Variety of possibilities for hybrid separations.

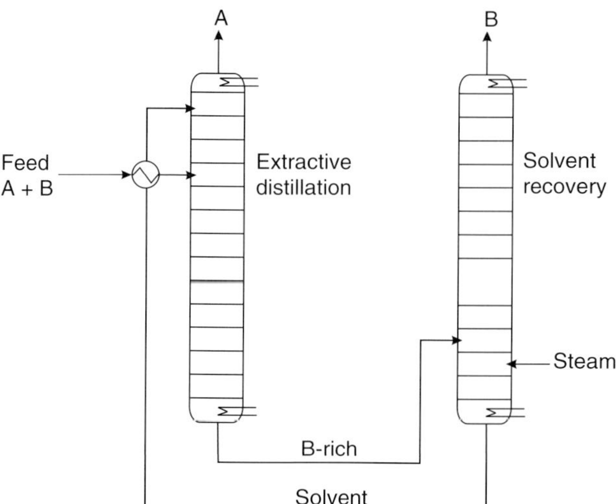

Figure 5.45 The two-column process for extractive distillation.

solvent alters the relative volatility of original feed components. It allows obtaining a pure component from the top of the distillation column while the other component leaves the column through the bottom together with the solvent and is separated in the secondary distillation column (Figure 5.45).

Originally, the extractive distillation was limited to two-component problems. However, recent developments in solvent technology enable applications of this hybrid separation also to multicomponent systems. For example, Lei et al. [105] discuss a four-column process for the separation of the C4 mixture. The system, shown in Figure 5.46, enables separation of four individual types of hydrocarbons and in this sense is superior to the traditional industrial two-column system using acrylonitrile as the solvent, which enabled only the separation of butadiene and vinylacetylene (VAC). Reduction in capital costs and energy savings present the most important advantages of extractive distillation and effects of up to c. 25% are

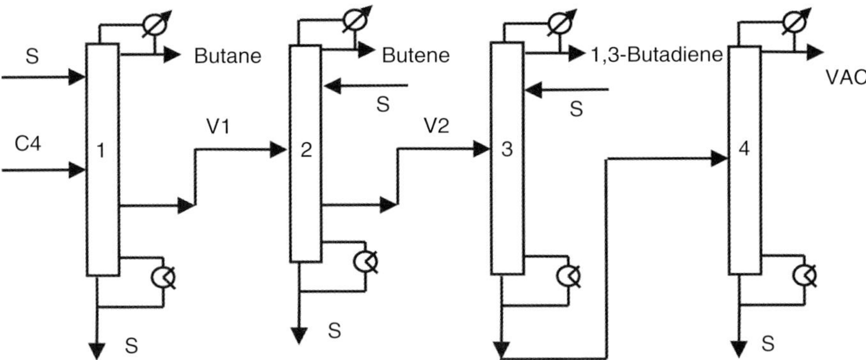

Figure 5.46 Four-column extractive distillation system for separation of the C4 mixture. Source: Lei and Chen 2003 [105]. Reproduced with permission of Taylor and Francis Group.

reported. More information on those effects and the existing commercial applications can be found in Ref. [106].

5.3.8.2 Adsorptive Distillation

Adsorptive distillation is a three-phase mass transfer operation, which is conceptually similar to extractive distillation. The role of the selective solvent in extractive distillation plays here a selective adsorbent. The adsorbent is usually a fine powder (particle size in 10 μm range), fluidized and circulated by an inert carrier. Also here, the process concept involves two columns: an adsorptive distillation column for the main separation and a distillative desorption column for the regeneration of the adsorbent (Figure 5.47). As it was in the case of extractive distillation, also adsorptive distillation can potentially be used for

Figure 5.47 The two-column process for adsorptive distillation in binary system.

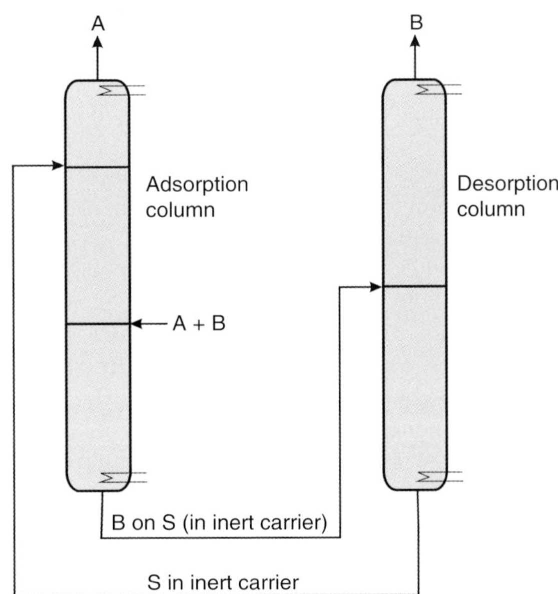

the separation of mixtures containing close boiling components or to bypass the azeotrope. However, in contrast to the extractive distillation and despite almost 50 years of research history, no large-scale commercial processes using adsorptive distillation have been reported. This is due both to the operational complexity of the processes involving moving fine solid materials and the insufficient separation/purification efficiency. Lei et al. [105] presented a direct experimental comparison between the extractive and adsorptive distillation for separation in a simple binary system consisting of ethanol and water. It has been shown that, in contrast to the extractive distillation, the adsorptive distillation failed to achieve high-purity (>99%) ethanol.

5.3.8.3 Membrane Distillation

Membrane distillation (MD) is considered a promising hybrid separation method, primarily applicable in water desalination and environmental engineering. In membrane distillation systems, the temperature difference between both sides of the membrane results in vapor pressure difference and is the main mechanism of mass transport through the (hydrophobic) membrane. Water molecules are transported through the pores of the membrane from the high vapor pressure to the low vapor pressure side. Here, they are coming in contact with a colder aqueous solution (the so-called *direct contact membrane distillation*), with a cold surface separated by an air gap (*air gap membrane distillation*), with a sweep gas (*sweep gas membrane distillation*), or with the vacuum (*vacuum membrane distillation*). These four basic types of membrane distillation are presented in Figure 5.48.

Membrane distillation units are energy-consuming, which makes them less attractive if the energy needs to be supplied in a conventional way, i.e. from fossil fuels. However, the fact that MD can be operated at low temperatures (much lower than in case of classical distillation) makes it worthwhile to consider the energy supply from sustainable sources, such as wind, solar, or geothermal. Figure 5.49 shows an exemplary concept of a wind/solar energy-operated membrane distillation unit [107].

The name "membrane distillation" is also sometimes used to describe *membrane-assisted distillation* systems. In those systems, a membrane (usually a pervaporation) unit is coupled to one or more distillation columns. Possible configuration functions of such a system are shown in Figure 5.50 [108]. In case (a), the membrane is used to break an azeotrope. The permeate is recycled to the first column while the retentate is sent to the second column for further purification. In case (b), the membrane is used to directly purify the azeotropic top product until the required product specification has been reached. In case (c), the membrane is used to prefractionate the mixture that relieves the distillation column. Finally, in case (d), the membrane is used to preseparate a three-component mixture.

Energy saving presents the most important effect of the incorporation of a membrane unit in a distillation system. Pribic et al. [109] analyzed, among other things, the energy consumption in a THF-water separation system using pressure-swing distillation and using a distillation/pervaporation hybrid distillation system (Figure 5.51). The reported energy savings on steam and

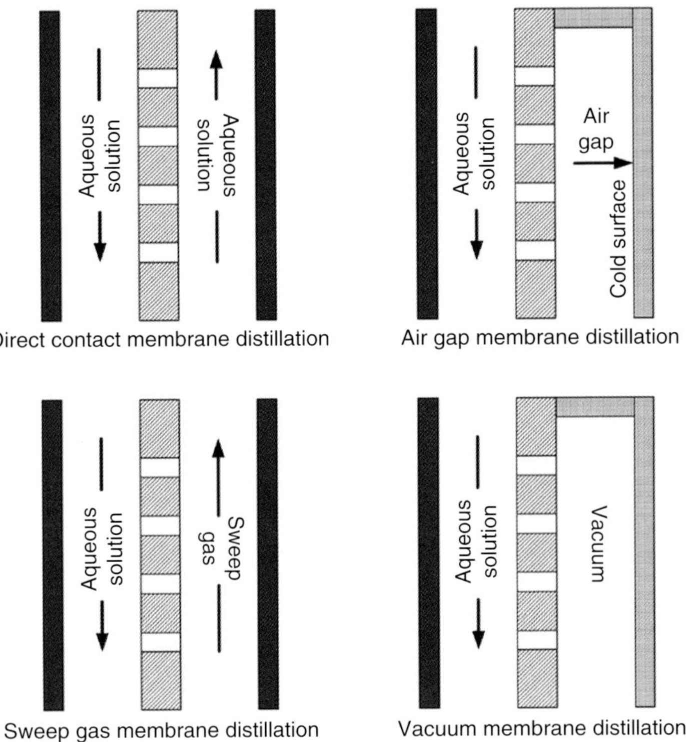

Figure 5.48 Four basic types of membrane distillation systems. Source: Curci and Drioli 2005 [152]. Reproduced with permission of Taylor and Francis.

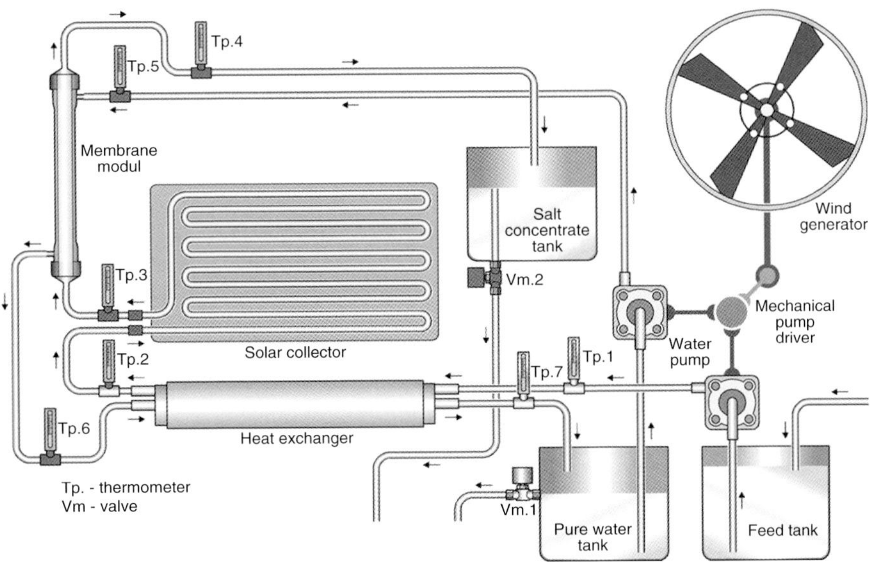

Figure 5.49 Solar/wind energy-operated membrane distillation system. Source: Susanto 2011 [107]. Reproduced with permission of Elsevier.

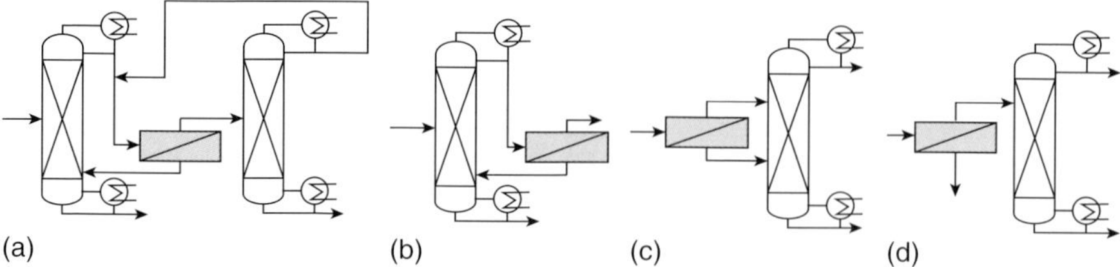

Figure 5.50 Various configuration on membrane-assisted distillation systems. Source: Adapted from Lutze and Górak 2013 [108].

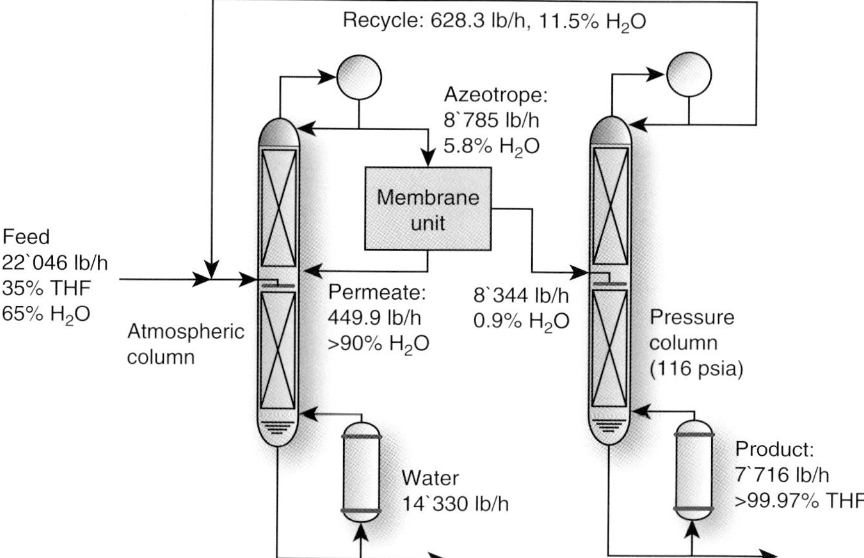

Figure 5.51 Membrane-assisted dual-pressure distillation system for THF water separation. Source: Pribic et al. 2006 [109]. Reproduced with permission of Taylor and Francis Group.

cooling water in the latter case exceeded 50%. Sommer and Melin [110] reported possible savings in energy consumption of 83% and reduction in investment and operational costs of 40% in the organics dehydration processes. Further analysis of the economics of different pervaporation-assisted distillation processes can be found in Ref. [111].

Pervaporation units can also be combined with reactive distillation. Most systems investigated in the literature concern esterification and etherification processes [108, 112]. Such hybrid systems are already offered commercially. An example here can be the fatty acid esterification technology developed by Sulzer and schematically shown in Figure 5.52.

5.3.8.4 Membrane Crystallization

Membrane crystallization is a relatively new hybrid separation technique that can potentially lead to an improved control of crystal morphology, structure, and purity [113]. Many conventional crystallization processes, carried out

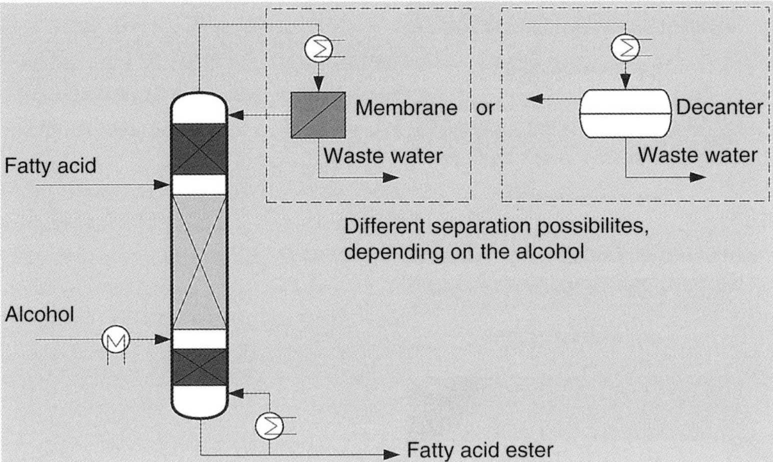

Figure 5.52 Sulzer's hybrid process for fatty acids esterification. Source: Courtesy of Sulzer Chemtech Ltd., Switzerland, www.sulzer.com.

either in evaporative or cooling crystallizers, suffer from poor reproducibility of product quality, which is partially due to limited control of partial phenomena occurring in the system. Additional problem is high energy consumption in conventional crystallization systems. A combination of crystallization and membrane separation can help the above issues and, depending on the case, a membrane can play three different roles in the systems [113]. It can be used for solvent removal only, where solvent is removed from the solution by the temperature gradient (Figure 5.53a). It can also be used for solvent/antisolvent demixing (Figure 5.53b), where selective removal of the solvent increases the antisolvent fraction and decreases solubility. Finally, it can be used for controlled antisolvent addition (Figure 5.53c). A separate role that a membrane could play in a crystallization system without a mass flux across the membrane is that of a template (mold) to control the crystal size or shape [114]. In that case, crystals are grown in the nanopores of the membrane and replicate the size and the morphology of the template. Most potential applications of membrane crystallization investigated so far concern production of (bio)pharmaceutical compounds [115–117]. More recently, membrane crystallization was also studied in the context of CO_2 sequestration [118].

5.3.8.5 Extractive Crystallization

Extractive crystallization can be used to bypass the eutectic barriers in binary or multicomponent systems. The driving force in this hybrid separation process is created by altering the solid–liquid phase relationships via the addition of a third component (usually liquid solvent) to the system. The solvent is chosen in such a way that it binds strongly at crystallization temperature but separates easily at another temperature, where it is usually regenerated via distillation. Examples of the so-defined extractive crystallization include separation of isomers, such as *m*- and *p*-cresols [119], *o*- and *p*-nitrochlorobenzenes [120], or *p*-xylene and *m*-xylene [121]. In the separation of chlorobenzoic acids by the dissociation

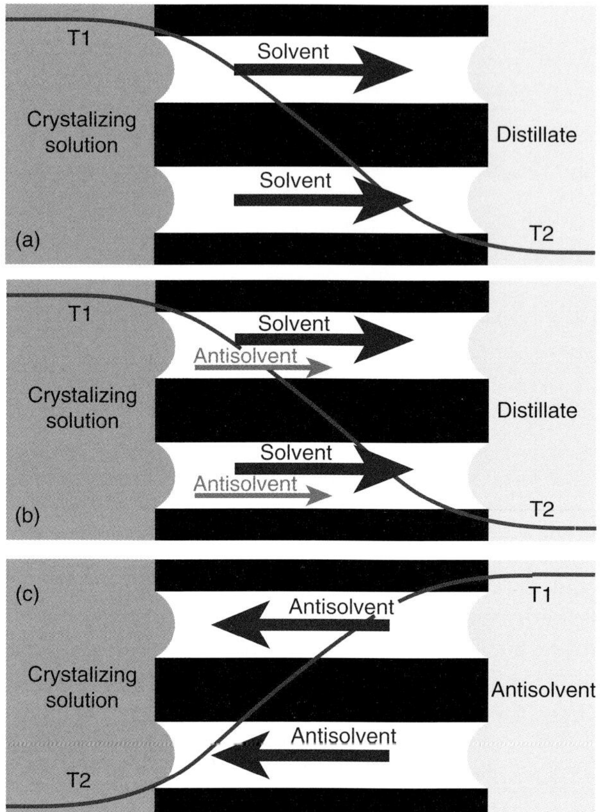

Figure 5.53 Three different forms of membrane crystallization: (a) solvent removal; (b) solvent/antisolvent demixing; and (c) antisolvent addition. Source: Drioli et al. 2012 [113]. Reproduced with permission of Elsevier.

extractive crystallization, good single-stage recovery of *o*-CBA (>90%) and very high separation factors (usually reaching infinity) have been reported [122].

A basic extractive crystallization system consists of two crystallizers operating at different temperatures, two filters, and a solvent recovery columns. Depending on chemicals involved, the solvent can be introduced to the first of second crystallizer. According to Rajagopal et al. [121], those two basic flow sheets depicted in Figure 5.54 can handle systems with a wide variety of solid–liquid-phase behaviors.

5.3.8.6 Membrane Absorption/Stripping

Membrane absorption and stripping are processes that Mother Nature had invented a long time before the engineers did. Human lungs and intestines are perfect examples of membrane absorption systems. In fact, medical care was the first application area of membrane absorption in the artificial lungs [123, 124]. In the simplest case of membrane absorption, a gaseous component is selectively transported via a membrane and dissolved in the absorbing liquid, as it is shown

Figure 5.54 Two basic flow sheets of extractive crystallization systems. Source: Adapted from Rajagopal et al. 1991 [121].

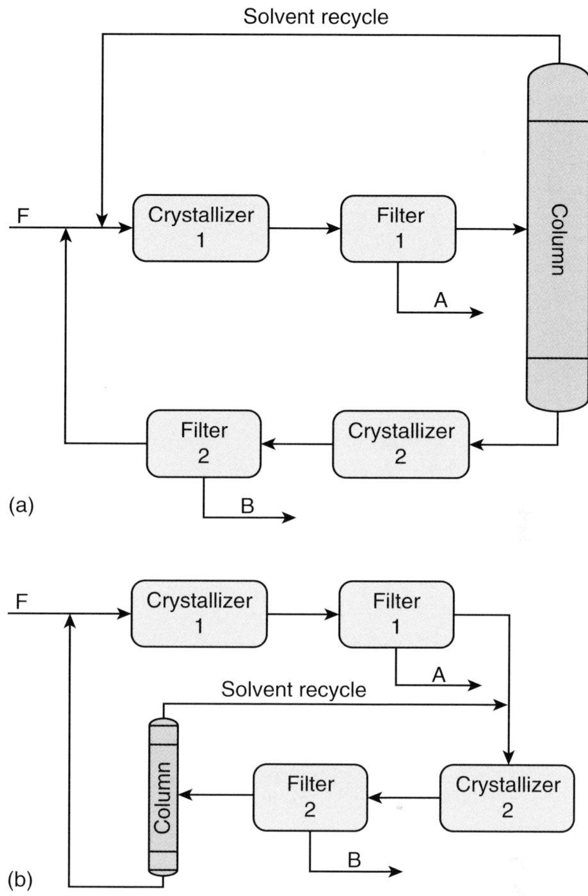

in Figure 5.55. In a membrane-stripping process in turn, selected components are removed from the liquid phase through the membrane by a stripping gas.

The most important application area of membrane absorption in chemical processing is the removal and capture of CO_2 from flue gas [125–128] or biogas [129]. Energy savings present the most often reported benefit from this technology. Okabe et al. [126] reported a 47% reduction of the energy consumption compared to the conventional chemical absorption method. Some years ago, Kværner presented a membrane absorption-based technology for the removal of CO_2 from turbine exhaust gases in offshore applications [128]. The process, based on the membrane-facilitated CO_2 absorption in amine, followed by a membrane-facilitated stripping with steam, is schematically shown in Figure 5.56. The expected cost reduction, in comparison with a conventional amine separation process, ranges between 30% and 40%, on both investment and operating costs. The new membrane-based process also offers a very significant reduction in the weight and size of equipment (on absorber and stripper 70–75% and 65%, respectively).

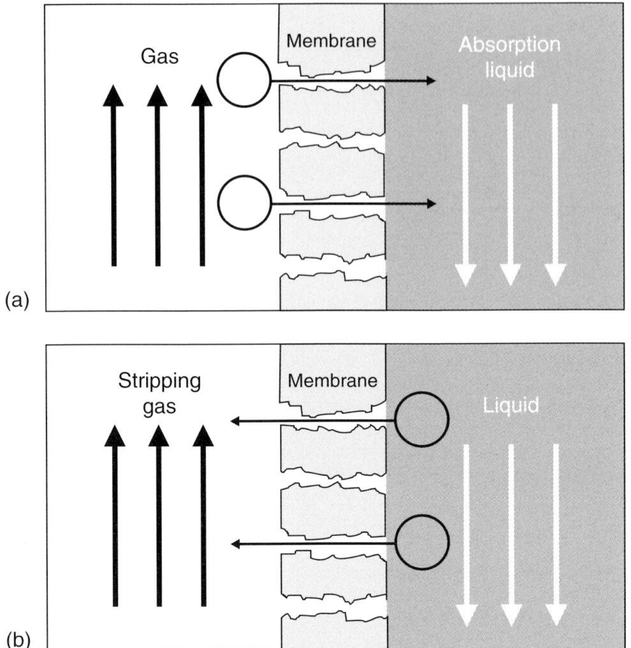

Figure 5.55 Membrane absorption (a) and stripping (b).

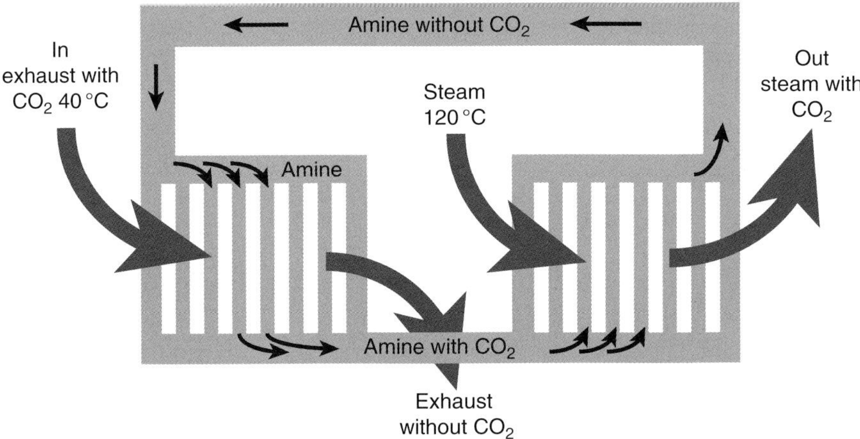

Figure 5.56 A scheme of Kværner's membrane absorption process for CO_2 removal from flue gases. Source: Adapted from Herzog and Falk-Pedersen 2000 [128].

An important characteristic feature of the membrane absorption is that it proceeds without creating any gas–liquid interface in the form of bubbles. Such a bubble-free mass transfer can be of advantage in certain processes, for instance, in shear-sensitive biological systems where numerous papers on membrane aeration or oxygenation in fermentation systems involving sensitive cell lines (e.g. mammalian cells for production of monoclonal antibodies) have been published (e.g. [130, 131]). Similar bubble-free systems have also been proposed in wastewater treatment, in cases when oxygen requirements are too high for conventional aeration systems or when bubbling of air would result in stripping of volatile organic compounds or foaming [132].

5.3.8.7 Membrane Chromatography (Adsorptive Membranes)

Membrane chromatography is a hybrid separation technique applied commercially in the biopharmaceutical sector for the downstream processing of proteins [133–135]. Traditionally, most chromatographic purification steps in the downstream processing of proteins take place in columns packed with adsorbent in form of beads. The main feature and advantage of membrane chromatography is the elimination of the *pore diffusion* in dead-ended pores of the adsorbent beads, which is the main transport resistance in traditional chromatography. This limitation becomes particularly important in the case of large molecules, such as proteins. In membrane chromatography, dissolved molecules are carried by the forced convective flow through the membrane pores, where the *film diffusion* (much faster than the pore diffusion) leads them to functional ligands attached to the inner pore structure, which act as adsorbents (Figure 5.57). This way, the process throughput can be significantly increased. Przybycien et al. [133] report dynamic capacities in membrane chromatography at least 1 order of magnitude greater than in traditional chromatography. Varadaraju et al. [136] compared membrane-based process for monoclonal antibody purification with a traditional three-packed bed-based chromatography and found

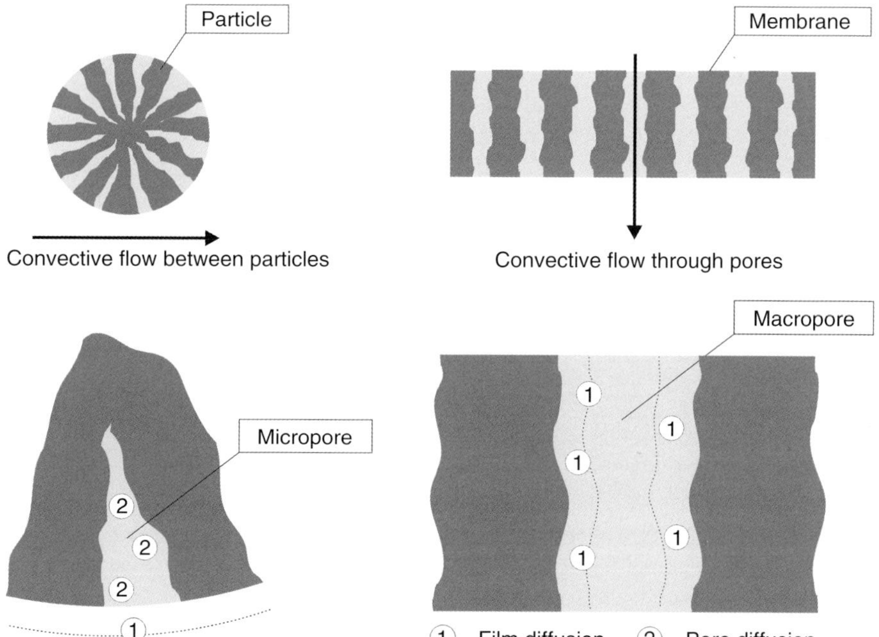

Figure 5.57 Mass transfer mechanisms in the conventional chromatography (a) and in membrane chromatography (b).

an up to 80% reduction in the adsorbent volume, 50% reduction in processing time, 40% reduction in labor costs, and 40% reduction in waste generated. Next to its established application area – biopharmaceutical processing – membrane chromatography/adsorption has also been investigated in the context of food processing [137, 138] and water purification [139].

5.3.8.8 Membrane Extraction

In *membrane extraction*, the treated solution and the extractant/solvent are separated from each other by means of a solid or liquid membrane. The technique is primarily applied in three areas: wastewater treatment (e.g. removal of heavy metals or recovery of trace components), biotechnology (e.g. removal of products from fermentation broths or separation of enantiomers), and analytical chemistry (e.g. on-line monitoring of pollutants concentrations in wastewater). It overcomes the limitations of the conventional liquid extraction (e.g. flooding, mixing-settling, and the requirement of the density difference) and provides large specific surface areas for mass exchange. The main form of membrane extraction used for the removal of metallic pollutants is *emulsion pertraction* [140, 141]. In this technique, hollow fiber membranes are often used. The water phase flows outside the membrane fibers, while the metal ions to be removed are bound by an extractant present in the membrane pores and inside the fibers. Through the fibers, a strip liquid is continuously flowing that regenerates the extractant (Figure 5.58).

Figure 5.59 shows schematically an industrial hollow fiber-based pertraction process for wastewater treatment, according to the TNO technology [142]. The membrane unit is integrated with a film evaporator to enable the release of pollutants in pure form. The reported purification costs per liter of the treated wastewater can be as low as a half of the cost of the air-stripping technique and one-third of the costs of the activated carbon filtration.

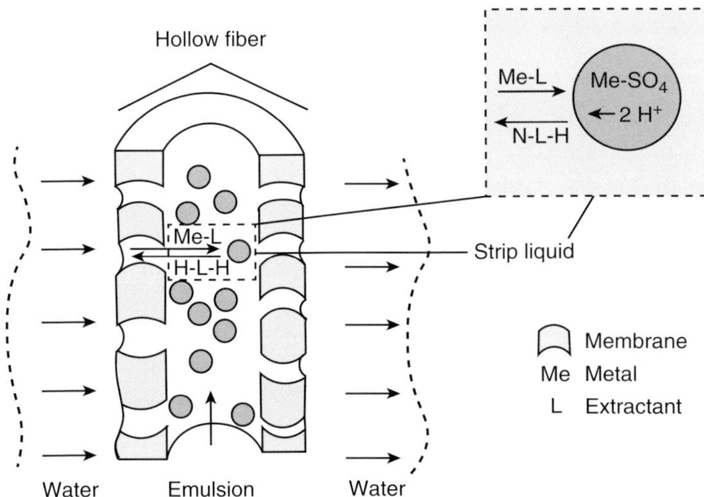

Figure 5.58 Scheme of emulsion pertraction. Source: Adapted from Klaassen et al. 2008 [140].

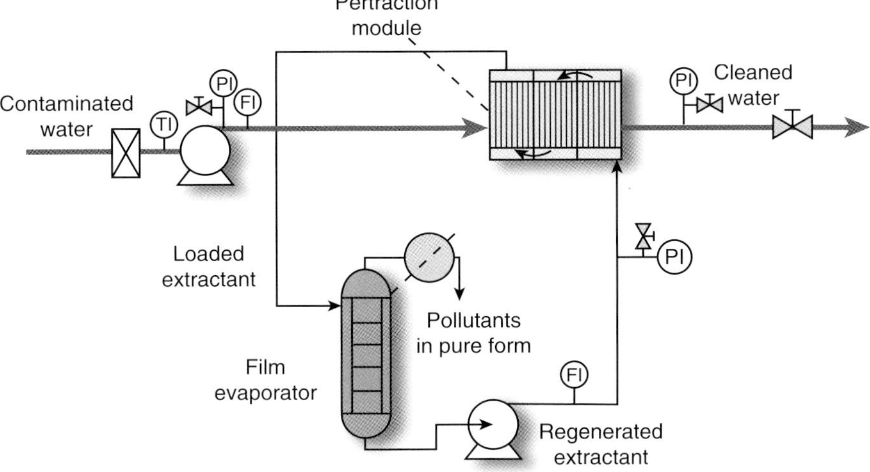

Figure 5.59 TNO pertraction–regeneration process for wastewater treatment, with pollutants released in pure form. Source: Adapted from Klaassen et al. 2008 [142].

References

1 Schembecker, G. and Tlatlik, S. (2003). Process synthesis for reactive separations. *Chem. Eng. Proc.* 42: 179–189.

2 Hesse, D. (2000). Multifunctional catalysts. In: *Catalysis from A to Z. A Concise Encyclopedia* (ed. B. Cornils, W.A. Herrmann, R. Schlogl and C.-H. Wong), 395. Weinheim: Wiley-VCH.

3 Weisz, P.B. (1962). Polifunctional heterogeneous catalysis. In: *Advances in Catalysis and Related Subjects*, vol. 13 (ed. D.D. Eley), 132–191. New York: Academic Press.

4 Gunn, D.J. and Thomas, W.J. (1965). Mass transport and chemical reaction in multifunctional catalyst system. *Chem. Eng. Sci.* 20: 89–100.

5 Bisio, C., Gatti, G., Marchese, L. et al. (2012). Design and applications of multifunctional catalysts based on inorganic oxides. In: *Nanoporous Materials for Energy and the Environment* (ed. G.C.G. Rios and N. Kanellopoulos), 13–53. Pan Stanford Publishing Pte.Ltd.

6 Iglesia, E., Barton, D.G., Biscardi, J.A. et al. (1997). Bifunctional pathways in catalysis by solid acids and bases. *Catal. Today* 38: 339–360.

7 Grotjahn, D.B. (2008). Bifunctionals catalysts and related complexes: structures and properties. *Dalton Trans.* 37: 6497–6508.

8 Climent, M.J., Corma, A., Iborra, S. et al. (2009). New one-pot multistep with multifunctional catalysts: decreasing the E factor in the synthesis of fine chemicals. *Green Chem.* 12: 99–107.

9 Cirujano, E.G., Llabrés i Xamena, F.X., and Corma, A. (2012). MOFs as multifunctional catalysts: one-pot synthesis of methanol form citronellal over a bifunctional MIL-101 catalysts. *Dalton Trans.* 41: 4249–4254.

10 Tanabe, K. and Hölderlich, W.F. (1999). Industrial application of solid acid-base catalysts. *Appl. Cat. A.: General* 181: 399–434.

11 Setoyama, T. (2006). Acid–base bifunctional catalysis: an industrial viewpoint. *Catal. Today* 116: 250–262.

12 Grünewald, M. and Agar, D. (2004). Enhanced catalyst performance using integrated structured functionalities. *Chem. Eng. Sci.* 59: 5519–5526.

13 Dietrich, W., Lawrence, P.S., Grünewald, M., and Agar, D. (2005). Theoretical studies on multifunctional catalysts with integrated adsorption sites. *Chem. Eng. J.* 107: 103–111.

14 Kapil, A., Bhat, S.A., and Sadhukhan, J. (2008). Multiscale characterization framework for sorption enhanced reaction processes. *AIChE J.* 54: 1028–1036.

15 Foley, H.C., Lafyatis, D.S., Mriwala, R.K. et al. (1994). Shape selective methylamines synthesis: reaction and diffusion in a $CMs\text{-}SiO_2\text{-}Al_2O_3$ composite catalyst. *Chem. Eng. Sci.* 49: 4771–4786.

16 Dautzenberg, F.M. and Mukherjee, M. (2001). Process intensification using multifunctional reactors. *Chem. Eng. Sci.* 56: 251–267.

17 Rakitzis, T.P., Van den Brom, A.L., and Janssen, M.H. (2004). Directional dynamics in the photodissociation of oriented molecules. *Science* 303: 1852–1854.

18 Edvinsson-Albers, R.K., Houterman, M.J.J., Vergunst, T. et al. (1998). Novel monolithic stirrer reactor. *AIChE J.* 44: 2459–2464.

19 Hoek, I. (2004). Towards the catalytic application of a monolithic stirrer reactor. PhD dissertation. Delft University of Technology.

20 De Lathouder, K.M., Bakker, J.J.W., Kreutzer, M.T. et al. (2006). Structured reactors for enzyme immobilization: a monolithic stirrer reactor for application in organic media. *Trans. IChemE, Part A* 84: 390–398.

21 Boldrini, D.E., Sánchez, J.F.M., Tonetto, G.M., and Damiani, D.E. (2012). Monolithic stirrer reactor: performance in the partial hydrogenation of sunflower oil. *Ind. Eng. Chem. Res.* 51: 12222–12232.

22 Wu, D., Guo, Y., Geng, S., and Xia, Y. (2013). Synthesis of propylene carbonate from urea and 1,2-propylene glycol in a monolithic stirrer reactor. *Ind. Eng. Chem. Res.* 52: 1216–1223.

23 Sulzer Chemtech (CH). (1997). Mixing and reaction technology. *Sulzer brochure*, 23.27.06.40-100.

24 Kumar, V., Mridha, M., Gupta, A.K., and Nigam, K.D.P. (2007). Coiled flow inverter as a heat exchanger. *Chem. Eng. Sci.* 62: 2386–2396.

25 Anxionnaz, Z., Cabassud, M., Gourdon, C., and Tochon, P. (2008). Heat exchanger/reactors (HEX reactors): concepts, technologies: state-of-the-art. *Chem. Eng. Proc. Proc. Intens.* 47: 2029–2050.

26 Santacesaria, E., Di Serio, M., Tesser, R. et al. (2009). Use of a corrugated plates heat exchanger reactor for obtaining biodiesel with very high productivity. *Energy Fuels* 23: 5206–5212.

27 Niedbalski, N., Johnson, D., Patnaik, S.S., and Banerjee, D. (2014). Study of a multi-phase hybrid heat exchanger-reactor (HEX reactor): part I – experimental characterization. *Int. J. Heat Mass Transfer* 70: 1078–1085.

28 Phillips, C. (1999). Development of a novel compact chemical reactor - heat exchanger. In: *Proceedings of 3rd International Conference on Process Intensification for the Chemical Industry*, vol. 38 (ed. A. Green), BHR Group Conference Series, 71–87. London: Professional Engineering Publishing Ltd.

29 Calder, R. (2000). HEX reactors reduce process time by 98.6%. *BHR Group News*, Summer, 4.

30 Despènes, L., Elgue, S., Gourdon, C., and Cabassud, M. (2012). Impact of the material on the thermal behavior of heat-exchangers-reactors. *Chem. Eng. Proc. Proc. Intens.* 52: 102–111.

31 Kiss, A.A. (2014). Distillation technology – still young and full of breakthrough opportunities. *J. Chem. Technol. Biotechnol.* 89: 479–498.

32 Stankiewicz, A. (2003). Reactive separations for process intensification. An industrial perspective. *Chem. Eng. Proc.* 42: 137–144.

33 Schoenmakers, H.G. and Bessling, B. (2003). Reactive and catalytic distillation from an industrial perspective. *Chem. Eng. Proc.* 42: 145–155.

34 Schmidt-Traub, H. and Górak, A. (2006). Introduction. In: *Integrated Reaction and Separation Opertions. Modelling and Experimental Validation* (ed. H. Schmidt-Traub and A. Górak), 1–6. Berlin: Springer VDI.

35 Hiwale, R.S., Bhate, N.V., Mahajan, Y.S., and Mahajani, S.M. (2004). Industrial applications of reactive distillation: recent trends. *Int. J. Chem. React. Eng.* 2: 1–52.

36 Harmsen, G.J. (2007). Reactive distillation: the front-runner of industrial process intensification. A full review of commercial applications, research, scale-up, design and operation. *Chem. Eng. Proc. Proc. Intens.* 46: 774–780.

37 Siirola, J.J. (1995). An industrial perspective on process synthesis. *AIChE Symp. Ser.* 91: 222–233.

38 Stein, E., Kienle, A., and Sundmacher, K. (2000). Separation using coupled reactive distillation columns. *Chem. Eng.* 107: 68–72.

39 Heils, R., Niesbach, A., Wierschem, M. et al. (2014). Integration of enzymatic catalysts in a continuous reactive distillation column: reaction kinetics and process simulation. *Ind. Eng. Chem. Res.* 53: 19612–19619.

40 Wierschem, M., Heils, R., Schlimper, S. et al. (2015). Enzymatic reactive distillation for the transesterification of ethyl butyrate: model validation and process analysis. *Comp. Aided Chem. Eng.* 37: 2135–2140.

41 Sirkar, K.K., Shanbhag, P.V., and Kovvali, A.S. (1999). Membrane in a reactor: a functional perspective. *Ind. Eng. Chem. Res.* 38: 3715–3737.

42 Westermann, T. and Melin, T. (2009). Flow-through catalytic membrane reactors – principles and applications. *Chem. Eng. Sci.* 48: 17–28.

43 Drioli, E., Brunetti, A., Di Profio, G., and Barbieri, G. (2012). Process intensification strategies and membrane engineering. *Green Chem.* 14: 1561–1572.

44 Rios, G.M., Belleville, M.P., Paolucci, D., and Sanchez, J. (2004). Progress in enzymatic membrane reactors – a review. *J. Membr. Sci.* 242: 189–196.

45 Kuczynski, M. (1987). The synthesis of methanol in a gas-solid-solid trickle flow reactor. PhD dissertation. University of Twente, Enschede.

46 Westerterp, K.R., Bodewes, T.N., Vrijland, M.S., and Kuczynski, M. (1988). Two new methanol converters. *Hydrocarbon Process.* 67: 69–73.

47 Fricke, J., Schmidt-Traub, H., and Schembecker, G. (2008). Chromatographic reactors. In: *Ullmann's Encyclopedia of Industrial Chemistry*, 103–129. Weinheim: Wiley-VCH Verlag GmbH & Co. KGaA.

48 Broughton, D.B. and Gerhold, C.G. (1961). Continuous sorption process employing fixed bed of sorbent and moving inlets and outlets. US Patent 2,985,589.

49 Carr, R.W. (1993). Continuous reaction chromatography. In: *Preparative and Production Scale Chromatography* (ed. G. Ganetsos and P.E. Barker), 421–447. New York: Marcel Dekker.

50 Molga, E., Lewak, M., and Ciach, M. (2006). Modeling of reactive chromatography carried out in a rotating annular reactor. *Chem. Proc. Eng.* 27: 1411–1429.

51 Sharma, M., Vyas, R.K., and Singh, K. (2013). A review on reactived adsorption for potential environmental applications. *Adsorption* 19: 161–188.

52 Pereira, C.S.M., Silva, V.M.T.M., and Rodrigues, A.E. (2012). Green fuel production using the PermSMBR technology. *Ind. Eng. Chem. Res.* 51: 8928–8938.

53 Pereira, C.S.M. and Rodrigues, A.E. (2013). Process intensification: new technologies (SMBR and PermSMBR) for the synthesis of acetals. *Catal. Today* 218–219: 148–152.

54 Viecco, G.A. and Caram, H.S. (2005). Use of a reverse-flow chromatographic reactor to enhance productivity in consecutive reaction systems. *Ind. Eng. Chem. Res.* 44: 3396–3401.

55 Behr, A. and Seuster, J. (2006). Reactive extraction. In: *Integrated Reaction and Separation Operations* (ed. H. Schmidt-Traub and A. Górak), 241–258. Berlin: Springer Verlag.

56 Stankiewicz, A. (2003). Reactive and hybrid separations: incentives, barriers, applications. In: *Re-engineering the Chemical Processing Plant: Process Intensification* (ed. A. Stankiewicz and J.A. Moulijn), 261–308. New York: Marcel Dekker Inc.

57 Gaidhani, H.K., Tolani, V.L., Pangarkar, K.V., and Pangarkar, V.G. (2002). Intensification of enzymatic hydrolysis of penicillin G: part 2. Model for enzymatic reaction with reactive extraction. *Chem. Eng. Sci.* 57: 1985–1992.

58 Cascaval, D. and Galaction, A.-I. (2004). New extraction techniques on bioseparations. 1. Reactive extraction. *Chem. Ind.* 58: 375–386.

59 Kelkar, V.V. and Ng, K.M. (1999). Design of reactive crystallization systems incorporating kinetics and mass-transfer effects. *AIChE J.* 45: 69–81.

60 Ulrich, J. and Heinrich, J. (2005). Reaktivkristallisation. *Chem.-Ing.-Tech.* 77: 1759–1772.

61 Alamdari, A. and Tabkhi, F. (2004). Kinetics of hexamine crystallization in industrial scale. *Chem. Eng. Proc.* 43: 803–810.

62 Alatalo, H., Hatakka, H., Kohonen, J. et al. (2010). Process control and monitoring of reactive crystallization of L-Glutamic acid. *AIChE J.* 56: 2063–2076.

63 Encarnación-Gómez, L.G., Bommarius, A.S., and Rousseau, R.W. (2015). Reactive crystallization of selected enantiomers: chemo-enzymatic stereoinversion of amino acids at supersaturated conditions. *Chem. Eng. Sci.* 122: 416–425.

64 Yildirim, Ö., Kiss, A.A., Hüser, N. et al. (2012). Reactive absorption in chemical process industry: a review on current activities. *Chem. Eng. J.* 213: 371–391.

65 Kenig, E., Górak, A., and Bart, H.-J. (2003). Reactive separations in fluid systems. In: *Re-engineering the Chemical Processing Plant: Process Intensification* (ed. A. Stankiewicz and J.A. Moulijn), 309–377. New York: Marcel Dekker Inc.

66 Baláž, P., Achimovičová, M., Baláž, M. et al. (2013). Hallmarks of mechanochemistry: from nanoparticles to technology. *Chem. Soc. Rev.* 42: 7571–7637.

67 Baláž, P. (2008). *Mechanochemistry in Nanoscience and Minerals Engineering*. Berlin, Heidelberg: Springer-Verlag.

68 Tan, Q. and Li, J. (2015). Recycling metals from wastes: a novel application of mechanochemistry. *Environ. Sci. Technol.* 49: 5849–5861.

69 Kumar, S., Alex, T.C. and Kumar, R. (2008). Mechanical activation of solids in extractive metallurgy. Proceedings of Extraction of Nonferrous Metals and their Recycling – A Training Programme, NML, Jamshedpur, India.

70 Ou, Z., Li, J., and Wang, Z. (2015). Application of mechanochemistry to metal recovery from second-hand resources: a technical overview. *Environ. Sci. Process. Impacts* 17: 1522–1530.

71 Takacs, L. (2013). The historical development of mechanochemistry. *Chem. Soc. Rev.* 42: 7649–7659.

72 Suryanarayana, C. (2001). Mechanical alloying and milling. *Progr. Mat. Sci.* 46: 1–184.

73 Frišcic, T., Halasz, I., Beldon, P.J. et al. (2012). Real-time and in situ monitoring of mechanochemical milling reactions. *Nat. Chem.* 5: 66–73.

74 Halasz, I., Kimber, S.A.J., Beldon, P.J. et al. (2013). In situ and real-time monitoring of mechanochemical milling reactions using synchrotron X-ray diffraction. *Nat. Protoc.* 8: 1718–1729.

75 Užarevic, K., Halasz, I., and Frišcic, T. (2015). Real-time and in situ monitoring of mechanochemical reactions: a new playground for all chemists. *J. Phys. Chem. Letters* 6: 4129–4140.

76 Ma, X., Yuan, W., Bell, S.E.J., and James, S.L. (2014). Better understanding of mechanochemical reactions: Raman monitoring reveals surprisingly simple 'pseudo-fluid' model for a ball milling reaction. *Chem. Comm.* 50: 1585–1587.

77 Baláž, P. (2003). Mechanical activation in hydrometallurgy. *Int. J. Miner. Process.* 72: 341–354.

78 Fokina, E.L., Budim, N.I., Kochnev, V.G., and Cherinik, G.G. (2004). Planetary mills of periodic and continuous action. *J. Mat. Sci.* 39: 5217–5221.

79 Hoffmann, U., Horst, C., and Kunz, U. (2005). Reactive comminution. In: *Integrated Chemical Processes* (ed. K. Sundmacher, A. Kienle and A. Seidel-Morgenstern), 407–436. Weinheim: Wiley-VCH Verlag GmbH & Co. KGaA.

80 Van Loy, S., Binnemans, K., and Van Gerven, T. (2018). Mechanochemical-assisted leaching of lamp phosphors: a green engineering approach for rare-earth recovery. *Engineering* 4: 398–405.

81 Sasikumar, C., Srikanth, S., Kumar, R. et al. (2011). Where does the energy go in high energy milling? In: *VI International Conference on Mechanochemistry and Mechanical Alloying (INCOME 2008), 1–4 December 2008*. India: NML Jamshedpur.

82 Ferencz, Z. (2016). Mechanochemical preparation and structural characterization of layered double hydroxides and their amino acid-intercalated derivatives. PhD thesis. University of Szeged.

83 James, S.L., Adams, C.J., Bolm, C. et al. (2012). Mechanochemistry: opportunities for new and cleaner synthesis. *Chem. Soc. Rev.* 41: 413–447.

84 Burmeister, C.F. and Kwade, A. (2013). Process engineering with planetary ball mills. *Chem. Soc. Rev.* 42: 7660–7667.

85 Mio, H., Kano, J., Saito, F., and Kaneko, K. (2002). Effects of rotational direction and rotation-to-revolution speed ratio in planetary ball milling. *Mater. Sci. Eng. A* 332: 75–80.

86 Mio, H., Kano, J., Saito, F., and Kaneko, K. (2004). Optimum revolution and rotational directions and their speeds in planetary ball milling. *Int. J. Miner. Process.* 74: S85–S95.

87 Schmidt, R., Fuhrmann, S., Wondraczek, L., and Stolle, A. (2016). Influence of reaction parameters on the depolymerization of H_2SO_4-impregnated cellulose in planetary ball mills. *Powder Technol.* 288: 123–131.

88 Gotor, F.J., Achimovicova, M., Real, C., and Baláž, P. (2013). Influence of the milling parameters on the mechanical work intensity in planetary mills. *Powder Technol.* 233: 1–7.

89 Ali, A.A., Baumlı, P., and Mucsi, G. (2004). Mechanical alloying and milling mechanical engineering. *Prog. Mater. Sci.* 46: 1–8.

90 Dolatmoradi, A., Raygan, S., and Abdizadeh, H. (2013). Mechanochemical synthesis of W-Cu nanocomposites via in-situ co-reduction of the oxides. *Powder Technol.* 233: 208–214.

91 Le Brun, P., Froyen, L., and Delaey, L. (1993). The modelling of the mechanical alloying process in a planetary ball mill: comparison between theory and in-situ observations. *Mat. Sci. Eng. A* 161: 75–82.

92 Baláž, P. and Dutková, E. (2009). Fine milling in applied mechanochemistry. *Min. Eng.* 22: 681–694.

93 McCormick, P.G. (1995). Application of mechanical alloying to chemical refining. *Mat. Trans. J.* 36: 161–169.

94 Suryanarayana, C. (2008). Recent developments in mechanical alloying. *Rev. Adv. Mat. Sci.* 18: 203–211.

95 Corrans, I.J. and Angove, E.J. (1991). Ultra fine milling for the recovery of refractory gold. *Min. Eng.* 4: 763–776.

96 Baláž, P., Sekula, F., Jakabský, Š., and Kammel, R. (1995). Application of attrition grinding in alkaline leaching of tetrahedrite. *Min. Eng.* 8: 1299–1308.

97 Jones, W. and Eddleston, M.D. (2014). Introductory lecture: mechanochemistry, a versatile synthesis strategy for new materials. *Faraday Disc.* 170: 9–34.

98 Khadka, P., Ro, J., Kim, H. et al. (2014). Pharmaceutical particle technologies: an approach to improve drug solubility, dissolution and bioavailability. *Asian J. Pharm. Sci.* 9: 304–316.

99 Kaupp, G. (2009). Mechanochemistry: the varied applications of mechanical bondbreaking. *CrystEngComm* 11: 388–403.

100 Xu, C., De, S., Balu, A.M. et al. (2015). Mechanochemical synthesis of advanced nanomaterials for catalytic applications. *Chem. Comm.* 51: 6698–6713.

101 Rosenkranz, S., Breitung-Faes, S., and Kwade, A. (2011). Experimental investigations and modelling of the ball motion in planetary ball mills. *Powder Technol.* 212: 224–230.

102 Michaud, D. (2015). Rod mill working principle and components. https://www.911metallurgist.com/blog/rod-mills (accessed 14 August 2018).

103 Bouvier, J.-M. and Campanella, O.H. (2014). *Extrusion Processing Technology: Food and Non-Food Biomaterials*. Hoboken, NJ.: Wiley-Backwell.

104 Brown, S.B. (1991). Chemical processes applied to reactive extrusion of polymers. *Annu. Rev. Mater. Sci.* 21: 409–435.

105 Lei, Z., Li, C., and Chen, B. (2003). Extractive distillation: a review. *Sep. Purif. Rev.* 32: 121–213.

106 Lee, F.-M. (2000). Extractive distillation. In: *Encyclopedia of Separation Science* (ed. I.D. Wilson, E.R. Adlard, M. Cooke and C.F. Poole), 1013–1022. San Diego: Academic Press.

107 Susanto, H. (2011). Towards practical implementations of membrane distillation. *Chem. Eng. Proc. Proc. Intens.* 50: 139–150.

108 Lutze, P. and Górak, A. (2013). Reactive and membrane-assisted distillation: Recent developments and perspective. *Chem. Eng. Res. Des.* 91: 1978–1997.

109 Pribic, P., Roza, M., and Zuber, L. (2006). How to improve the energy savings in distillation and hybrid distillation-pervaporation systems. *Sep. Sci. Technol.* 41: 2581–2602.

110 Sommer, S. and Melin, T. (2004). Design and optimization of hybrid separation processes for the dehydration of 2-propanol and other organics. *Ind. Eng. Chem. Res.* 43: 5248–5259.

111 Lipnizki, F., Field, R.W., and Ten, P.-K. (1999). Pervaporation-based hybrid process: a review of process design, applications and economics. *J. Mem. Sci.* 153: 183–210.

112 Aiouache, F. and Goto, S. (2003). Reactive distillation–pervaporation hybrid column for tert-amyl alcohol etherification with ethanol. *Chem. Eng. Sci.* 58: 2465–2477.

113 Drioli, E., Di Profio, G., and Curcio, E. (2012). Progress in membrane crystallization. *Curr. Opin. Chem. Eng.* 1: 178–182.

114 Charcosset, C., Kieffer, R., Mangin, D., and Puel, F. (2010). Coupling between membrane processes and crystallization operations. *Ind. Eng. Che. Res.* 49: 5489–5495.

115 Di Profio, G., Curcio, E., Cassetta, A. et al. (2003). Membrane crystallization of lysome: kinetic aspects. *J. Crystal Growth* 257: 359–369.

116 Di Profio, G., Stabile, C., Caridi, A. et al. (2009). Antisolvent membrane crystallization of pharmaceutical compounds. *J. Pharm. Sci.* 98: 4902–4913.

117 Brito Martínez, M., Jullok, N., Rodríguez Negrína, Z. et al. (2014). Membrane crystallization for the recovery of a pharmaceutical compound from waste streams. *Chem. Eng. Res. Des.* 92: 264–272.

118 Luis, P., Van Aubel, D., and Van der Bruggen, B. (2013). Technical viability and exergy analysis of membrane crystallization: closing the loop of CO_2 sequestration. *Int. J. Greenhouse Gas Control* 12: 450–459.

119 Chivate, M.R. and Shah, S.M. (1956). Separation of m-Cresol and p-Cresol by Extractive crystallization. *Chem. Eng. Sci.* 5: 232–241.

120 Dikshit, R.C. and Chivate, M.R. (1970). Separation of ortho and para nitrochlorobenzenes by extractive crystallization. *Chem. Eng. Sci.* 25: 311–317.

121 Rajagopal, S., Ng, K.M., and Douglas, J.M. (1991). Design and economic trade-offs of extractive crystallization processes. *AIChE J.* 37: 437–447.

122 Lashanizadegan, A., Newsham, D.M.T., and Tavare, N.S. (2001). Separation of chlorobenzo acids by dissociation extractive crystallization. *Chem. Eng. Sci.* 56: 2335–2346.

123 Federspiel, W., Sawzik, P., Borovetz, H. et al. (1996). Temporary support of the lungs – the artificial lung. In: *The Transplantation and Replacement of Thoracic Organs: The Present Statos of Biological and Mechanical Replacement of the Hear and Lungs* (ed. D.K.C. Cooper, L.W. Miller and G.A. Patterson), 717–728. Berlin: Springer Verlag.

124 Mockros, L.F. and Gaylor, J.D.S. (1975). Artificial lung design: tubular membrane units. *Med. Biol. Eng.* 13: 171–181.

125 Li, J.-L. and Chen, B.-H. (2005). Review of CO_2 absorption using chemical solvents in hollow fiber membrane contactors. *Sep. Purif. Technol.* 41: 109–122.

126 Okabe, K., Nakamura, M., Mano, H. et al. (2006). CO_2 separation by membrane/absorption hybrid method. *Stud. Surf. Sci. Catal.* 159: 409–412.

127 Ahmada, A.L., Sunarti, A.R., Lee, K.T., and Fernando, W.J.N. (2010). CO_2 removal using membrane gas absorption. *Int. J. Greenhouse Gas Control* 4: 495–498.

128 Herzog, H. and Falk-Pedersen, O. (2000). The Kvaerner membrane contactor. Lessons from a case study in how to reduce capture costs. In: *Proceedings of 5th International Conference on Greenhouse Gas Control Technologies*, Cairns, Australia, August 13–16 (ed. D.J. Williams, R.A. Dutie, P. McMullan, et al.). Collingwood: CSIRO Publishing.

129 Shalygin, M.G., Yakovlev, A.V., Khotimskii, V.S. et al. (2011). Membrane contactors for biogas conditioning. *Petrol. Chem.* 51: 601–609.

130 Ozturk, S.S. (1996). Engineering challenges in high cell density culture systems. *Cytotechnology* 22: 3–16.

131 Frahm, B., Brod, H., and Langer, U. (2009). Improving bioreactor cultivation conditions for sensitive cell lines by dynamic membrane aeration. *Cytotechnology* 59: 17–30.

132 Bersillon, J.-L. and Huyard, A. (1989). Bubble-free aeration using membranes: mass transfer analysis. *J. Mem. Sci.* 47: 91–106.

133 Przybycien, T.M., Pujar, N.S., and Steele, L.M. (2004). Alternative bioseparation operations: life beyond packed-bed chromatography. *Curr. Opinions Biotech.* 15: 469–478.

134 Charcosset, C. (2006). Membrane process in biotechnology: an overview. *Biotech. Adv.* 24: 482–492.

135 Orr, V., Zhong, L., Moo-Young, M., and Chou, C.P. (2013). Recent advances in bioprocessing application of membrane chromatography. *Biotech. Adv.* 31: 450–465.

136 Varadaraju, H., Schneiderman, S., Zhang, L. et al. (2011). Process and economic evaluation for monoclonal antibody purification using a membrane-only process. *Biotechnol. Prog.* 27: 1297–1305.

137 Kreuß, M. and Kulozika, U. (2009). Separation of glycosylated caseino-macropeptide at pilot scale using membrane adsorption in direct-capture mode. *J. Chromatogr. A.* 1216: 8771–8777.

138 Voswinkel, L. and Kulozik, U. (2011). Fractionation of whey proteins by means of membrane adsorption chromatography. *Procedia Food Sci.* 1: 900–907.

139 Liu, L., Zheng, G., and Yang, F. (2010). Adsorptive removal and oxidation of organic pollutants from water using a novel membrane. *Chem. Eng. J.* 156: 553–556.

140 Klaassen, R., Feron, P., and Jansen, A. (2008). Membrane contactor applications. *Desalination* 224: 81–87.

141 Ortiz, M. and Irabien, J.A. (2009). Membrane-assisted solvent extraction for the recovery of metallic pollutants: process modeling and optimization. In: *Handbook of Membrane Separations* (ed. A.K. Pabby, S.S.H. Rizvi and A.M. Sastre Requena), 1023–1039. London: Taylor and Francis.

142 Klaassen, R., Feron, P., and Jansen, A. (2005). Membrane contactors in industrial applications. *Chem. Eng. Res. Des.* 83: 234–246.

143 Agar, D.W. (1999). Multifunctional reactors: old preconceptions and new dimensions. *Chem. Eng. Sci.* 54: 1299–1305.

144 Górak, A. and Stankiewicz, A. (2011). Intensified reaction and separation systems. *Ann. Rev. Chem. Biomol. Eng.* 2: 431–451.

145 Stichlmair, J. and Frey, T. (1999). Reactive distillation processes. *Chem. Eng. Technol.* 22: 95–103.

146 Au-Yeung, P.H., Resnick, S.M., Witt, P.M. et al. (2013). Horizontal reactive distillation for multicomponent chiral resolution. *AIChE J.* 59: 2603–2620.

147 Barbieri, G., Brunetti, A., Caravella, A., and Drioli, E. (2011). Pd-based membrane reactors for one-stage process of water gas shift. *RSC Adv.* 1: 651–661.

148 Diban, N., Aguayo, A.T., Bilbao, J. et al. (2013). Membrane reactors for in situ water removal: a review of applications. *Ind. Eng. Chem. Res.* 52: 10342–10354.

149 Kiss, A.A., Bildea, C.S., and Grievink, J. (2010). Dynamic modeling and process optimization of an industrial sulfuric acid plant. *Chem. Eng. J.* 158: 241–249.

150 Bernotat, S. and Schonert, K. (1998). Size reduction. In: *Ullmann's Encyclopedia of Industrial Chemistry*, vol. B2, 5.1–5.39. Weinheim: VCH Verlagsgesellschaft.

151 Tzoganakis, C. (1989). Reactive extrusion of polymers: a review. *Adv. Polymer Technol.* 9: 321–330.

152 Curci, E. and Drioli, E. (2005). Membrane distillation and related operations – a review. *Sep. Purif. Rev.* 34: 35–86.

153 Sekiguchi, K., Sasaki, C., and Sakamoto, K. (2011). Synergistic effects of high-frequency ultrasound on photocatalytic degradation of aldehydes and their intermediates using TiO_2 suspension in water. *Ultrason. Sonochem.* 18: 158–163.

154 Torres, R.A., Nieto, J.I., Combet, E. et al. (2008). Influence of TiO_2 concentration on the synergistic effect between photocatalysis and high-frequency ultrasound for organic pollutant mineralization in water. *Appl. Cat. B: Environ.* 80: 168–175.

155 Shirgaonkar, I.Z. and Pandit, A.B. (1998). Sonophotochemical destruction of aqueous solution of 2,4,6-trichlorophenol. *Ultrason. Sonochem.* 5: 53–61.

156 Cravotto, G., Beggiato, M., Penoni, A. et al. (2005). High-intensity ultrasound and microwave, alone or combined, promote Pd/C-catalyzed aryl-aryl couplings. *Tetrahedron Lett.* 46: 2267–2271.

157 Peng, Y. and Song, G. (2001). Simultaneous microwave and ultrasound irradiation: a rapid synthesis of hydrazides. *Green Chem.* 3: 302–304.

158 Peng, Y. and Song, G. (2002). Combined microwave and ultrasound assisted Williamson ether synthesis in the absence of phase-transfer catalysts. *Green Chem.* 4: 349–351.

159 Wu, Z.-L., Ondruschka, B., and Cravotto, G. (2008). Degradation of phenol under combined irradiation of microwaves and ultrasound. *Environ. Sci. Technol.* 42: 8083–8087.

160 Chemat, S., Lagha, A., Amar, H.A., and Chemat, F. (2004). Ultrasound assisted microwave digestion. *Ultrason. Sonochem.* 11: 5–8.

161 Chemat, F., Poux, M., Di Matino, J.-L., and Berlan, J. (1996). An original microwave-ultrasound combined reactor for organic synthesis: application to pyrolysis and esterification. *J. Microwave Power Electromagn. Energy* 31: 19–22.

6

TIME – PI Approaches in Temporal Domain

In this chapter, process intensification in the TIME domain will be discussed. Here, the underlying manipulative parameter for process intensification is *time*. In this domain, two approaches can be distinguished and will be further analyzed in the subsequent sections in the context of specific unit operations and processes. The first approach concerns *introduction of purposeful periodic or cyclic operation of process units.* Oscillatory flow reactors (OFRs), reverse flow reactors, periodic operation of trickle bed reactors (TBRs), cyclic distillation, pulse combustion, pressure swing adsorption (PSA), desorptive cooling, and variable volume operation of stirred tank reactors are some prominent examples in this category that are discussed in Sections 6.1–6.8.

It is important to realize that there are many cases of natural periodicity in common chemical processes and unit operations. Indicatively, on the active sites of catalytic surfaces, molecules continuously undergo a sequence of adsorption, reaction, and desorption. In bubble columns, gaseous feed is dispersed in the form of small bubbles in a liquid phase; then, bubbles may grow, coalesce, and break up several times. In fluidized bed reactors, solid particles may recurrently form clusters that subsequently disintegrate. In catalytically coated microreactors and monolithic reactors that process gas–liquid Taylor flows, a differential catalytic surface at any given point along the reactor length will alternatingly contact gas and liquid volumes that are conveyed from inlet to outlet. Chemical processes are characterized by a broad range of time scales. Phenomena related to active sites and molecular level show a time duration shorter than nanoseconds, whereas mechanisms involving bubbles, drops, and film transfer have periodicity between 0.001 and 0.1 s. On the other hand, time scales of 0.1–1 s are typical of monoliths and pulsed trickle beds, whereas risers operate in time interval of seconds. Furthermore, chemical processes involved in fluidized beds last from seconds to minutes, whereas operational times in moving beds are between minutes and hours. Lastly, bioreactors and many other batch processes run in time scales of hours to days. Inspired by such natural periodic (cyclic) phenomena, the goal of process intensification in the TIME domain is to introduce controlled and optimized periodicity in an inherently steady-state process in order to attain operating conditions that can significantly improve the process.

The Fundamentals of Process Intensification, First Edition.
Andrzej Stankiewicz, Tom Van Gerven, and Georgios Stefanidis.
© 2019 Wiley-VCH Verlag GmbH & Co. KGaA. Published 2019 by Wiley-VCH Verlag GmbH & Co. KGaA.

The second approach is based on the combined application of novel reactor designs and operating windows and aims at *manipulation (usually extreme shortening) of targeted reaction or transport events* within a reactive process in order to control the product distribution as the main goal. In Section 6.9, this approach is highlighted through examples of transformation of gaseous and solid (oxygenated) hydrocarbon fuels to higher value products.

6.1 Oscillatory Flow Reactors

In its standard configuration, an OFR is a tubular reactor (>15 mm diameter) with equidistantly spaced internal baffles that define a number of segments "in-series" along the reactor length. Two types of flow are discerned in this type of reactor, namely the bulk flow that is pumped from inlet to outlet and an oscillatory flow, which is superimposed on the bulk flow by means of a reciprocating piston. The result of the flow superposition is the generation of vortices on the backside of the baffles, which propagate toward the inner side of each segment creating intense mixing and mass and heat transfer in the segment itself. Eventually, the OFR operates as a system of "well-mixed" isothermal reactors in series that enable near plug flow reactor (PFR) operation, given a sufficient number of baffles and segments. Figure 6.1 presents in a schematic the operating principle of the OFR as well as visualization of the flow in an OFR segment [1].

Aside from effective multiphase mixing due to eddies formation, particularly when immiscible phases are involved, another major advantage of the technology is that it allows linear scale up by increasing the number of baffles and segments in the flow direction without influencing the surface-to-volume ratio and the hydrodynamics of the flow. This is conceptually superior to three-dimensional scale-up of traditional batch or continuous stirred tank reactors that result in modifications of heat and mass flow patterns upon increase in the reactor size and inevitably in increased development times and costs. More importantly, OFRs allow carrying out relatively slow chemistries under near plug flow conditions and low net flow rates; this is impossible to attain in smoothed wall reactors. Further, owing to the combination of plug flow and low net flow rates (ml/min), mesoscale OFRs (4.4–5 mm diameter) can be used as a process-screening platform in continuous mode with low feedstock utilization and waste production [2]. The mesoscale OFR in particular is compatible with different baffle configurations presented in Figure 6.2 and aimed at different applications.

Transport phenomena in the OFR are determined by a number of dimensionless groups, which also form geometry and operation parameters amenable to optimization. These are (i) the baffle spacing ratio defined as the ratio of baffle spacing to reactor diameter, (ii) the open baffle flow area defined as the ratio of orifice diameter to OFR diameter, (iii) the net flow Reynolds number ($\rho u D/\mu$), (iv) the oscillatory flow Reynolds number ($2\pi f x \rho D/\mu$), where f and x are the oscillation frequency and amplitude, respectively, and (v) the Strouhal number ($D/4\pi x$) [2].

OFRs have been applied to various gas–liquid, liquid–liquid, and liquid–solid systems in which effective phase dispersion and good mass transfer are required.

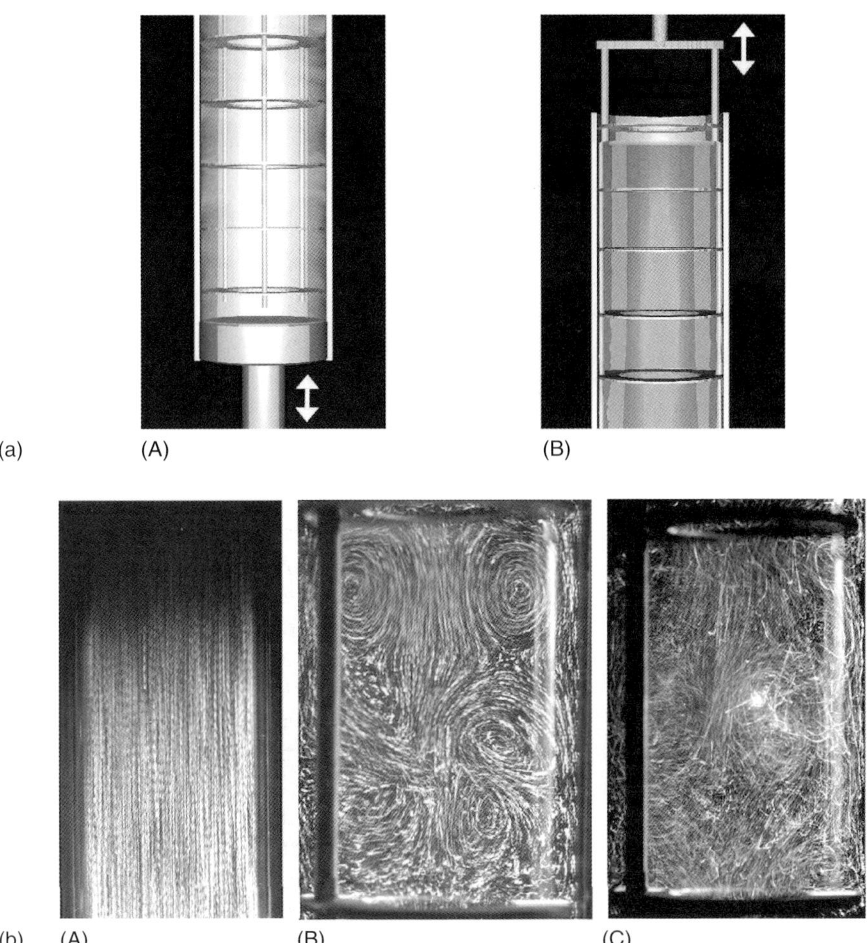

Figure 6.1 (a) The oscillatory baffled column; (b) visualization of the flow in a smoothed wall column (A) and in an oscillatory baffled column (B and C). Source: Ni et al. 2003 [1]. Reproduced with permission of Elsevier.

Relevant applications include biodiesel synthesis, polymerization, heterogeneous catalysis, and crystallization. Indicatively, Mazubert et al. [3] performed waste cooking oil transesterification with methanol in a NiTech continuous OFR and in a batch reactor. Nearly equilibrium conversion (92.1 wt%) of esters was obtained in 6 min at 44 °C in the NiTech OFR compared to 94.8 wt% of esters in 10 min at 60 °C in the batch reactor. Phan et al. [4] performed alkali-catalyzed transesterification of rapeseed oil with methanol in a mesoscale OFR comprising four connected glass tubes with a total length of 320 mm (Figure 6.3). Three different baffle configurations, namely integral baffles, wire wool, and sharp-edged baffles with a central rod, were used. More than 95% conversion at 60 °C and 5 min residence time was obtained using the integral baffled configuration. Low (1.5 wt%) catalyst (KOH, potassium hydroxide) concentration and a methanol:rapeseed oil

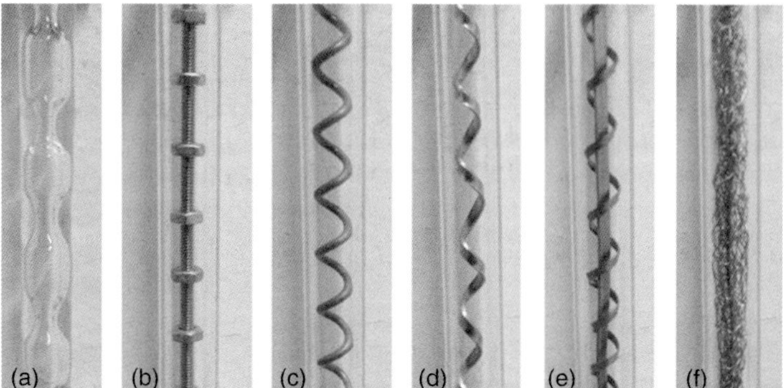

Figure 6.2 Mesoscale baffle configurations: (a) integral baffles, (b) central axial, (c) round-edged helical baffles, (d) sharp-edged helical baffles, (e) sharp-edged helical baffles with a central insert, and (f) wire wool baffles. Source: McDonough et al. 2015 [2]. Reproduced with permission of Elsevier.

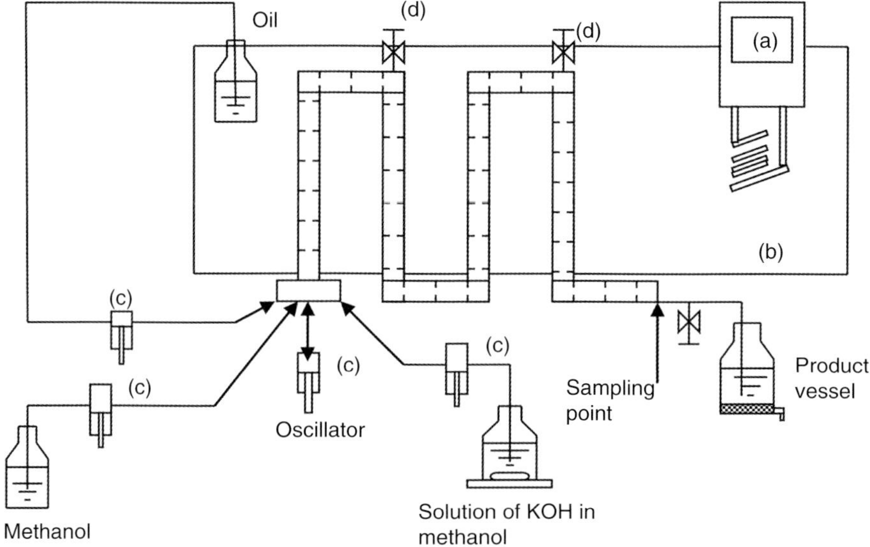

Figure 6.3 Schematic of the experimental setup for continuous biodiesel production. (a) Temperature controller, (b) water bath, (c) syringe pumps, and (d) valves. Source: Phan et al. 2012 [4]. Reproduced with permission of John Wiley and Sons.

ratio of 6 : 1 were applied. These results indicate the potential for an order of magnitude reduction in process time compared to conventional biodiesel processing in continuously stirred tank reactors (CSTRs).

Lobry et al. [5] used an OFR to perform suspension polymerization for the production of poly(vinyl acetate), PVAc. The authors demonstrated the ability of the OFR technology to (i) control the droplet size and distribution in the initial liquid–liquid dispersion step and (ii) control the final particle size during the agglomeration step. In the last step, a 30% mass conversion was reached. Both aforementioned elements were bottlenecks in continuous flow mode. Therefore, the results provide an encouraging step toward the transition from stirred batch reactors to continuous flow OFRs for industrial processes.

Eze et al. [6] demonstrated the applicability of the technology to heterogeneously catalyzed liquid phase reactions. In particular, $PrSO_3H$-SBA-15 in powder form was used to catalyze the esterification of hexanoic acid with methanol. The excellent semiquantitative agreement between the OFR and the stirred batch reaction kinetics that the authors report is indicative of the efficient mixing that OFR can enable, as well as of its capacity to be used for processing and kinetic investigations of liquid–solid catalytic systems.

In the field of crystallization, effective mixing and controlled residence time in the OFR allows for continuous production of small particles with narrow size distribution. As an example, Teixeira and coworkers [7] demonstrated the ability of a mesoscale OFR (35 cm long, 4.4 mm internal diameter) and an upscaled meso-OFR, comprising eight meso-OFRs in series, for continuous production of hydroxyapatite (HAp) nanoparticles. Specifically, nanoparticles with $d_{50} = 77$ nm, narrow size distribution, and uniform, mostly rod-shaped, morphology were produced from both reactors at short residence times, namely 0.4 and 3.3 min for the meso-OFR and the upscaled meso-OFR, respectively. A recent comprehensive review on application of OFRs to continuous manufacturing and crystallization is given by McGlone et al. [8].

Finally, an example of an industrial application of the technology was given by Ni et al. [9]. The company James Robinson Ltd, a manufacturer of dyes for different applications, replaced a system of two large batch reactors in series for the production, through wetting, diazotization, and cyclization, of an intermediate chemical for a photographic chemical product with a two-section OFR unit that allowed continuous production of the intermediate chemical. Significant reduction in process time and consequently in equipment footprint was gained for approximately the same yield, purity, and production capacity (Figure 6.4 and Table 6.1; "NiTech – Continuous Oscillatory Baffled Reactor™" [10]).

Table 6.1 Chart showing the benefits of replacement of the James Robinson batch reactor system with the OFR.

Parameter	James Robinson reactor	Oscillatory flow reactor
Reactor volume (l)	16 000	270
Floor area used (m^2)	1 200	45
Reaction wetting time (h)	12	<0.1
Diazotization reaction time (h)	2	<0.1
Cyclization reaction time (h)	0.3	<0.1
Yield (%)	83	89
Purity (%)	99	99
Production capacity (kg/d)	180	180–205

Source: Adapted from "NiTech – Continuous Oscillatory Baffled Reactor™" [10].

Figure 6.4 Comparison of space requirements of OFR and conventional batch process equipment. Source: Courtesy: NiTech Solutions Ltd., Scotland. www.nitechsolutions.co.uk.

6.2 Reverse Flow Reactors

The reverse flow reactor technology enables one of the four thermal management strategies in chemical reactor engineering, namely thermal management through heat regeneration. The other three approaches are heat recuperation, direct thermal coupling between exothermic and endothermic reactions, and convection by addition or withdrawal of side streams. Flow reversal is typically applied to adiabatic packed bed reactors. If the catalyst of the reactor is preheated above the ignition temperature of the reaction under investigation, then it is possible to set the inlet of the reactor at the ambient temperature, thus eliminating the need for preheating. In this case, a moving reaction front will start developing at the position where the cold feed with high reactants concentration meets the hot spot of the reactor, where the catalyst temperature is higher than the ignition temperature (first cycle). The reaction front will propagate toward the back end of the reactor, increasing the catalyst temperature near the reactor outlet with a concomitant decrease in the catalyst temperature near the inlet. The cold reactor inlet can be re-ignited by reversing the flow direction and consequently the propagation direction of the reaction front and heat wave from the "outlet" to the "inlet" (second cycle). Collectively, periodic flow reversal will lead to establishment of periodic "steady-state" temperature and conversion profiles along the reactor length. A schematic of an adiabatic fixed bed reactor with flow reversal is shown in Figure 6.5. [11]. The principle of reverse flow operation and axial temperature and conversion profiles in the two cycles described above are shown in Figure 6.6 [11].

A major feature of the temperature profiles in reverse flow reactors is that the difference between the maximum temperature in the reactor bed and the inlet

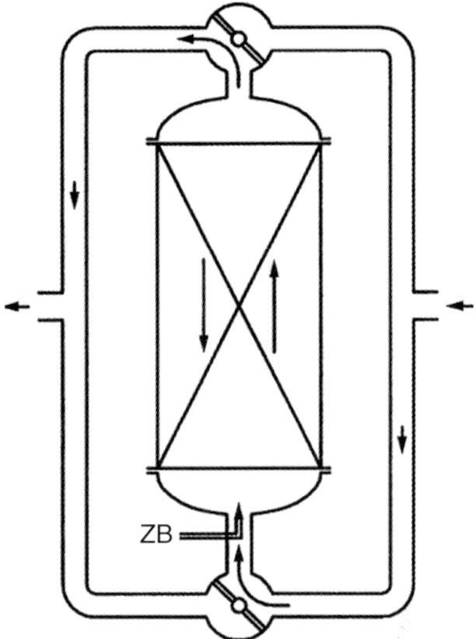

Figure 6.5 Adiabatic fixed bed reactor with switching inlet valves for periodic flow reversal. Source: Kolios et al. 2000 [11]. Reproduced with permission of Elsevier.

temperature exceeds the adiabatic temperature increase in an exothermic process as determined by the inlet composition and the thermophysical properties of the solid structure inside the reactor. Provided that the maximum temperature can be effectively controlled to avoid catalyst compromise or thermal runaway conditions, the advantage of this feature is twofold: (i) the technology can be used to attain autothermal operation for oxidation of combustible streams with low calorific value and low temperature rise and (ii) the equipment can essentially function as a compact reactor heat exchanger that can efficiently store and utilize heat for an endothermic reaction.

Three commercial processes that use periodic flow reversal have been reported in a key critical review of the technology by Matros and Bunimovich [12]. These concern (i) oxidation of volatile organic compounds (VOC) in industrial gaseous waste streams, (ii) SO_2 oxidation for sulfuric acid production, and (iii) NO_x reduction to ammonia in an industrial exhaust gas. In all cases, the technology is applied to lean gaseous streams with low concentrations of VOCs, SO_2, and NO_X. In the same context, Gosiewski et al. [13] have presented a research and demonstration plant (Figure 6.7) based on two thermal flow reversal reactors (TFRR; inert ceramic structures are used for thermal energy storage) for volatile air methane (VAM) combustion from hard coal seams. Depending on the location of the coal bed, VAM emissions may contain up to ~1 vol% CH_4 contributing to the greenhouse effect apart from wasting valuable fuel and chemical resources. Stable autothermal operation for VAM concentrations >0.2 vol% and operation with stable heat recovery for VAM concentrations >0.4 vol% is possible with the TFRR technology.

Glöckler et al. [14] demonstrated experimentally the thermal coupling of a methane steam reforming process with a fuel gas combustion (PSA-off gas) process in a packed bed reactor operated with flow reversal in a quartz glass laboratory reformer (Figure 6.8). Nearly complete methane conversion and respective steady hydrogen production rate were obtained in the reforming

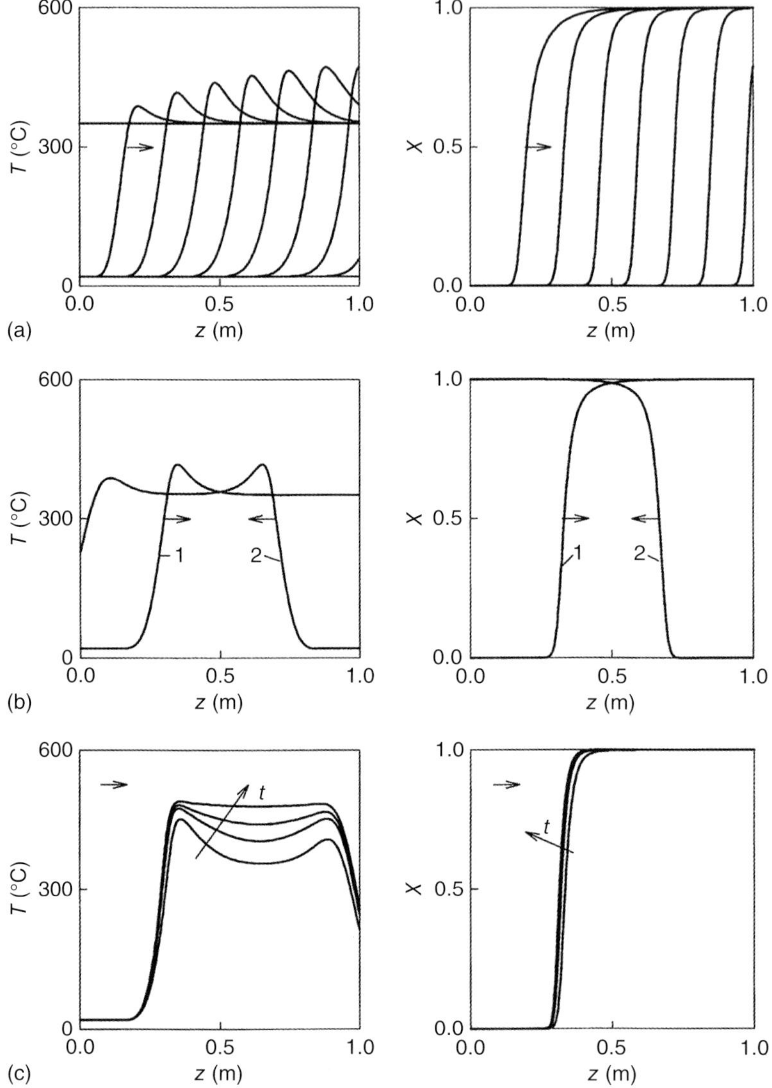

Figure 6.6 Principles of reverse-flow operation: (a) temperature (T) and conversion (X) profiles of a moving reaction front in an adiabatic fixed bed reactor; (b) temperature and conversion profiles at the end of the two first semicycles; (c) transient to periodic steady state. Source: Kolios et al. 2000 [11]. Reproduced with permission of Elsevier.

cycle, which corresponded to a thermal power of 5.1 kW$_{LHV,H2}$ at appropriately limited maximum temperature (Figure 6.9).

Kolios and Eigenberger [15] performed experimental and modeling studies for styrene synthesis through ethylbenzene dehydrogenation ($C_8H_{10} \rightarrow C_8H_8 + H_2$) in a reverse flow reactor schematically shown in Figure 6.10. The fixed bed reactor (50 mm internal diameter, 1200 mm length) is divided into inert and catalytic zones. A catalytic combustor, placed at the center of the reactor

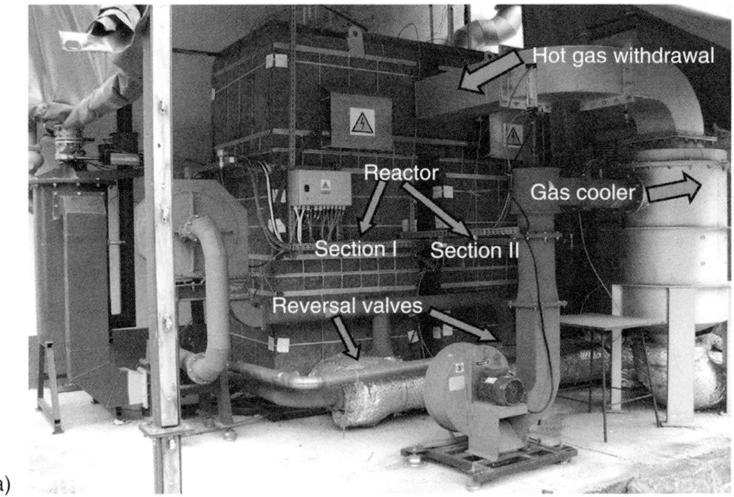

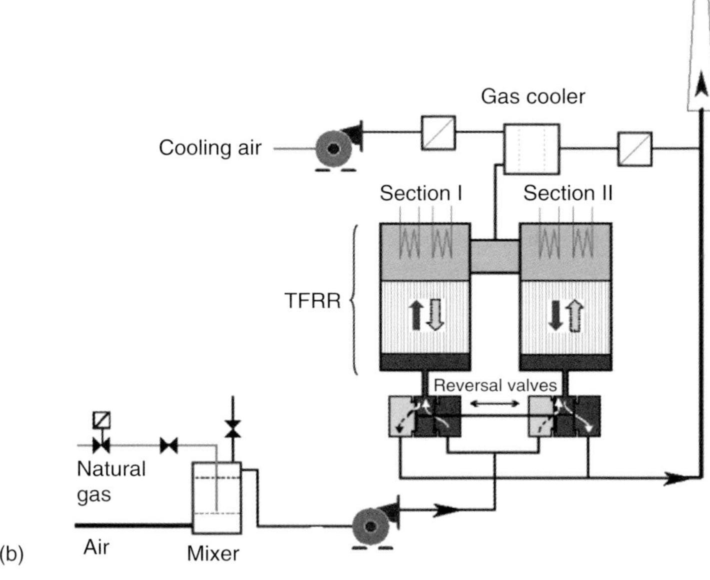

Figure 6.7 (a) General view of the research and demonstration of TFRR (thermal flow reversal reactors) plant; (b) main TFRR elements, comprising two ceramic monolith beds arranged horizontally in sections I and II and connected by a duct at the top, and flow sheet of the plant. Source: Gosiewski et al. 2015 [13]. Reproduced with permission of Elsevier.

(symmetric operation), is used for combustion of H_2 while air is fed into it through separate tubes. The effluent gases exit through radial holes and transfer heat directly to the styrene synthesis. The direction of feed flow for styrene synthesis (ethylbenzene and steam) changes periodically. Figure 6.11 shows the measured periodic steady-state temperature profiles in the reactor in two successive cycles of flow reversal. The experimental investigation revealed the potential for autothermal operation (the hydrogen generation and consumption

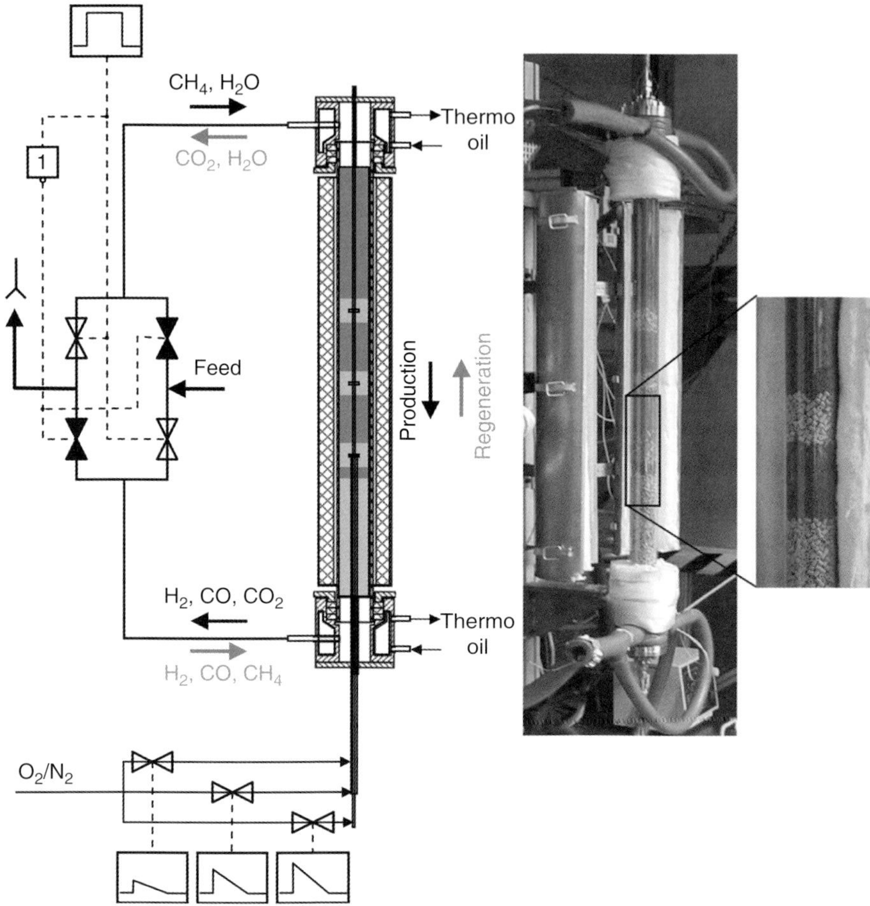

Figure 6.8 Photo of a quartz glass laboratory reformer operated under flow reversal. The tube is insulated with a ceramic fiber mat in a demountable insulation casing (opened in the photo). Both ends of the glass tube are connected to heated stuffing boxes, providing the gas supply. The fixed bed has a diameter of 50 mm (ID of the glass tube) and a length of 900 mm. Source: Glöckler et al. 2007 [14]. Reproduced with permission of Elsevier.

rates are equal), high ethylbenzene conversion (64–68%), and high styrene selectivity (91–96%).

At the microreactor level, Kaisare and Vlachos [16] presented a "proof-of-concept" simulation study on homogeneous combustion of propane in a microburner operated with flow reversal (RF burner; Figure 6.12). In this case, heat release due to combustion in one cycle is stored in the microburner walls that preheat the cold combustible stream in the next cycle. Unlike meso- and macroscale reverse flow reactors, where inert or catalytic solid structures inside the reactor volume exchange heat with a gas stream, in microreactors, there is strong PI thermal coupling between the reactor walls and the gas stream due to the small (<1 mm) characteristic diameter of the reactor. It was also shown that depending on the frequency of flow reversal, the upper stability limit of the

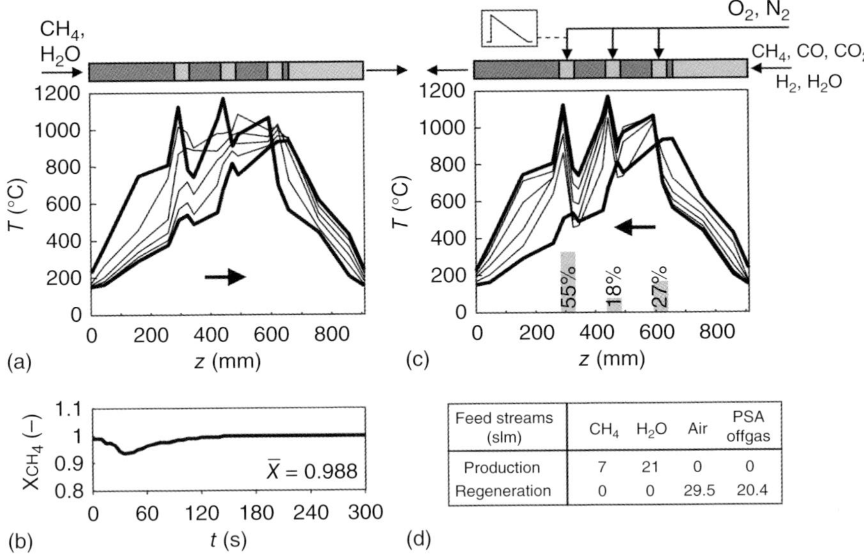

Feed streams (slm)	CH$_4$	H$_2$O	Air	PSA offgas
Production	7	21	0	0
Regeneration	0	0	29.5	20.4

Figure 6.9 Periodic steady state in the quartz glass laboratory reformer (Figure 6.8). (a) Temperature evolution during production (reforming) time (time steps: 60 s); (b) methane conversion during production (reforming) and average conversion; and (c) temperature evolution during regeneration (combustion) (time steps: 60 s) and air feed distribution. The dark-shaded zones in the reactor schematics represent spaces filled with inert γ-Al$_2$O$_3$ rings. The light-shaded zones represent spaces filled with γ-Al$_2$O$_3$ pellets coated with noble metal shell-type catalyst. Because of the slower evolution of the temperature profiles in the inert zones compared to the catalytic ones during the reforming period, small temperature peaks created in the reforming cycle are magnified at the air feed side ports during the combustion cycle. Source: Glöckler et al. 2007 [14]. Reproduced with permission of Elsevier.

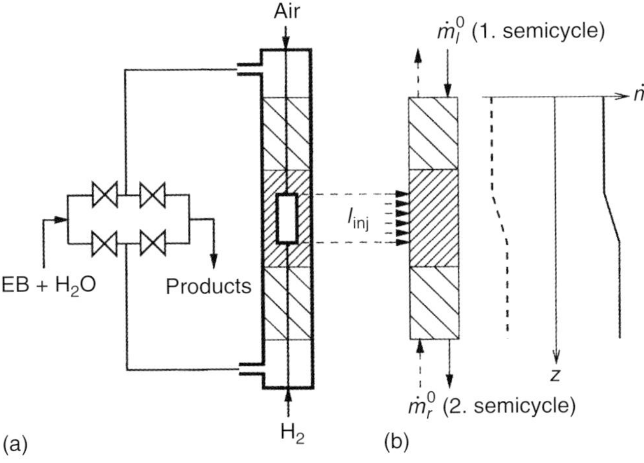

Figure 6.10 (a) sketch of lab-scale reactor and (b) reactor model. Source: Kolios and Eigenberger 1999 [15]. Reproduced with permission of Elsevier.

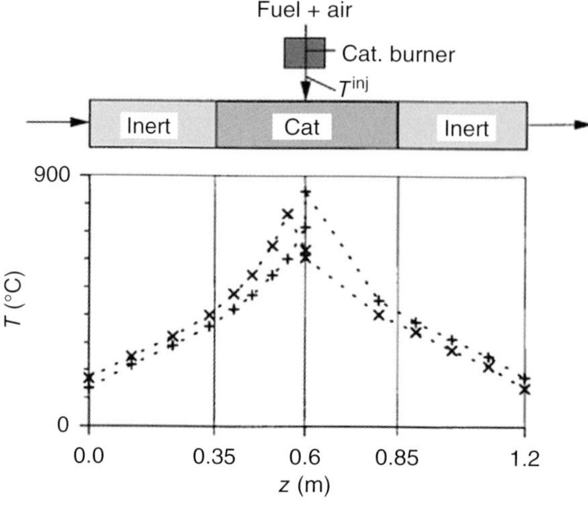

Figure 6.11 Temperature profiles at the end of two successive semicycles in the periodic steady state showing heat accumulation in the center zone of the reactor where combustion takes place; (+) measured temperature profiles with feed from the left; (×) measured temperature profiles with feed from the right. Source: Kolios and Eigenberger 1999 [15]. Reproduced with permission of Elsevier.

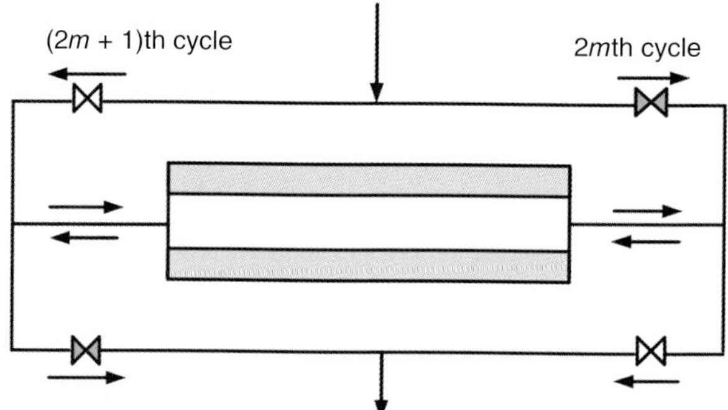

Figure 6.12 Schematic illustrating the working principle of reverse flow operation in a homogeneous microburner. The flow direction is reversed by alternately switching on/off the shaded and unshaded valves. Source: Kaisare and Vlachos 2007 [16]. Reproduced with permission of Elsevier.

reactor corresponding to the maximum velocity or flow rate at which combustion is sustained, the so-called blowout limit, can be substantially extended compared to unidirectional operation (Figure 6.13). Extended operating windows imply higher heat production rates and higher throughputs.

A reverse flow fixed bed reactor concept has been explored via simulations for application to chemical looping combustion (CLC) by Han and Bollas [17]. In a fixed bed CLC process, successive cycles of reduction and oxidation of a metal oxide (O_2 source) take place. In the reduction phase, a fuel enters in a fully oxidized bed and converted to CO_2 and H_2O with simultaneous reduction of the metal oxide. In the oxidation cycle, air is fed to the reactor to regenerate the oxygen carrier. Han and Bollas [17] studied the effect of flow reversal in the reduction cycle considering methane and syngas as fuels and CuO and NiO as oxygen carriers. Simulations showed that, in comparison to unidirectional flow, periodic flow reversal could enhance the contact of the fuel with the oxygen carrier, which results in suppression of side catalytic reactions and coke formation and improves the uniformity of the spatial conversion and temperature fields. At system level, it has been shown that combination of the reverse-flow CLC reactor with a gas turbine downstream results in higher power generation efficiency compared to the combination of a packed bed reactor, operated in one direction, with the gas

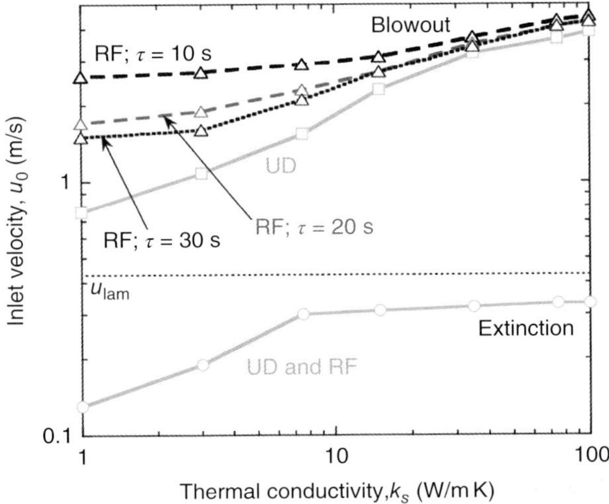

Figure 6.13 Critical inlet velocity for self-sustained combustion versus wall thermal conductivity. The blowout limit of the reverse flow (RF) operation expands as the frequency of switching increases. The dotted line indicates the laminar burning velocity, u_{lam}, for propane. The extinction limit, minimum combustible velocity, or flow rate at which combustion is self-sustained is not affected by flow reversal. UD, unidirectional operation. Source: Kaisare and Vlachos 2007 [16]. Reproduced with permission of Elsevier.

turbine. Finally, use of a fixed bed, instead of a circulating fluidized bed, addresses the issues of solid attrition and gas–solid separation [17].

6.3 Periodic Operation of Trickle Bed Reactors

TBRs are three-phase flow reactors in which mobile gas and liquid phases flow down a reactor filled with catalyst pellets. TBRs are commonly used for hydroprocessing in oil refineries (hydrogenation, hydrodesulphurization, and hydrocracking). Periodic (cyclic) operation has been investigated experimentally and by means of computational fluid dynamics (CFD) simulations as one way to improve reactor performance, which is mass-transfer limited. In the case of liquid cyclic operation, the liquid flow rate varies between a minimum and a maximum value (base-pulse) model. When the minimum flow rate is zero, this is called "on–off mode" [18].

Lange et al. [19] studied periodic operation of a trickle bed for the hydrogenation of α-methylstyrene to cumene $[C_6H_5(CH_3) = CH_{2(L)} + H_{2(g)} \rightarrow C_6H_5CH(CH_3)_{2(L)}]$ over a Pd catalyst (0.7% Pd/γ-alumina). It was shown that periodic liquid flow variation strongly affects liquid holdup and catalyst wetting and improves time-average conversion (Figures 6.14 and 6.15).

Most experimental works on the hydrodynamics of periodic flow operation of TBRs have been studied in the so-called slow mode (duration of periods in the order of few minutes). Aydin et al. [20] studied the influence of temperature and pressure on fast mode (duration of periods in the order of few seconds) cyclic

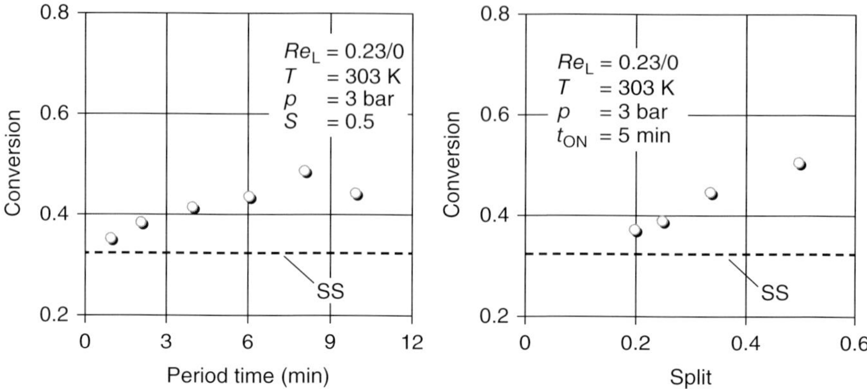

Figure 6.14 Time-average α-methylstyrene conversions depending on two-level volumetric liquid flow rate, period time, and cycle split (*S*) in comparison to the conversion under equivalent steady-state (SS) conditions ($U_{SS} = 0.33$). Source: Lange et al. 2004 [19]. Reproduced with permission of Elsevier.

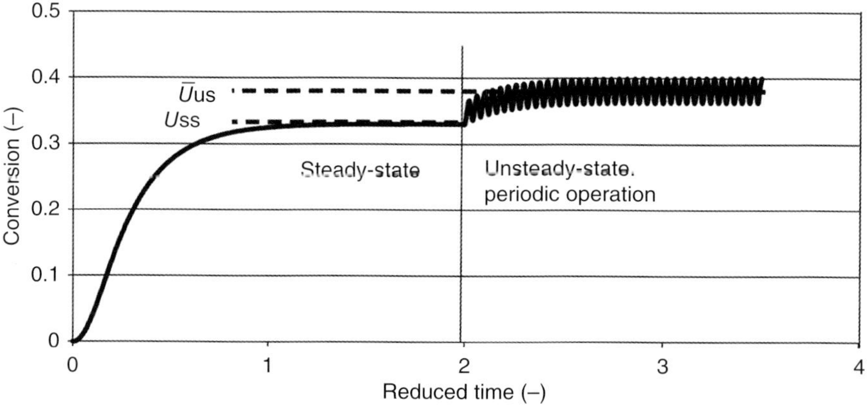

Figure 6.15 Simulation results of the time-average conversion for a periodically liquid flow rate (on–off control modes, $Re_L = 0.23/0$) at a constant cycle split of $S = 0.5$ and a period time of 10 min. Experimental data: steady-state (SS) condition $U_{SS} = 0.33$, unsteady-state (US) condition. Average $U_{US} = 0.44$. Source: Lange et al. 2004 [19]. Reproduced with permission of Elsevier. Although there is a mismatch between the simulations (Figure 6.14) and experimental predictions (Figure 6.14), the qualitative trend is the same: Periodic liquid flow operation results in an increase in α-methylstyrene conversion.

operation hydrodynamics in trickle-bed reactors. Experiments were done with an air–water system flowing through a stainless steel column filled with glass spheres. The effect of temperature and pressure on liquid holdup and pressure drop were studied. In Figure 6.16, holdup deviation indices are defined and distinction is made between the holdup patterns of cyclic operation and nonforced constant throughput operation for a given base superficial liquid velocity (u_{Lb}) and peak superficial liquid velocity (u_{Lp}) set.

Figure 6.17 shows that the pulse and base holdup deviation indices decrease with increasing temperature and pressure both for the fast and slow operation

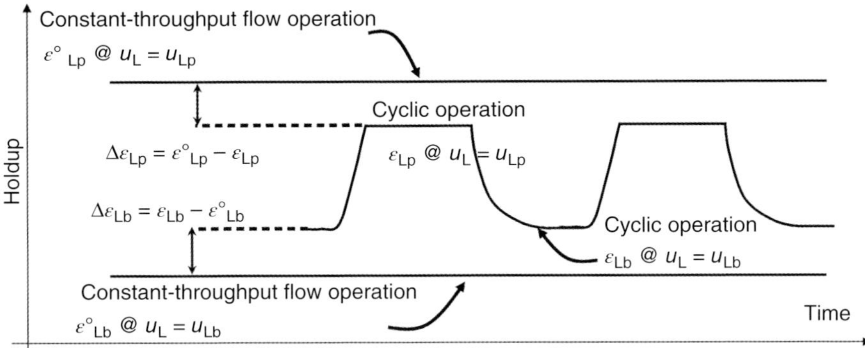

Figure 6.16 Definition of the deviation indices and distinction between the holdup patterns of cyclic operation and nonforced constant-throughput operation for a given (u_{Lb}, base superficial liquid velocity; u_{Lp}, peak superficial liquid velocity) set. Source: Aydin et al. 2008 [20]. Reproduced with permission of Elsevier.

modes. However, the decrease is more pronounced for the fast mode case. This implies that under the same conditions of temperature, pressure, fluid throughputs, and split ratio, the fast mode better preserves the pulse holdup content than the slow mode (~2 times according to the deviation indices). This creates an advantage for the fast operation mode in terms of more efficient removal of reaction product heat from the catalyst surface [20].

Recent modeling works on the hydrodynamics of TBRs under slow and fast periodic operation modes have been reported in Gancarczyk et al. [18] and Hamidipour et al. [21].

6.4 Cyclic Distillation

Cyclic distillation refers to an alternative operating mode of a distillation column that is based on the temporal isolation of liquid and vapor movements, contrary to the conventional counter-current operation where liquid–vapor phases flow simultaneously through the column. This type of operation can lead to distinct advantages over the conventional operation because of maximization of the driving force between gas and liquid phase in each separation stage, as well as minimization of mixing liquids of different compositions in each stage. Cyclic operation can be achieved by numerous methods. The simplest operation is controlled cycling, which is based on applying repeated cycles with periods of vapor flow upward the column, in which liquid remains stationary, followed by periods where the vapor flow is interrupted, feed and reflux are supplied, and the liquid flows down the column due to gravitation. This type of operation is shown in Figure 6.18 [22]. During vapor flow periods, the liquid flow is prevented by the thrust of the upcoming vapor, whereas in the liquid flow period, the liquid flows downward the column passing through each tray. A time-controlled valve in the vapor line that connects the reboiler to the column allows for the cyclic operation. This configuration is the most advantageous one as it can be directly

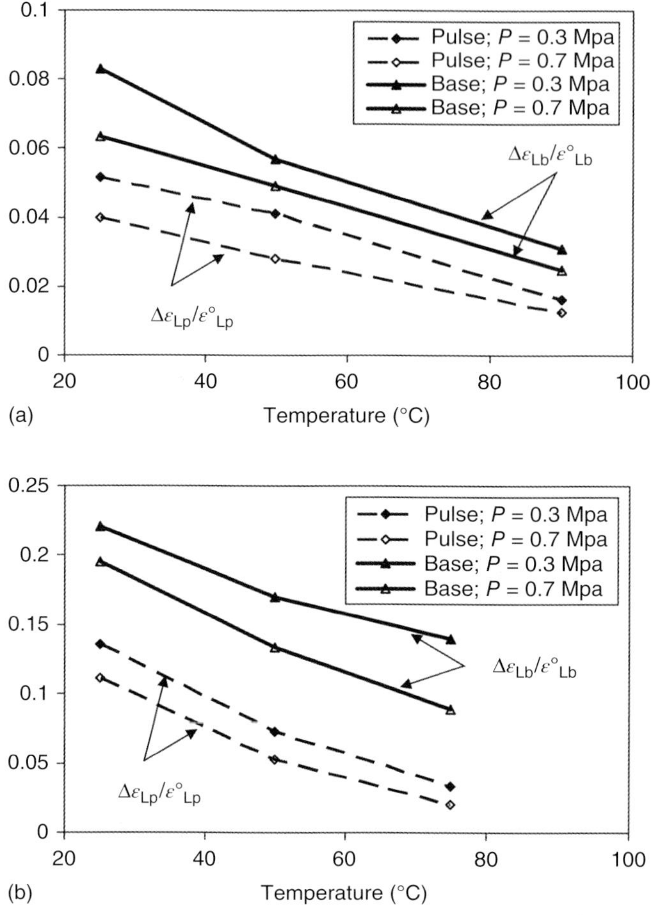

Figure 6.17 Liquid holdup deviation indices as a function of temperature and pressure obtained at 40 cm depth. $u_{Lb} = 0.0035$ m/s, $u_{Lp} = 0.014$ m/s, $u_G = 0.2$ m/s: (a) fast mode operation ($t_b = 4$ s, $t_p = 4$ s) and (b) slow mode operation ($t_b = 60$ s, $t_p = 60$ s). Source: Aydin et al. 2008 [20]. Reproduced with permission of Elsevier.

applied to already existing columns and alleviates the use of downcomers, making the plate design simpler and less expensive. Other advantages of cyclic operation over conventional distillation include superior capacity and tray efficiency. This in turn enables the same separation to be carried out at lower energy input. Main disadvantage of this configuration is that the separate movement of liquid and vapor is hindered as the number of trays increases (>10).

A solution to this limitation is possible by special tray design as seen in Figure 6.19 [23]. These trays feature valves and sluice chambers under the trays. During vapor flow, the valves below the trays are closed and the liquid remains on the tray. When liquid flow is applied, the valves open directing the liquid from the tray to the chamber below. When the next vapor flow period begins, the sluice chamber opens and the liquid flows to the empty tray below.

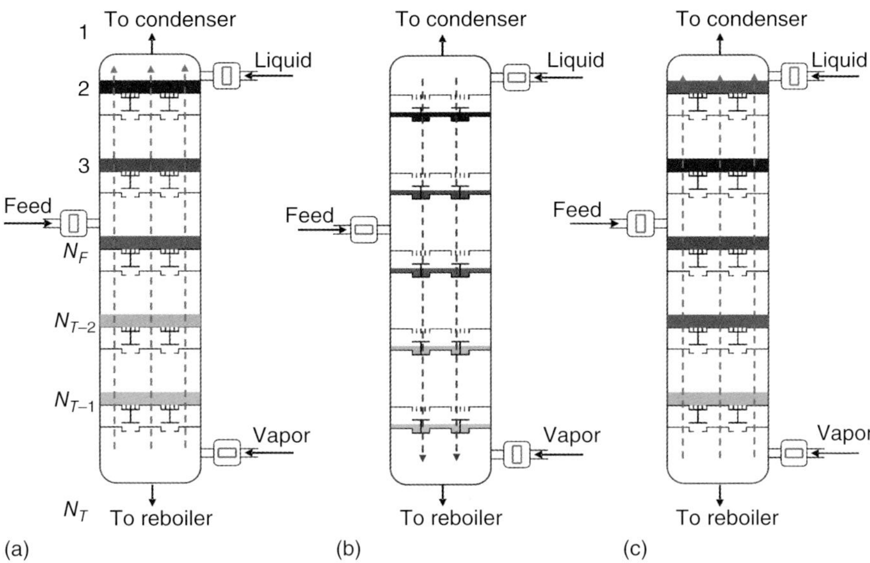

Figure 6.18 Operation of cyclic distillation. Alternation of a vapor flow period (a) with a liquid flow period (b) followed by another vapor flow period (c). Darkening colors from bottom to the top indicate the concentration gradient in the liquid phase present across the column. Source: Pătruț et al. 2014 [22]. Reproduced with permission of Elsevier.

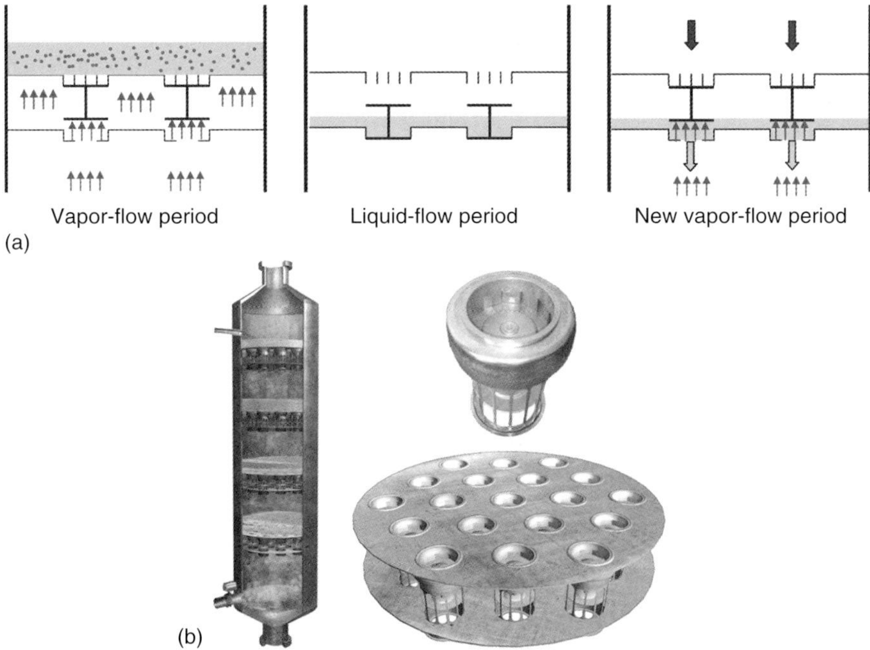

Figure 6.19 Trays designed for cyclic distillation. Source: Bildea et al. 2016 [23]. Reproduced with permission of John Wiley and Sons.

Other configurations for circumventing the hydrodynamic problems of early designs are stepwise periodic operation and stage-switching mode. In stepwise periodic operation, trays are equipped with an inlet and outlet together with a side reservoir that can contain one tray holdup. No downcomers are needed and the reservoir can be emptied into the reboiler directly by the outlet. During the vapor flow period, the reservoir collects the condensate. Before the beginning of a new cycle, the bottom product is removed from the reboiler, all trays are emptied to the reboiler, and the reservoir content is transferred to the corresponding tray. In the stage-switching mode, nozzle stages and complicated connections between the reboiler and condenser are used.

The modeling of cyclic distillation processes can be thought of as analogous to modeling of conventional distillation with liquid-phase concentration gradients across the plates of the column. In conventional distillation, the distance across the plate of the column is the independent variable, whereas for cyclic distillation, the independent variable can be substituted with time. With regard to column design, for the vapor period, the mass and energy balances for each tray are employed in the dynamic form without the terms that represent liquid flow in and out of the tray. In turn, for the liquid flow period, hydrodynamic models are used to describe the downward movement of the liquid from the trays, taking into account the mixing that occurs in the feed tray and reboiler. The vapor and liquid equations are then solved simultaneously. Besides the design parameters relevant to conventional distillation, important design parameters for cyclic distillation are the weeping limit (vapor velocity above which liquid will overflow from tray to tray during the vapor flow period) and the duration of vapor- and liquid flow periods [24].

Cyclic distillation offers several advantages over conventional distillation such as higher tray efficiency, higher throughput, reduced energy requirements, and increased quality of the product because of higher separation efficiency. As an example, Pătruţ et al. [22] modeled the case of benzene–toluene separation using cyclic distillation. Figure 6.20 shows that the energy requirements (expressed as vapor to feed flow rates ratio, V/F) as a function of the product purity (X_D) are significantly lower for cyclic distillation compared to conventional (classic) distillation [22].

Cyclic distillation has already been applied for pilot-scale separation of methanol/water, methyl-cyclohexane/n-heptane, acetone/water, and benzene/toluene mixtures. Indicatively, Schrodt et al. [25] were the first to compare the performance of cyclic operation versus conventional operation for a semi-plant-scale separation of acetone and water. They found that in cyclic operation (2.3–8 s liquid flow periods and 5–10 s vapor flow periods) of a 15-stage column, capacity increased two- to threefold compared to conventional operation while maintaining the same degree of separation. However, the authors noticed that the maximum efficiency was lower for the cyclic distillation. Higher efficiency is attained in the cyclic distillation process only if the liquid flow from the trays (during the liquid flow period) is uniform and no mixing of the liquid flowing from a tray to the tray below occurs, leading to near plug flow behavior. It was surmised by the same authors that in practice, only columns with <10–12 trays can be operated without flow problems. However, after the

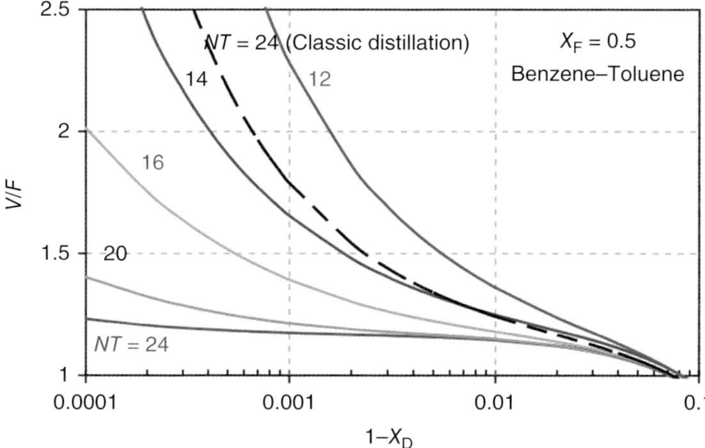

Figure 6.20 Comparison of energy requirements expressed as vapor to feed flow rate ratio (V/F) in cyclic distillation versus classic distillation for the ideal mixture benzene/toluene. The energy requirements are lower for cyclic distillation compared to classic distillation for the same number of trays, NT. Source: Pătruţ et al. 2014 [22]. Reproduced with permission of Elsevier.

advent of better tray designs, cyclic distillation was applied in the industry. Since 2005, MaletaCD has built and installed numerous cyclic distillation columns comprising 5–42 trays with diameters up to 1.7 m. A notable example is a cyclic distillation column with simultaneous tray drainage developed for the Lipnitsky Commercial Alcohol Plant in Ukraine [26] (Figure 6.21).

6.5 Pulse Combustion

Pulse combustion (P-C) is an intermittent combustion technique that can burn diverse-quality fuels achieving high combustion efficiency and low pollutant emissions. The operating principle of the technique is based on oscillatory mass flow rate conditions associated with a periodic variation in operating parameters (pressure, temperature, and velocity). It is a combustion system characterized by complicated dynamics; three-dimensional turbulent flow, resonant pressure field conditions, transient heat transfer, and chemical reactions are concurrently involved. An extensive overview regarding the principle, modeling approach, and applications of P-C has been published by Meng et al. [27]. Although a P-C device generally comprises inlet valves, a combustion chamber, and a tailpipe, two main design configurations prevail: (i) the Schmidt, Helmholtz, and Rijke types that are based on the operating system geometry (Figure 6.22) and (ii) the valved and aerovalved (valveless) types that are distinguished by the way fuel and air enter the combustor (Figure 6.23). The key features and the characteristics of each design configuration are presented in Tables 6.2 and 6.3.

The operating principle of each pulse combustor is based on the following four-phase cyclic operation described below and schematically presented in Figure 6.24.

Figure 6.21 Industrial implementation of cyclic distillation. Source: Courtesy of MaletaCD [26].

- *Phase 1 – ignition and combustion* (A–B): Fuel and air enter the combustion chamber. A spark plug or the residual flue gases from the previous cycle initiates the ignition. A rapid pressure rise occurs (from point A to B) because of the combustion of the fuel and air mixture.
- *Phase 2 – expansion* (B–C): The high-temperature flue gases are expanded and escape from the combustion chamber through the tailpipe causing a pressure decrease toward below atmospheric pressure (from point B to C).
- *Phase 3 – purge and recharge* (C–D): Pressure decrease results in suction of a certain amount of the flue gases back to the combustion chamber and concurrently admission of fresh fuel and air into the combustion chamber (from point C to D).
- *Phase 4 – recharge and compress* (D–A): The mixture of the fresh fuel with the air and the sucked flue gases is autoignited because of the high temperature of the sucked flue gases. Pressure increases up to the initial value (from

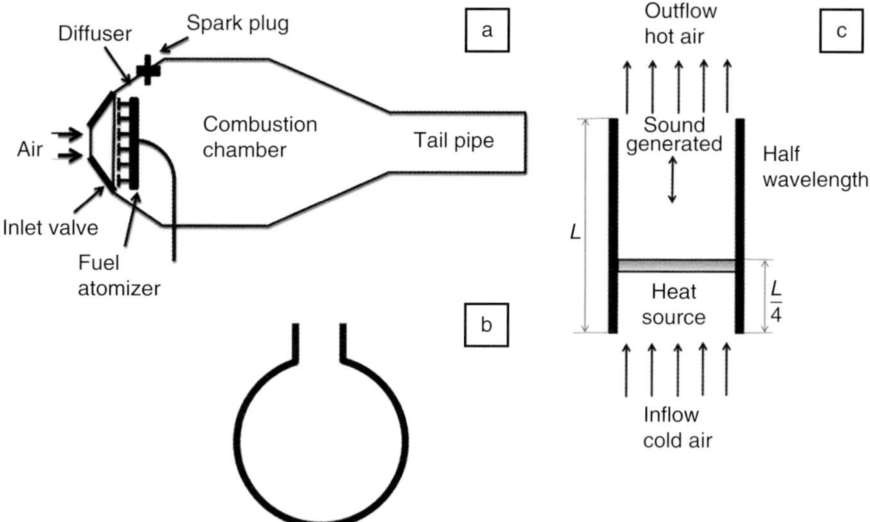

Figure 6.22 Simplified schemes of (a) Schmidt tube pulsejet, (b) Helmholtz resonator, and (c) Rijke tube. Source: Adapted from Entezam et al. 1997 [28].

point D to A). Once self-sustention is achieved, no spark plug is required to ignite the fuel/air mixture. Combustion cycles repeat themselves for a definite time period.

The P-C devices are very complex systems, which make their control and optimization a demanding task. Parameters such as air/fuel supply rate, air/fuel mixing, fuel types, and tailpipe length influence the performance of a pulse combustor. Pulse combustors operate within a temperature range of 810–1470 K and flue gases oscillation frequency varies between 20 and 250 Hz (typically 125–150 Hz). Pressure oscillation in the combustion chamber is approximately 710 kPa, resulting in velocity oscillation in the tailpipe of about 7100 m/s and gas velocity range at the tailpipe exit of 0–100 m/s. Pulse combustors have many advantages over the conventional steady continuous combustion such as higher combustion efficiency that leads to lower energy consumption, up to 10 times higher combustion intensity, two to five times higher heat and mass transfer rates, up to three times lower NOx and CO pollutant emissions, better temperature uniformity in the combustor, and lower volumetric footprint. The main advantages of P-C devices over conventional steady continuous combustion are summarized and presented in Table 6.4.

P-C systems face significant limitations as well. The major downside of pulse combustors is the high operational noise levels (in the range of 130–180 dB). The high noise levels mainly come from three sources: (i) the detonating nature of pulse combustion, (ii) the vibrations of the metal walls, and (iii) other mechanical parts of the pulse combustor, which may cause intense mechanical vibrations. Moreover, the noise can be of mechanical origin; in particular, the operation of flapper valves, the blower that supplies air for combustion, and the principal

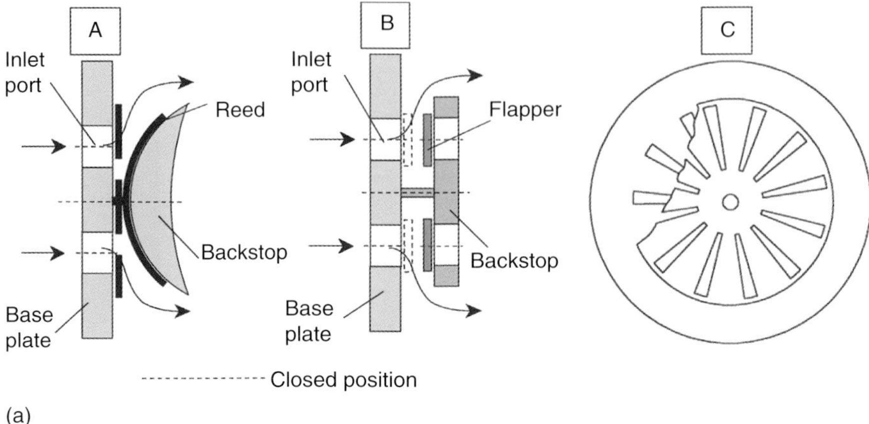

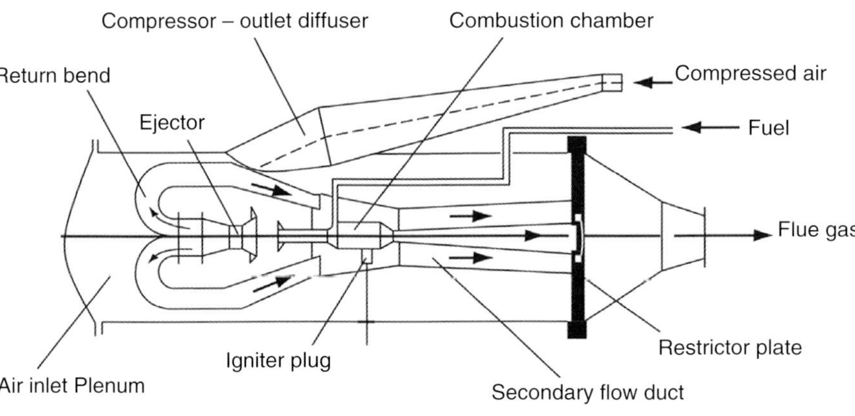

Figure 6.23 (a) Mechanical valves used in valved combustors; A, reed valve; B, flapper valve; C, rotary valve; (b) valveless pulse combustor. Source: (a) (A), (B), and (b) Kudra and Mujumdar 2010 [29]. Reproduced with permission of Taylor and Francis Group LLC Books. (a) (C) Zbicinski 2002 [30]. Reproduced with permission of Elsevier.

blower that delivers air to reduce the temperature of drying gases may be a source of noise.

P-C has been a combustion technique with high commercial interest as it has been widely tested in numerous industrial and domestic applications. The potential of P-C has recently been investigated in applications such as drying, calcination, incineration, water heating, steam production, and heating air in open field to protect plants from freezing [32]. The different applications that the P-C technique has been proposed as encouraging alternative for are presented in Table 6.5.

One of the most important applications of P-C is the drying process. P-C can be used to dry a broad range of materials including minerals, chemicals, food, agriculture products, heat-sensitive biomaterials, and high-viscosity suspensions. The main advantages that the P-C drying offers compared to the conventional dryers are

Table 6.2 Key features of Schmidt tube pulsejet, Helmholtz resonator, and Rijke tube.

	Schmidt-type pulse combustor	Helmholtz-type pulse combustor	Rijke-type pulse combustor
Basics	Closed–open, quarter-wave Schmidt tube	Closed–open, open–open Helmholtz resonator	Open–open, half-wave Rijke tube
Geometry	An inlet valve, a combustion chamber, and a tailpipe with general identical diameter	An inlet, a separated combustion chamber, and a long tailpipe with a smaller diameter	A vertical tube that comprises a heat source and two isobaric ends
Fuel used	Liquid fuels	Gaseous fuels	Solid/liquid fuels
Combustion	Near the vicinity of the inlet valve	Inside the combustion chamber	At around the distance 1/4 length of the vertical tube
Remarks	Fuel mixing/firing is essentially important to sustain pulsation and must be completed within half a period of pulsation	Frequency determination via both the combustion chamber volume and the length and cross-sectional area of the tailpipe	The length of the tube equals half the wavelength of the oscillation

Source: Meng et al. 2016 [27]. Reproduced with permission of Elsevier.

Table 6.3 Key features of valved and valveless pulse combustors.

	Mechanical valve pulse combustors		Valveless pulse combustors
Valve type	**Flapper and reed types**	**Rotary types**	**Aerodynamic diode**
Advantages	Easy design, unidirectional flow	Durable, flexible to use, unidirectional flow	Lack of valves and moving parts, no fatigue failure problem
Disadvantages	Fatigue failure	Fatigue failure	Backflow

Source: Meng et al. 2016 [27]. Reproduced with permission of Elsevier.

- high heat transfer because of high velocity fluctuations;
- higher drying rate because of high heat transfer rate;
- higher product quality because of elimination of material distribution within the dryer;
- lower temperature of gases and product;
- lower air consumption;
- atomization of slurries and suspensions without the need of disk atomizers or high-pressure nozzles;
- lower energy consumption;
- no requirement for air blowing systems in the combustion system;

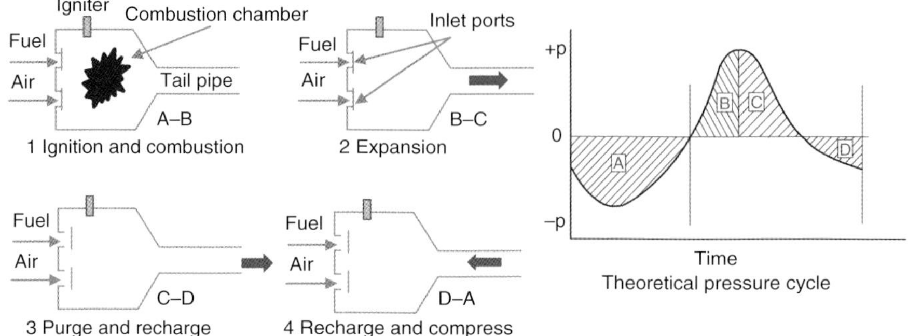

Figure 6.24 Schematic representation of a complete combustion cycle (the current cycle refers to the combustion cycle of a Helmholtz-type pulse combustor). Source: Wu 2007 [31].

Table 6.4 Comparison of steady continuous and pulse combustion.

Features of combustion	Continuous	Pulse
Combustion intensity (kW/m^3)	100–1 000	10 000–50 000
Efficiency of burning (%)	80–96	90–99
Losses due to chemical under-burning (%)	0–3	0–1
Losses due to mechanical under-burning (%)	0–15	0–5
Temperature level (K)	2 000–2 500	1 500–2 000
CO concentration in exhaust (%)	0–2	0–1
NOx concentration in exhaust (mg/m^3)	100–7 000	20–70
Convective heat transfer coefficient (W/m^2 K)	50–100	100–500
Residence time (s)	1–10	0.01–0.5
Excess air ratio	1.01–1.2	1.00–1.01
Noise produced (dB)	85–100	110–130

Source: Reproduced with permission of Royal Society of Chemistry.

- destructive effect on odor, toxic compounds, and VOCs; and
- high compatibility with different processed materials.

P-C has been largely used in spray dryers and flash dryers and to a lesser extent in fluidized bed dryers. Minneapolis United States, Sonodyne Industries Inc., Portland (United States), Pulse Combustion Systems (PCS), and San Rafael (United States) are some well-known contributors in P-C spray drying development. Two configurations of spray dryer have been developed: the vertical and the horizontal P-C spray dryer (Figure 6.25). The operating principle of the vertical P-C spray dryer is as follows: fuel and air enter the combustion chamber, mix, and get ignited. The high-temperature flue gases rush down the tailpipe toward the orifice. Air and fuel are recharged into the hot combustion chamber to repeat the cycle. Just above the orifice, quench/diluent air is blended with the hot flue gasses to attain the desired product contact temperature. The feed materials such as liquids and dilute suspensions are atomized by the hot

Table 6.5 Commercial P–C applications.

General purpose	Specific techniques	Activity level
Thrust generation	Propulsion	a)
	Torque production	b)
	Vertical lift devices	
	Surface cleaning and descaling	
Pressure gain and fluid pumping	Pumping fog oil and insecticide	a)
	Gas turbine combustors	b)
	Fan/blower replacement	
	Flue gas recirculation	
Heating of liquids	Hydronic space heating (hot water)	a)
	Potable water heating	b)
	Steam raising with gas, liquid, and solid fuels	b)
	Deep fat fryers	b)
	Chemical processing	
	Petroleum processing	
	Viscosity control of liquids	
Indirect heating of air	Space heating in vehicles	a)
	Space heating in homes	a)
	Space heating in industry	
Miscellaneous uses	Snow and ice melting	a)
	Griddles	b)
	Coagulation of materials	b)
	Refuse incineration	
	Gasification	
	Power generation	
	Heat pumps	
	Sawdust incinerator-boiler	
	Pilot burners and igniters for low grade fuels	

a) Significant applications are in use.
b) Active development has been underway (currently or recently).
Source: Putnam et al. 1986 [33]. Reproduced with permission of Elsevier.

pulsating gases and come into a conventional drying chamber. The dried product is mainly collected downstream of the cyclone, while exhaust gases are vented to the atmosphere.

In a horizontal P-C spray dryer, a valveless pulse combustor is used as the source of a gaseous drying agent. Air and fuel enter the combustion chamber; they are mixed and ignited by a spark plug. Raw material is injected directly into the drying cone, which is situated downstream of the tailpipe where the hot flue gases pass by. The heavier dry particles are separated by gravity in the primary

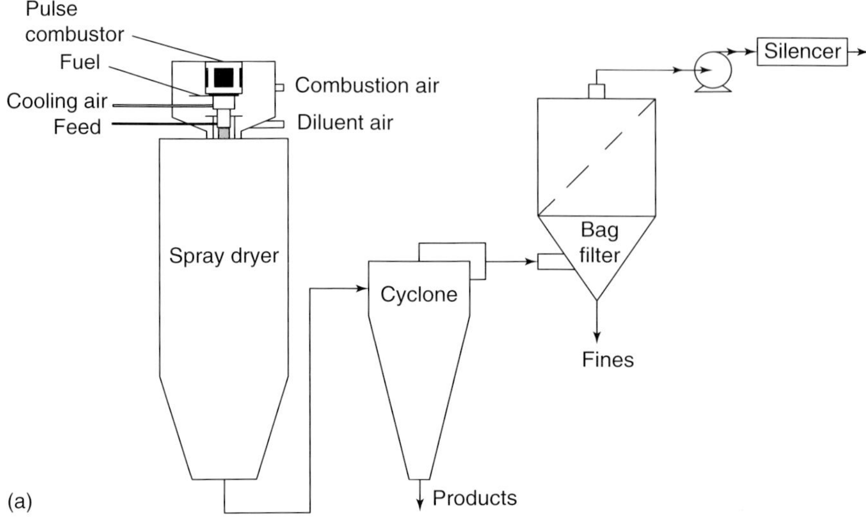

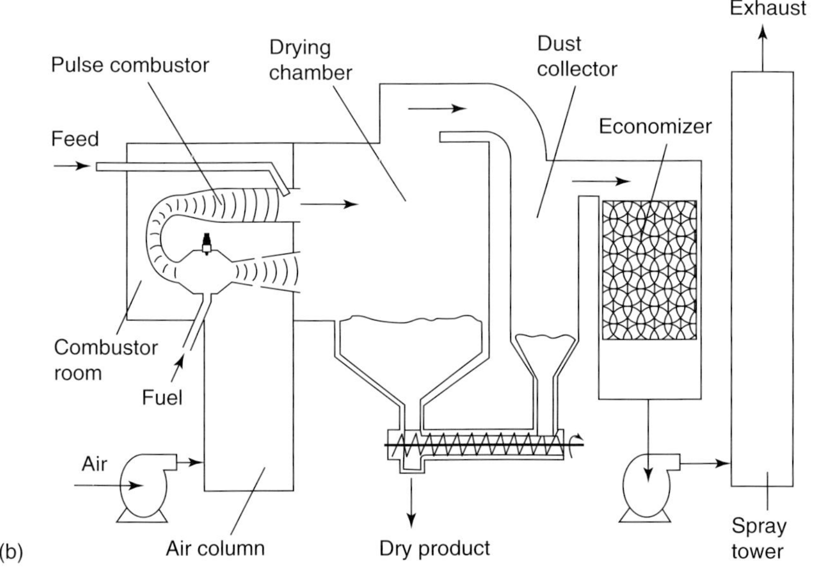

Figure 6.25 P-C spray dryer configurations; (a) vertical P-C spray dryers. (b) Horizontal P-C spray dryer [34]. Source: (a) Adapted from Kudra and Mujumdar 2010 [29]. (b) Kudra 2008 [34]. Reproduced with permission of Taylor & Francis.

cyclone while fine particles are centrifugally removed from the gas stream in the dust collector.

Novadyne Ltd., Canada is the lead company in the development of the P-C flash dryer. Its configuration is presented in Figure 6.26a. The P-C flash dryer comprises a pulse combustor with a flapper valve, a pneumatic duct extended from the tailpipe of the combustor, and equipment for material feed and collection.

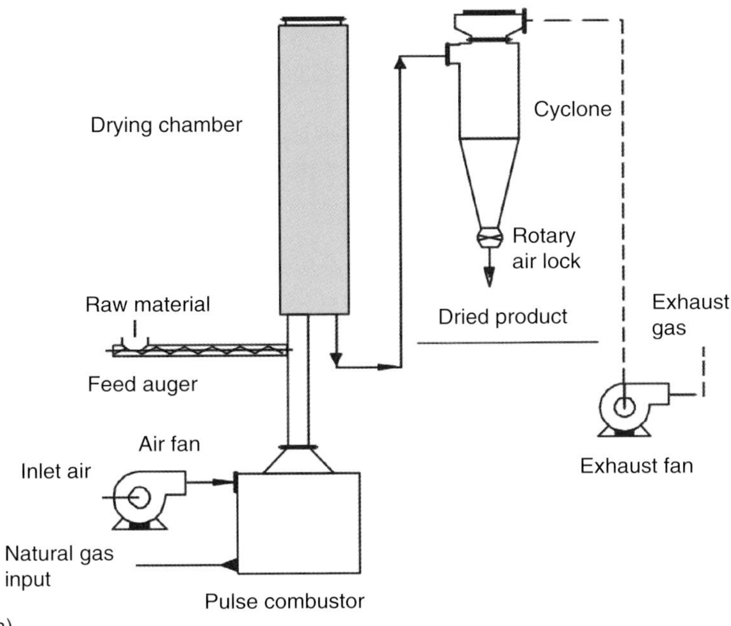

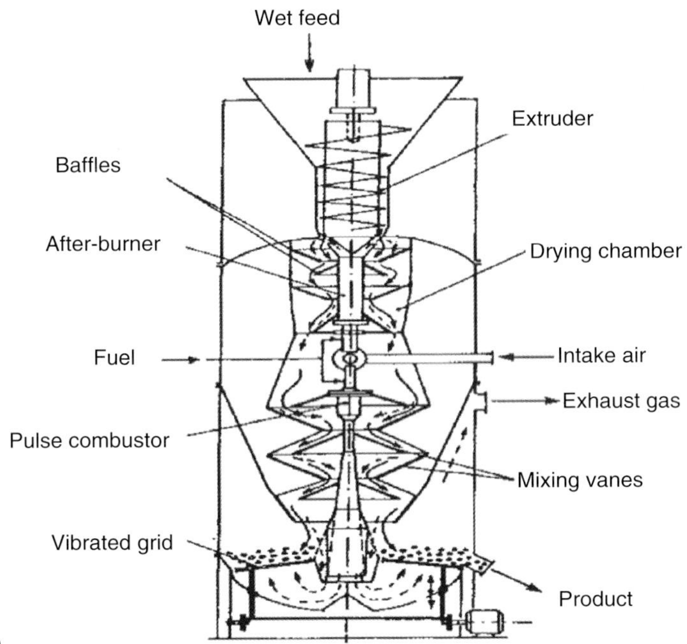

Figure 6.26 Different P-C dryers: (a) P-C flash dryer; (b) P-C fluidized bed dryer. Source: (a) Courtesy of Novadyne Ltd., Canada. (b) Zbicinski 2002 [30]. Reproduced with permission of Elsevier.

The material is fed directly into the tailpipe where it is dispersed and dried by the hot flue gases. The drying process is finalized in the pneumatic duct. It has been proven that the thermal efficiency of a P-C flash dryer is similar to that of conventional flash dryers, but the electricity consumption is lower (40–50% less) as the motors of the material handling equipment are smaller [29].

In Figure 6.26b, the P-C fluidized bed dryer configuration is presented. The drying process takes place in different sections of the pulse combustor chamber. First, the wet material is transferred to the upper part of the drying chamber. There, the heat from the afterburner eliminates the water. Then, the partially dried solids are mixed with the hot pulsating flue gases in the middle part of the dryer. Eventually, the particles, containing $\sim 10\%$ moisture on dry basis, are removed through a vibrating bed at the bottom part of the fluidized bed [34].

Aside from drying applications, P-C systems have also been developed in the areas of heaters, propulsion engine, calcination, incineration, and cement making.

6.6 Pressure Swing Adsorption

The principle of adsorption is based on the affinity of a gas for an adsorbent material. In principle, gases with high polarity have high affinity to create bonds with an adsorbent. The type of adsorbent material, the partial pressure of the gas component, and the operating temperature are parameters, which affect the affinity of a gas for an adsorbent and consequently the efficiency of separation. Adsorption processes are employed to separate binary gas mixtures and to purify or to enrich gaseous streams. In the case of binary mixtures, the gas with the higher affinity (adsorbate) adheres to the adsorbent surface, whereas the gas with the lower affinity breaks through the column faster than the adsorbate, thus enabling the gas separation. In the case of gas purification, the impurity adheres to the adsorbent surface. Similarly, in gas enrichment, the adsorbate is removed by the adsorbent. Consequently, the concentration of the other gas compounds increases in the effluent gas. However, after a certain contact time between the gas feed and the adsorbent, an equilibrium state is achieved, which defines the thermodynamic limit of the adsorbent loading for given operating conditions (fluid phase composition, temperature, and pressure).

To achieve efficient separation, the adsorbate should not break through the column. Therefore, the adsorbent should be regularly regenerated by desorbing the adsorbate. By changing one of the process parameters (composition, temperature, and pressure), it is possible to regenerate the adsorbent. In the case of adsorbent regeneration by reducing the total pressure of the system, the process is called PSA as the total pressure of the system "swings" between high pressure in feed and low pressure in regeneration. A schematic of PSA unit is presented in Figure 6.27.

The operational principle of a PSA is rather simple. The basic phases of a complete cycle are as follows: gas is fed to the first column (C1) at a pressure higher than atmospheric and at constant temperature. The adsorbent selectively adsorbs

Figure 6.27 Schematic design of a two-column pressure swing adsorption unit. Source: Grande 2012 [35]. ISRN Chemical Engineering – open access journal.

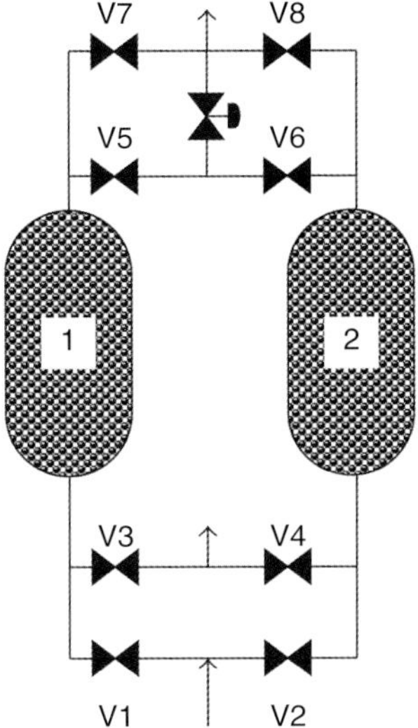

the gas that needs to be removed, rendering the exiting stream (after valve V7) richer in the gas that needs to be reclaimed. Concurrently, heat is released as adsorption is an exothermic process. When the adsorbent packed in C1 is saturated and cannot adsorb more gas, the feed is directed to the second column (C2). To release the adsorbate from C1, the flow direction is reversed and the total pressure of the column is reduced by venting (opening valve V3). This step is called the blowdown. In the blowdown step, the adsorbate is desorbed from the adsorbent and released. After the adsorbent is regenerated, the overall pressure of the system is restored. That takes place in the pressurization step using the feed stream. At the end of this sequence of steps, a full cycle is completed.

The separation efficiency of a PSA unit is strongly determined by the adsorbent employed for the separation. Therefore, it is essential to select the adsorbent very carefully. Adsorbents are mostly porous materials with high specific surface area. They are shaped into spherical pellets or extruded. They can also be shaped into honeycomb monolithic structures resulting in reduced pressure drop along the bed. Zeolites, inorganic mesoporous materials, activated carbon, and metal–organic frameworks (MOFs) are the most known materials that are used as effective adsorbents [36].

PSA is one of the most known and established industrial processes for gas separation. Low volumetric footprint of the equipment, low energy requirements, low capital investment cost, safety, and simplicity of operation are considerable strengths of the technology [37]. Over the years, it has been demonstrated that the PSA technology can be used in a large variety of applications; Riboldi and Bolland [38] investigated the potential of PSA as a CO_2 separation technique in coal-fired power plants. It was reported that PSA required lower energy compared to amine absorption. Augelletti et al. [37] studied PSA for biogas upgrading. Methane recovery greater than 99% was obtained with very low energy consumption (~1250 kJ/kg of biomethane). Other applications of PSA are hydrogen purification, air separation, OBOGS (on-board gas generation system), and noble gases (He, Xe, and Ar) purification [39].

6.7 Desorptive Cooling

The high enthalpies often associated with phase changes make the utilization of the latter a feasible method to store thermal energy. Examples of such recuperative processes are evaporative cooling and liquid injection. For reactor cooling applications (exothermic reactions), the process of evaporation can be replaced by desorption with several distinct advantages. Desorption processes involve enthalpies of a similar order of magnitude (or higher) compared to evaporation, while avoiding the technical difficulties that arise because of the presence of a liquid phase to evaporate. The application of desorptive cooling involves the selection of an inert adsorbate and a suitable adsorbent. An appropriate combination can in several instances extend the feasible range of operating pressures and temperatures of a reactor [40].

An integrated catalytic reaction/desorption scheme (Figure 6.28) can be established by employing a conventional catalytic fixed bed diluted with adsorbent particles. The catalyst/adsorbent ratio can be varied along the length of the reactor. Before the reaction, an inert (to both reaction and catalyst) adsorbate is adsorbed homogeneously onto the adsorbent. The adsorption process may initially result in some heat generation that has to be removed conventionally from the system before the reaction. Care must be taken in order to make sure that the catalyst can withstand periodic temperature cycling. Once the adsorption cycle ends and the system attains a uniform temperature profile, the reactants are loaded and the reactive cycle begins.

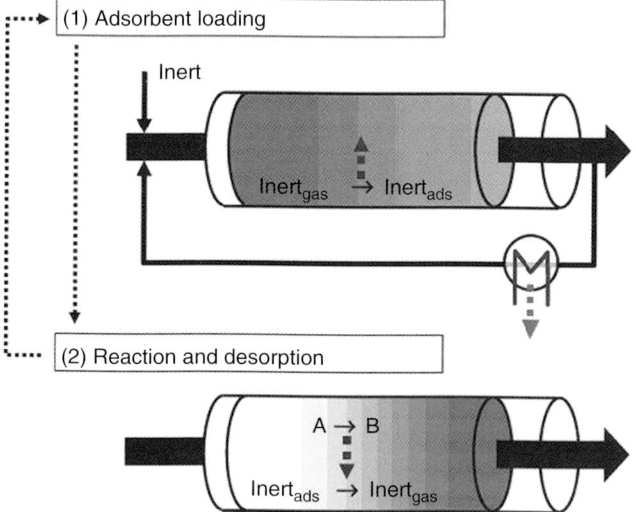

Figure 6.28 Operating principle of an integrated reaction/desorption process. Source: Grünewald and Agar 2004 [41]. Reproduced with permission of American Chemical Society.

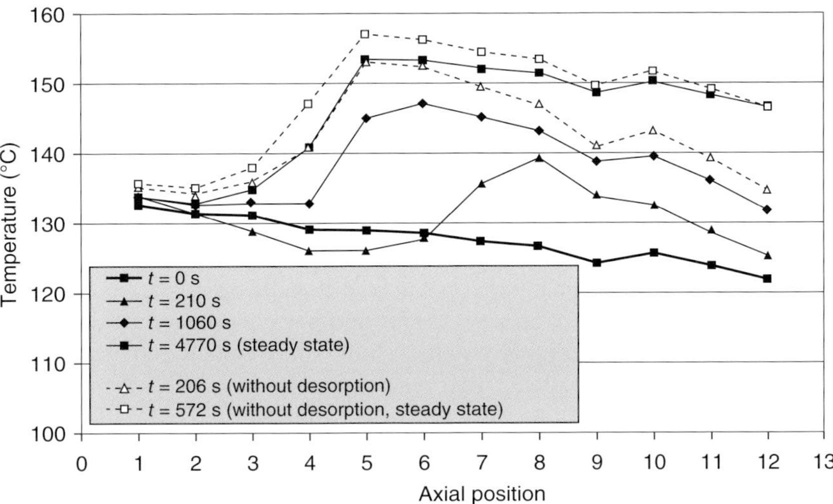

Figure 6.29 Experimental results for the temperature profile dynamics of reaction with and without integrated desorption. Source: Grünewald and Agar 2004 [41]. Reproduced with permission of American Chemical Society.

The heat generated by the exothermic reaction occurring onto the catalyst particles is conducted and adsorbed by the nearby adsorbent particles through desorption of the inert material in the gas phase. After a certain amount of time, the adsorbate is depleted and the reaction temperature cannot be maintained any further. At that time, a new adsorption cycle must begin.

The desorptive cooling process offers several advantages as the reactor design becomes simpler without the need of extensive heat transfer area. The heat removal rate can be tuned by the adjustment of the adsorbent/catalyst ratio along the axial dimension. Grünewald and Agar [41] studied desorptive cooling for CO oxidation on a supported palladium shell catalyst using 3A zeolite as the adsorbent and water vapor as the adsorbate. Their results in Figure 6.29 reveal the experimental temperature profile measured in the axial dimension by 12 thermocouples distributed along the catalyst/adsorbent fixed bed. It was found that the integrated reaction–desorption scheme results in lower temperatures compared to the case without desorption. However, some disadvantages of desorptive cooling were also evident, such as difficulties in the attainment of steady state and lower space–time yield as volume is less because of the presence of the adsorbent. Additional disadvantages may include difficulties in finding a compatible inert adsorbate/adsorbent system with respect to the reaction/catalyst system.

Besides the choice of adsorbent/adsorbate, the main parameters influencing a desorptive cooling process are the local rate of reactive heat liberation, which in turn depends on temperature, reactant concentration, and adsorbent mass fraction. Furthermore, the desorption rate depends on the partial pressure of the inert gas phase and the mass transfer resistances influencing the diffusion of the adsorbate into the gas stream [42]. Development of strategies to control desorption rates would lead to a more controlled and ideal temperature profile in the reactor. Through computer simulations, several effective strategies have been suggested such as local variation of the adsorbent fraction in the fixed bed or temporal variation of the inlet level for the inert component over the cycle.

6.8 Variable Volume Operation of Stirred Tank Reactors

Design and analysis of chemical reactors are traditionally based on ideal reactor models, i.e. ideal batch, PFR, and CSTR. PFR and perfectly mixed (ideal) batch reactor have the same performance equation (only space time in PFR is replaced by the residence time in the ideal batch reactor). In the case of a set of parallel reactions, or a combination of parallel and serial reactions, either an ideal batch or a CSTR reactor may be more desirable depending on the reaction kinetics. Variable volume operation of a stirred tank reactor represents a mixing regime in-between the two limits above and may improve reactor performance, for certain reaction systems, compared to the two ideal reactor models. A schematic showing the general principle of variable volume operation of a stirred tank reactor is given in Figure 6.30. Each cycle starts at a chosen minimum volume level (V_{min}); then, a certain feeding rate is applied that brings the reactor volume to a chosen maximum level (V_{max}). In a third step, the reactor contents are left to react for a batch time without any inflow or outflow and then the reactor volume is brought to the original level according to a chosen discharge rate. Finally, the reactor contents are left to rest for a given time before the next cycle starts. In this operating context, the minimum and maximum volume levels, the feeding rate, the batch time, the discharge rate, and the rest time are degrees of freedom.

To exemplify the merit of variable volume operation of stirred tank reactors, the Van de Vusse reaction scheme has been examined by Lund and Seagrave [44, 45]

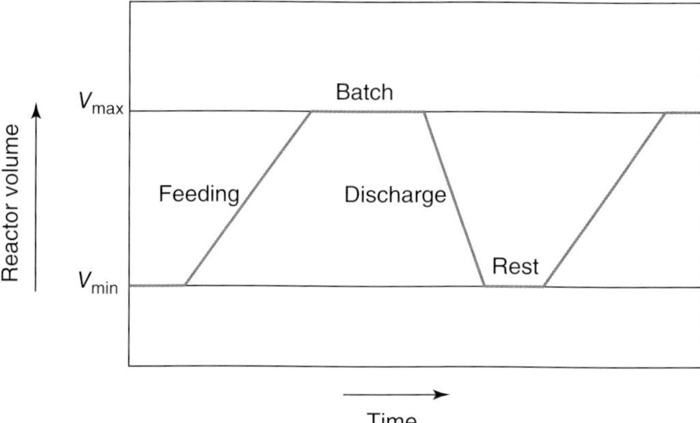

Figure 6.30 Principle of cyclic operation of a stirred tank reactor operated between a minimum and a maximum volume limit. Source: Zwijnenburg et al. 1998 [43]. Reproduced with permission of American Institute of Chemical Engineers Publication.

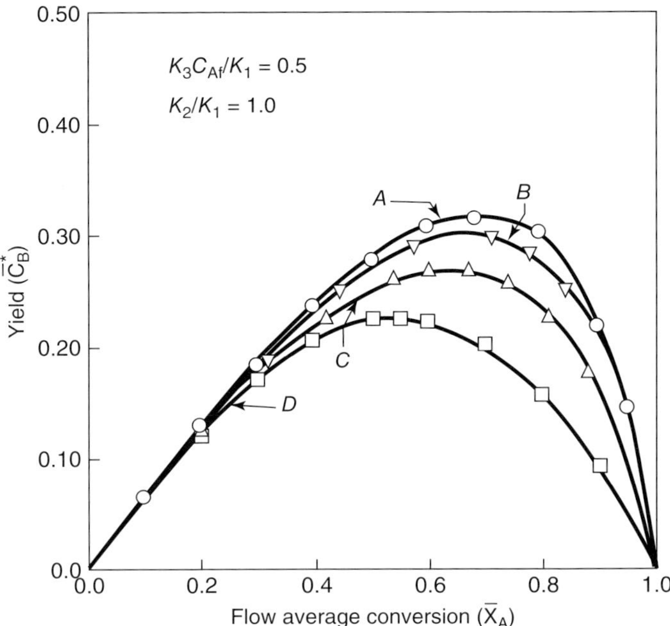

Figure 6.31 Yield of intermediate product *B* in different reactor systems with Van de Vusse reactions vs. conversion of feed component: curve A, plug flow reactor; curves B and C, semibatch operation of the stirred tank reactor; curve D, continuous stirred tank reactor. Source: Lund and Seagrave 1971 [44, 45]. Reproduced with permission of American Chemical Society.

and Riedlehoover and Seagrave [46].

$$A \xrightarrow{K1} B \xrightarrow{K2} C$$

$$2A \xrightarrow{K3} D$$

Figure 6.31 presents simulation results in terms of the yield of intermediate product *B* in different reactor systems with Van de Vusse reactions versus conversion of feed component *A*. The relative kinetics are determined by the equations $K_3C_{Af}/K_1 = 0.5$ and $K_2/K_1 = 1.0$. The simulation results show that PFR operation gives the highest yield in *B* across the entire conversion range; CSTR operation gives the lowest yield and variable volume (semibatch) operation has a performance in-between the two ideal reactor models.

Figure 6.32 shows simulation results for the same reactor systems and reaction scheme, but with different kinetics compared to Figure 6.31. Specifically, K_3C_{Af} is now big compared to K_1, $K_3C_{Af}/K_1 = 10.0$, while $K_2/K_1 = 1.0$. The simulation results show that now the CSTR model gives the highest yield in *B* for conversions X_A up to ~75%. At higher conversions of *A*, variable volume (semibatch) operation gives the highest yield in *B*. Collectively, for certain reaction schemes and kinetics, optimized variable volume operation of stirred tank reactors may give better yields compared to the PFR/ideal batch and CSTR operating limits.

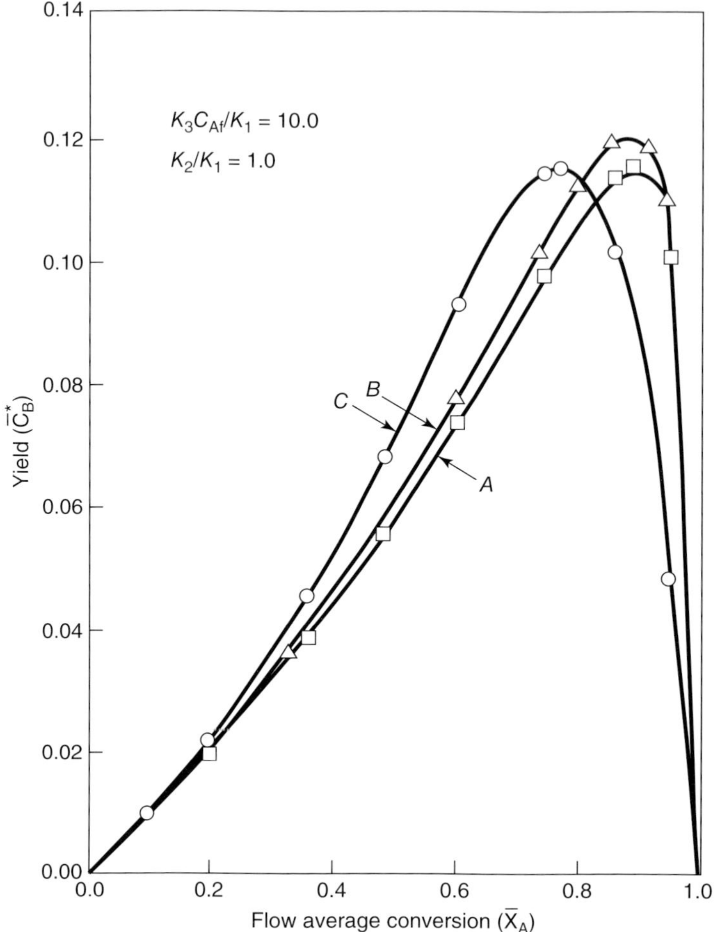

Figure 6.32 Yield of intermediate product *B* in different reactor systems with Van de Vusse reactions versus conversion of feed component: curve A, plug flow reactor; curve B, semibatch operation of a stirred tank reactor; curve C, continuous stirred tank reactor. Source: Lund and Seagrave 1971 [44, 45]. Reproduced with permission of American Chemical Society.

6.9 Short Contact Time Reactors

Short contact time reactors can be considered the catalytic reactors in which the superficial contact time between the reactants and the catalyst is less than 100 ms. Given appropriate thermal management and use of very active catalyst, high (intermediate) product yields can be attained, mainly through fast catalytic partial oxidation reactions. A good study dedicated to short contact time reactors is written by Veser [47]. Here, two prominent examples from the literature will be discussed: the catalytic partial oxidation of light alkanes to olefins and the catalytic partial oxidation of biomass to syngas without tar formation.

6.9.1 Catalytic Partial Oxidation of Alkanes

In the first example [48], catalytic partial oxidation of light alkanes (ethane, propane, *n*-butane, and isobutene) takes place in a short contact time reactor by rapid chemical heating (~ microseconds). The process is then followed by instant quenching to avoid product decomposition and overoxidation. The catalyst plays a very important role in the selectivity of the reaction toward the desired products and its configuration ultimately determines the effectiveness of the whole process.

In order to achieve the desired residence time and temperature profile, different chemical catalyst configurations and types have been tested, namely low-activity metal oxide ceramic foams, high-activity metal-coated monolithic reactors, and metal gauzes. In the first case, a preheating step is necessary because of the low activity of the catalyst employed. However, preheating is undesirable because it allows homogeneous reactions to take place before the reactants interact with the catalyst and leads to the production of CO_2 and H_2O, hindering the partial oxidation processes (Figure 6.33a). In the second case, the monolithic reactor rapidly heats up the reactants but slowly cools down the products via radial heat transfer through the reactor walls. This is also undesirable as slow cooling leads to intermediate product (olefins and oxygenates) decomposition (Figure 6.33b). In the advent of highly active catalysts, e.g. Pt–10%Rh in gauze reactor, preheating was essentially eliminated allowing selective catalytic partial oxidation to determine product distribution. In particular, chemical heating occurred in ~five microseconds, whereas the contact time over the gauze (wire diameter divided by gas velocity) was 8–500 μs. The goal is to avoid complete oxidation and to form intermediate products by rapid reactants heating and product cooling. To this end, a grid-shaped (gauze) reactor configuration was used (Figure 6.34). Aside from

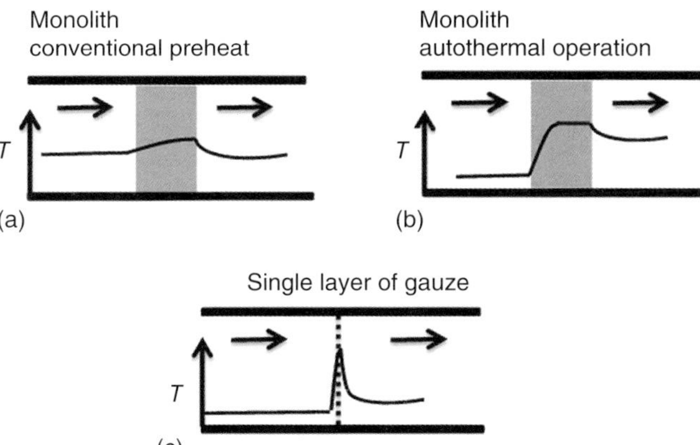

Figure 6.33 (a) Reactor that uses low-activity catalysts such as metal oxides and thus requires preheating. (b) Reactor that uses high-activity catalysts such as metal-coated monolith, which rapidly heats the reactants to the reaction temperature but slowly cools the products by heat transfer through the reactor walls. (c) Reactor with a single layer of metal gauze that allows rapid heating of the reactants and then rapid quenching of the products after microsecond partial oxidation of light alkanes to olefins. Source: Adapted from Goetsch and Schmidt 1996 [48].

Figure 6.34 Single gauze reactor. (a) Photograph of an ignited 1.8 cm-diameter Pt–10% Rh gauze reactor catalyzing the partial oxidation of butane. Although the reactants are supplied at 25 °C, reaction on the surface of the gauze heats it to >800 °C. (b, c) Scanning electron micrographs of the gauze before reaction show a smooth surface. (d) Scanning electron micrograph of the Pt gauze used to catalyze the partial oxidation of butane, which shows significant roughening and facet formation after exposure of the gauze to partial oxidation conditions for ∼50 h. (e) Higher magnification of this surface shows extensive faceting, which promotes mass transfer mixing. Source: Goetsch and Schmidt 1996 [48]. Reproduced with permission of AAAS.

rapid heating and reaction mentioned above, this grid-type configuration also allows for rapid mixing of the hot product gas with the cold unreacted gas that passes between the gauze wires. The mixing process results in rapid quenching from ~800 to ~400 °C in ~200 μs. That way, it is possible to avoid homogeneous decomposition reactions. A schematic of the catalytic partial oxidation process in the gauze reactor is shown in Figure 6.33c.

Using this gauze reactor, high selectivity to olefins and oxygenate products, mainly formaldehyde and acetaldehyde, were attained at oxygen conversions in excess of 90%. Comparison of carbon atom selectivity shows that olefin selectivity decreases, whereas oxygenate selectivity increases with increasing alkyl chain length (Table 6.6) [48].

6.9.2 Catalytic Partial Oxidation of Cellulose

Lignocellulosic biomass, which comes from agricultural waste or energy crops, is nowadays an alternative and sustainable raw material for the production of fuels and chemicals. There are many different routes for biomass utilization. One of the promising routes is gasification, in which biomass is thermally converted to synthesis gas (syngas – mixture of hydrogen and carbon monoxide). The produced syngas can be catalytically converted to liquid products (after cleaning and conditioning), such as alkanes and alcohols through the Fischer–Tropsch process.

One of the main technological limitations of biomass gasification processes is the formation of tars (a mixture of high molecular weight polyaromatic hydrocarbons). The presence of tars in the product gas significantly complicates

Table 6.6 Effect of chain length on alkane oxidation over a single layer of Pt–10% Rh gauze.

	Alkane			
	C_2H_6	C_3H_8	C_4H_{10}	iso-C_4H_{10}
Conversion (%)				
Alkane	34	16	10	3
Oxygen	100	99	90	70
Carbon selectivity (%)				
Olefins	62	58	36	<1
Oxygenates	7	9	40	<0.2
Alkanes	13	10	<1	Trace
$CO + CO_2$	18	23	24	99
Temperature (°C)				
Surface	900	850	800	800
Gas	580	504	400	395

The conditions were as follows: superficial velocity, 25 cm/s; pressure, 1.4 atm; C/O = 7; and inlet gas temperature, 25 °C.
Source: Goetsch and Schmidt 1996 [48]. Reproduced with permission of The American Association for the Advancement of Science.

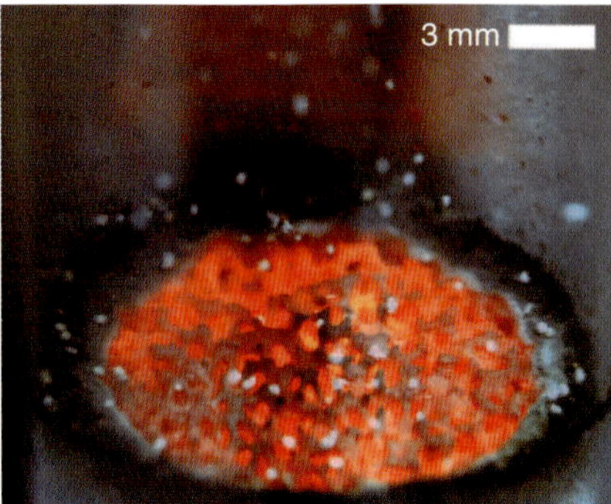

Figure 6.35 The face (0 mm) of the catalytic foam reactor during millisecond autothermal reforming. Source: Dauenhauer et al. 2007 [50]. Reproduced with permission of John Wiley and Sons.

downstream processing causing blockage of particle filters and of fuel lines. Furthermore, if a catalytic conversion process follows the gasification process, tars can deposit on the surface of the catalyst causing deactivation. The formation of tars is favored at relatively low temperatures and long processing times.

Millisecond catalytic reforming (catalytic partial oxidation) of solid biomass is a method that helps overcome the aforementioned problem [49–51]. According to this method, pretreated biomass particles are brought in very short contact (milliseconds) with a glowing hot catalytic surface (Figure 6.35) at 700–800 °C to rapidly heat up and pyrolyze into VOCs without tar formation [50]. Overall, at least three conventional biomass process steps (gasification, tar removal, and water–gas shift reaction) are integrated into a single autothermal catalytic reforming reactor, which is able to produce high-quality syngas at very short residence times (~30 ms).

In this process, short residence times result in better utilization of biomass for synthetic fuels. Specifically, biomass particles (cellulose) are impinged upon a hot catalytic bed in the presence of steam flow, O_2, and N_2. By using a ceramic foam monolith as the catalyst in a quartz tube, the total surface area of the catalyst is increased. Consequently, biomass particles undergo fast endothermic volatilization upon impact with the catalyst, resulting in the production of gases and VOCs. The volatile species flow into the catalyst (continuous flow of particles) and produce combustion products (CO_2, H_2O, and heat) by highly exothermic partial oxidation under oxygen-lean conditions, which leads to rapid increase in temperature. Finally, the thermal energy that is produced by the exothermic reaction in the oxidation zone is conducted upstream to the volatilization zone and downstream to the reforming zone. The VOCs that remain after partial oxidation undergo rapid steam reforming (SR: $VOC + H_2O \rightarrow H_2 + CO$), water gas shift (WGS: $H_2O + CO \rightarrow H_2 + CO_2$), and cracking reactions (Figure 6.36).

The advantages of millisecond catalytic reforming for production of clean syngas include lower residence times compared to conventional methods and a reactor effluent that is essentially free of tars and organics. Finally, the process can be implemented without the demand of external heating and is easily scalable and relatively simple.

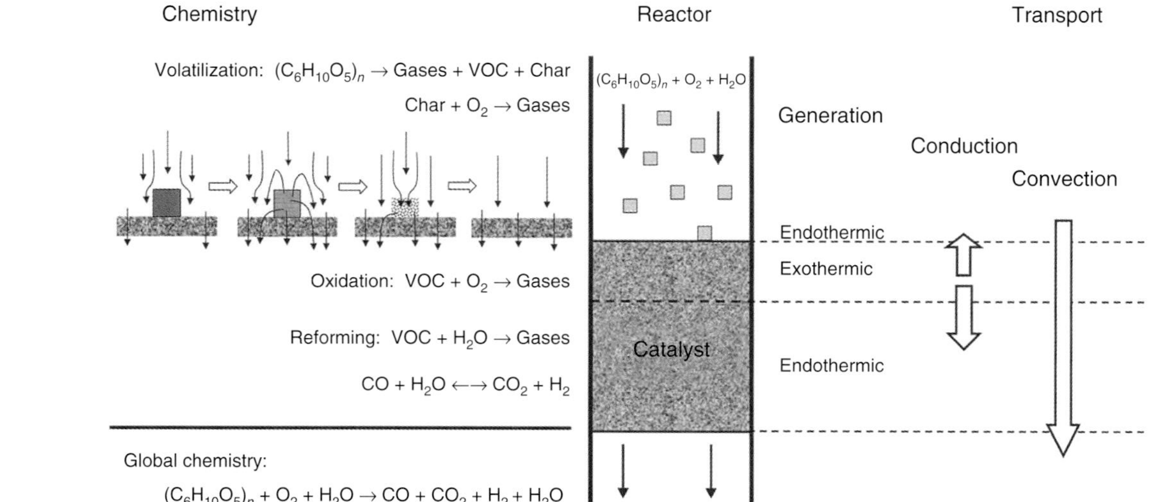

Figure 6.36 The chemically and thermally integrated steps of the millisecond autothermal reforming process. Source: Colby et al. 2008 [49]. Reproduced with permission of Royal Society of Chemistry.

References

1 Ni, X., Mackley, M.R., Harvey, A.P. et al. (2003). Mixing through oscillations and pulsations—a guide to achieving process enhancements in the chemical and process industries. *Chem. Eng. Res. Des.* 81 (3): 373–383.

2 McDonough, J.R., Phan, A.N., and Harvey, A.P. (2015). Rapid process development using oscillatory baffled mesoreactors–A state-of-the-art review. *Chem. Eng. J.* 265: 110–121.

3 Mazubert, A., Aubin, J., Elgue, S., and Poux, M. (2014). Intensification of waste cooking oil transformation by transesterification and esterification reactions in oscillatory baffled and microstructured reactors for biodiesel production. *Green Process. Synth.* 3 (6): 419–429.

4 Phan, A.N., Harvey, A.P., and Eze, V. (2012). Rapid production of biodiesel in mesoscale oscillatory baffled reactors. *Chem. Eng. Technol.* 35 (7): 1214–1220.

5 Lobry, E., Lasuye, T., Gourdon, C., and Xuereb, C. (2015). Liquid–liquid dispersion in a continuous oscillatory baffled reactor–Application to suspension polymerization. *Chem. Eng. J.* 259: 505–518.

6 Eze, V.C., Phan, A.N., Pirez, C. et al. (2013). Heterogeneous catalysis in an oscillatory baffled flow reactor. *Catal. Sci. Technol.* 3 (9): 2373–2379.

7 Castro, F., Ferreira, A., Rocha, F. et al. (2013). Continuous-flow precipitation of hydroxyapatite at 37 C in a meso oscillatory flow reactor. *Ind. Eng. Chem. Res.* 52 (29): 9816–9821.

8 McGlone, T., Briggs, N.E., Clark, C.A. et al. (2015). Oscillatory flow reactors (OFRs) for continuous manufacturing and crystallization. *Org. Process Res. Dev.* 19 (9): 1186–1202.

9 Ni, X.W., Fitch, A., and Webster, P. (2003). From maximum to most efficient production using a continuous oscillatory baffled reactor. *PIN News-The Newsletter of Process Intensification Network* 9: 12–15.

10 Ni, X.-W. (2008). NiTech – continuous oscillatory baffled reactor™. https://connect.innovateuk.org/documents/3255552/3747706/12+-+NiTech+-+Continuous+Oscillatory+Baffled+Reactor%E2%84%A2.pdf/ad122d0a-184d-4531-b16d-e0209ab51c71 (accessed 20 February 2019).

11 Kolios, G., Frauhammer, J., and Eigenberger, G. (2000). Autothermal fixed-bed reactor concepts. *Chem. Eng. Sci.* 55 (24): 5945–5967.

12 Matros, Y.S. and Bunimovich, G.A. (1996). Reverse-flow operation in fixed bed catalytic reactors. *Catal. Rev.* 38 (1): 1–68.

13 Gosiewski, K., Pawlaczyk, A., and Jaschik, M. (2015). Energy recovery from ventilation air methane via reverse-flow reactors. *Energy* 92: 13–23.

14 Glöckler, B., Dieter, H., Eigenberger, G., and Nieken, U. (2007). Efficient reheating of a reverse-flow reformer—An experimental study. *Chem. Eng. Sci.* 62 (18): 5638–5643.

15 Kolios, G. and Eigenberger, G. (1999). Styrene synthesis in a reverse-flow reactor. *Chem. Eng. Sci.* 54 (13–14): 2637–2646.

16 Kaisare, N.S. and Vlachos, D.G. (2007). Extending the region of stable homogeneous micro-combustion through forced unsteady operation. *Proc. Combust. Inst.* 31 (2): 3293–3300.

17 Han, L. and Bollas, G.M. (2016). Chemical-looping combustion in a reverse-flow fixed bed reactor. *Energy* 102: 669–681.

18 Gancarczyk, A., Janecki, D., Bartelmus, G. and Burghardt, A. (2014). Analysis of the hydrodynamics of a periodically operated trickle-bed reactor—A shock wave velocity. *Chem. Eng. Res. Des.* 92 (11): 2609–2617.

19 Lange, R., Schubert, M., Dietrich, W., and Grünewald, M. (2004). Unsteady-state operation of trickle-bed reactors. *Chem. Eng. Sci.* 59 (22): 5355–5361.

20 Aydin, B., Cassanello, M.C., and Larachi, F. (2008). Influence of temperature on fast-mode cyclic operation hydrodynamics in trickle-bed reactors. *Chem. Eng. Sci.* 63 (1): 141–152.

21 Hamidipour, M., Chen, J., and Larachi, F. (2013). CFD study and experimental validation of trickle bed hydrodynamics under gas, liquid and gas/liquid alternating cyclic operations. *Chem. Eng. Sci.* 89: 158–170.

22 Pătruţ, C., Bîldea, C.S., Liţă, I., and Kiss, A.A. (2014). Cyclic distillation–Design, control and applications. *Sep. Purif. Technol.* 125: 326–336.

23 Bildea, C.S., Pătruţ, C., Jorgensen, S.B. et al. (2016). Cyclic distillation technology – a mini-review. *J. Chem. Technol. Biotechnol.* 91: 1215–1223.

24 Maleta, V.N., Kiss, A.A., Taran, V.M., and Maleta, B.V. (2011). Understanding process intensification in cyclic distillation systems. *Chem. Eng. Process.* 50 (7): 655–664.

25 Schrodt, V.N., Sommerfeld, J.T., Martin, O.R. et al. (1967). Plant-scale study of controlled cyclic distillation. *Chem. Eng. Sci.* 22 (5): 759–767.

26 MaletaCD (2012). Innovation energy-saving technology of mass transfer in the tray columns. http://www.maletacd.com/index.php/portfolio/photo/distillation-system-turnkey.

27 Meng, X., de Jong, W., and Kudra, T. (2016). A state-of-the-art review of pulse combustion: Principles, modeling, applications and R&D issues. *Renewable Sustainable Energy Rev.* 55: 73–114.

28 Entezam, B., Van Moorhem, W., and Majdalani, J. (1997). Modeling of a Rijke-tube pulse combustor using computational fluid dynamics. In: *33rd Joint Propulsion Conference and Exhibit*, 2718. American Institute of Aeronautics and Astronautics.

29 Kudra, T. and Mujumdar, A.S. (2010). *Pulse Combustion Drying. Advanced Drying Technologies*, 211–237. Boca Raton, FL: CRC press.

30 Zbicinski, I. (2002). Equipment, technology, perspectives and modeling of pulse combustion drying. *Chem. Eng. J.* 86 (1): 33–46.

31 Wu, Z. (2007). Mathematical modeling of pulse combustion and its applications to innovative thermal drying techniques. Doctoral thesis. National University of Singapore, Singapore.

32 Li, P., Mi, J., Dally, B.B. et al. (2011). Progress and recent trend in MILD combustion. *Sci. China Technol. Sci.* 54 (2): 255–269.

33 Putnam, A.A., Belles, F.E., and Kentfield, J.A.C. (1986). Pulse combustion. *Prog. Energy Combust. Sci.* 12 (1): 43–79.

34 Kudra, T. (2008). Pulse-combustion drying: status and potentials. *Drying Technol.* 26 (12): 1409–1420.

35 Grande, C.A. (2012). Advances in pressure swing adsorption for gas separation. *ISRN Chemical Engineering* 982934: 13 pages.

36 Ilić, B. and Wettstein, S.G. (2017). A review of adsorbate and temperature-induced zeolite framework flexibility. *Microporous Mesoporous Mater.* 239: 221–234.

37 Augelletti, R., Conti, M., and Annesini, M.C. (2017). Pressure swing adsorption for biogas upgrading. A new process configuration for the separation of biomethane and carbon dioxide. *J. Cleaner Prod.* 140: 1390–1398.

38 Riboldi, L. and Bolland, O. (2015). Evaluating pressure swing adsorption as a CO_2 separation technique in coal-fired power plants. *Int. J. Greenhouse Gas Control* 39: 1–16.

39 Voss, C. (2005). Applications of pressure swing adsorption technology. *Adsorption* 11 (1): 527–529.

40 Yongsunthon, I. and Alpay, E. (1999). Design of periodic adsorptive reactors for the optimal integration of reaction, separation and heat exchange. *Chem. Eng. Sci.* 54 (13–14): 2647–2657.

41 Grünewald, M. and Agar, D.W. (2004). Intensification of regenerative heat exchange in chemical reactors using desorptive cooling. *Ind. Eng. Chem. Res.* 43 (16): 4773–4779.

42 Richrath, M., Lohse, S., Grünewald, M., and Agar, D.W. (2005). Particle-scale heat removal in fixed-bed catalytic reactors: modelling and optimisation of a desorptive cooling process. *Comput. Aided Chem. Eng.* 20: 673–678.

43 Zwijnenburg, A., Stankiewicz, A., and Moulijn, J.A. (1998). Dynamic operation of chemical reactors: friend or foe? *Chem. Eng. Prog.* 94 (11): 39–47.

44 Lund, M.W. and Seagrave, R.C. (1971). Variable volume operation of a stirred tank reactor. *Ind. Eng. Chem. Fundam.* 10 (3): 494–504.

45 Lund, M.M. and Seagrave, R.C. (1971). Optimal operation of a variable-volume stirred tank reactor. *AIChE J.* 17 (1): 30–37.

46 Ridlehoover, G.A. and Seagrave, R.C. (1973). Optimization of Van de Vusse reaction kinetics using semibatch reactor operation. *Ind. Eng. Chem. Fundam.* 12 (4): 444–447.

47 Veser, G. (2008). Short contact-time reactors. In: *Handbook of Heterogeneous Catalysis* (ed. G. Ertl, H. Knözinger, F. Schüth and J. Weitkamp). Wiley-VCH, 10:10.5:2174–2188.

48 Goetsch, D.A. and Schmidt, L.D. (1996). Microsecond catalytic partial oxidation of alkanes. *Science* 271 (5255): 1560.

49 Colby, J.L., Dauenhauer, P.J., and Schmidt, L.D. (2008). Millisecond autothermal steam reforming of cellulose for synthetic biofuels by reactive flash volatilization. *Green Chem.* 10 (7): 773–783.

50 Dauenhauer, P.J., Dreyer, B.J., Degenstein, N.J., and Schmidt, L.D. (2007). Millisecond reforming of solid biomass for sustainable fuels. *Angewandte Chemie-International Edition* 46 (31): 5864–5867.

51 Salge, J.R., Dreyer, B.J., Dauenhauer, P.J., and Schmidt, L.D. (2006). Renewable hydrogen from nonvolatile fuels by reactive flash volatilization. *Science* 314 (5800): 801–804.

Part III

Fundamentals in Practice – Designing a Sustainable, Intensified Process

7

Process Intensification and Sustainable Processing

7.1 Introduction

7.1.1 Sustainable Earth?

Chemical industry has brought enormous wealth and prosperity to the society and everyday life. Chemical products are omnipresent throughout our entire lives. Babies use disposable diapers, kids play with toys, sick people need pharmaceuticals, men and women use cosmetics, and all need food and clothing. All these elements of our daily lives, along with houses, cars, high-tech equipment, etc., are based on chemical products. In addition, consumption of chemical products is still increasing: new types of chemical and biochemical products are brought to the market every day and new markets open up in different parts of the world for already existing products. The underlying causes are related to the rapid growth in the world population, growth in consumer wealth, and growth in consumer needs.

For now, all chemical production takes place on the Earth. If we consider the Earth to be a giant chemical reactor, then input and output of this reactor is marginally small. Solar energy is obviously essential to natural life in terms of both heat and photosynthesis, but in the chemical and materials industry, this energy source is (up to now) still scarcely used. Output from this giant reactor to the surrounding space is negligible, except for some space crafts and electromagnetic wave pollution. Most of the organic chemicals used in this reactor are based on fossil fuels: oil, coal, and gas. These fossil fuels took more than 100 million years to form. However, we are rapidly consuming them now, with consumption drastically increasing since the industrial revolution, which occurred less than 200 years ago. From a mass balance perspective, this means that we are in a highly unsustainable situation: the consumption rate is higher than the generation rate by a factor of c. 500 000.

The energy and material inefficiency of our current manufacturing model is clearly illustrated in various reports. To pick only two, the US Energy Information Administration predicts that the decline of oil production after we have reached the peak oil moment is going to be massively quick because of the high oil consumption and the low availability at that moment [1]. With regard to our material consumption, the United Nations Environment Program has reported that improvements in material productivity are substantially smaller

The Fundamentals of Process Intensification, First Edition.
Andrzej Stankiewicz, Tom Van Gerven, and Georgios Stefanidis.
© 2019 Wiley-VCH Verlag GmbH & Co. KGaA. Published 2019 by Wiley-VCH Verlag GmbH & Co. KGaA.

than those in labor and energy productivity and that the absolute global material productivity has even declined since 2000 [2].

To solve the problem of resource availability problem, there are three conceptual solutions. The first one is to start exploitation of extraterrestrial resources, which is basically trying to increase the input component of our giant Earth reactor. As this option is for now still very much science fiction, we will not consider it further here. The second option is to develop technically and economically feasible processes based on renewable feedstock, the so-called biomass-based processes. Although this is a very reasonable option, it is clear that there are also many challenges associated with this route. There are ethical questions such as whether we will have enough arable land to feed mankind, provide energy, supply raw materials simultaneously without skyrocketing prices. Another ethical question deals with the genetic manipulations that are applied for increasing the yield of biomass plantations. Furthermore, reconversion of agricultural sector in such a biomass-based economy requires huge investments as well as sufficient time. Finally, renewable resources are not the raw materials for inorganic chemical products such as metals and minerals. Therefore, the third option should not be forgotten. That is, the option to develop innovative methods and technologies that drastically increase the efficiency of chemical and biochemical processes. The term "drastic" is somewhat abstract, but policy makers have suggested that we should be thinking in the order of a factor of 4 (Ernst von Weizsäcker, founder of the Wuppertal Institute for Climate, Environment & Energy), factor of 10 (Friedrich Schmidt-Bleek, from the same Institute, in his dematerialization studies in the early 1990s), or even a factor of 20 (Jan Venselaar, chairman of the European Federation of Chemical Engineering [EFCE] Working Party on Environmental Protection) [3–5]. It is this option where process intensification can play its role. Most probably, the solution to the resource problem will have to come from a combination of the biomass-based economy – where appropriate – and processing efficiency.

7.1.2 Sustainable Processing and the Position of PI

Because of the above-described process inefficiencies and resource scarcity, the concept of sustainability has also spread in the domain of industrial processing and that of process engineering in particular [6–8]. A wide range of terms have been used (e.g. sustainable engineering, green engineering, green manufacturing, and cleaner production), and it is often difficult to rise above the abstract definition of, for example, "A sustainable product or process is one that constrains resource consumption and waste generation to an acceptable level, making a positive contribution to the satisfaction of human needs, and provides enduring economic value to the business enterprise" [7].

Within the domain of sustainable chemical manufacturing, three approaches can be distinguished. Firstly, there is the field of *green chemistry* where chemists are studying green synthetic routes, green solvents, atom economy, biocatalysis, etc. Secondly, engineers come in the picture and worry about *green process technology* including waste and energy minimization in the unit processes, optimization of process schemes, development of end-of-pipe technologies, etc.

Process intensification can also be found in this class. Finally, there is the level of *green procedures and operations*, with focus on environmental management systems (e.g. ISO 14001, EU Eco-Management and Audit Scheme [EMAS]), environmentally conscious plant maintenance, safe transport and storage, etc.

Within the green process technology area, Figure 7.1 gives a view on where process intensification is positioned from a variety of perspectives. Process intensification is business-driven. Most major companies have process intensification (PI) in their research and development portfolio, with clear impact on the economic viability of a new process and product. PI is also applied in the process itself and not as an end-of-pipe technology (although these techniques may also benefit from process intensification principles). This also explains why it is not dedicated toward one single environmental compartment (atmosphere, water, soil, and waste) but has potential impact on all of them. Finally, as its name implies, process intensification focuses on the process, not on the whole lifecycle or system, neither on the single pollutant nor on the chemical of concern.

It is interesting to recall the 12 principles of green engineering [10] and compare them with the fundamentals of process intensification. The 12 principles of green engineering were articulated as an answer to the 12 principles of green chemistry [11]. More recently, these principles were abbreviated into sets with the acronyms PRODUCTIVELY for the green chemistry principles [12] and IMPROVEMENTS for the green technology principles [13]. Table 7.1 shows that process intensification clearly addresses many of the 12 principles of green engineering. The relations between process intensification and sustainability (illustrated by the "T" and "S" in IMPROVEMENTS) on the one hand and safety (the "I" in IMPROVEMENTS) on the other hand are more complex. They are further examined in the following sections.

7.1.3 Sustainability Assessment Tools Applied to Process Intensification

Process intensification is often advocated as one of the routes for the chemical process industry to become more sustainable. However, this statement is rarely supported by a quantitative and reliable proof. One reason is that often process intensification technologies are still immature and not yet fully developed industrially to allow for ecological assessment. Another reason is that ecological assessment tools themselves are not yet fully developed, let alone for application to an emerging field as process intensification. The fact that process intensification technologies cannot necessarily be classified in the conventional unit operation concepts does not make it easier to define the boundaries of the system to assess.

Literature concerning the sustainability evaluation of process intensification is very limited. A large number of scientific articles and review reports mention the advantages of process intensification in a number of domains, but in most cases, all remain qualitative in nature. The 10 most mentioned advantages of process intensification are decreased energy consumption, waste generation, required residence time in the process, size of the equipment, increased mass efficiency, financial cost, safety, flexibility, simplicity of the process, and the quality of the end product (Table 7.2). Disadvantages of process intensification are less easy to

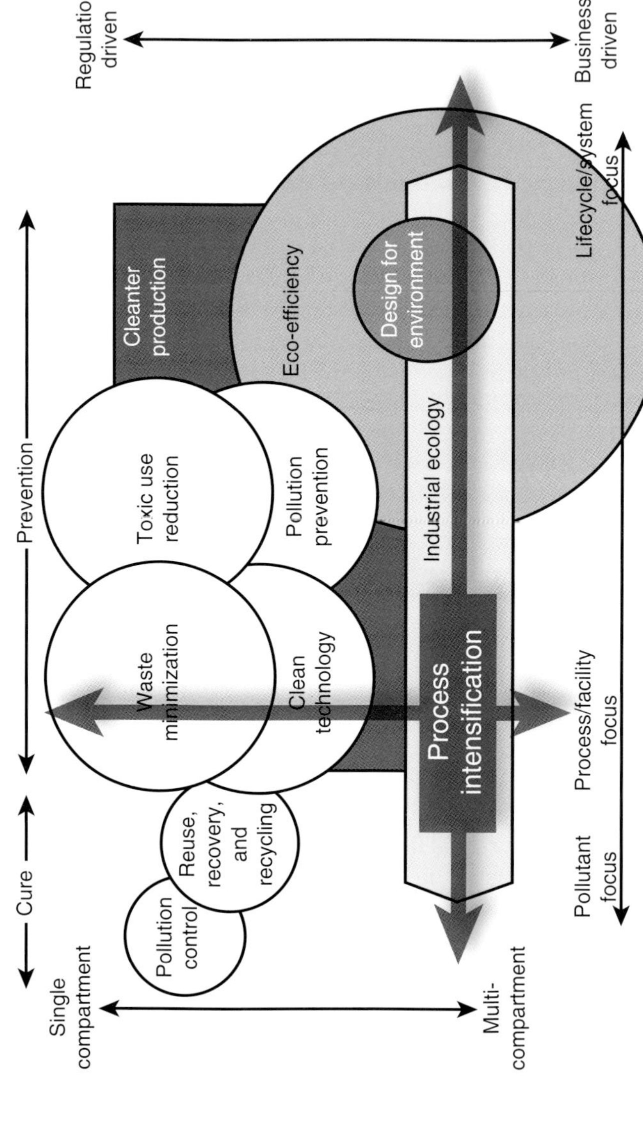

Figure 7.1 Position of process intensification in the field of green technology. Source: Adapted from van Berkel, 2006 [9].

Table 7.1 The principles of green engineering and their relation with process intensification.

Principles of green engineering		Process intensification
I	Inherently nonhazardous and safe	(see Section 7.3)
M	Minimize material diversity	… advocates the principle of "giving each molecule the same processing history"
P	Prevention instead of treatment	… focuses on in-process technologies rather than end-of-pipe technology
R	Renewable material and energy inputs	
O	Output-pulled design	
V	Very simple	
E	Efficient use of mass, energy, space, and time	… applies novel approaches in the spatial (*structure*), thermodynamic (*energy*), and temporal (*time*) domain
M	Meet the need	
E	Easy to separate by design	… applies novel approaches in the multifunctional (*synergy*) domain
N	Networks for exchange of local mass and energy	… multifunctional and hybrid technologies
T	Test the life cycle of the design	(see Section 7.2)
S	Sustainability throughout product life cycle	(see Section 7.2)

find in the literature. An example of a disadvantage discussed in the literature is more difficult control of some intensified systems leading to their reduced safety [26]. Another limitation regularly discussed is the propensity of microreactors for fouling and subsequent clogging [22, 27], which reduces their flexibility. In the field of reactive separations, the loss of degrees of freedom because of the integration of unit operations is reported [28], leading to a more complex process design and control and therefore affecting the simplicity and safety. Ten criteria reported in the literature and listed in Table 7.2 can be seen as an exhaustive set of criteria to assess the sustainability of intensified technologies.

Increased **energy efficiency** is often reported to be a prime consequence of process intensification because of an improvement in the transport phenomena and/or synergistic effects (integration of unit operations requiring similar energy forms). Reduced energy consumption increases the sustainability of a process because of lower energy costs and lower indirect CO_2 emissions (depending on the applied energy mix). Increased **mass efficiency** is related to improved reaction selectivity as well as reduced waste generation and results in lower operational costs [19]. The **waste generation** component should be understood in its broadest sense by incorporating solid waste as well as liquid and gaseous emissions. Waste reduction can occur in two ways: decrease of the amount of waste or reduction of the toxicity of the waste. This parameter is related to the energy efficiency through its indirect CO_2 emission. It is also related to the

Table 7.2 Criteria for the assessment of intensified technologies, as given in a variety of reports.

	Energy efficiency	Mass efficiency	Waste generation	Safety	Cost	Product quality	Flexibility	Simplicity	Residence time	Size
Stankiewicz and Moulijn (2000) [14]	X		X		X				X	X
Bakker (2003) [15]	X		X		X		X	X		X
Richardson et al. (2003) [16]	X		X	X		X	X		X	X
Harmsen et al. (2003) [17]	X	X	X	X	X					X
Charpentier (2007) [18]		X	X	X	X			X		X
Criscuoli and Drioli (2007) [19]	X	X	X	X					X	X
European Roadmap for PI (2008) [20]	X	X	X	X	X				X	X
Platform Keten Efficientie (2008) [21]	X		X	X	X	X	X	X		X
Reay et al. (2008) [22]	X		X	X	X	X	X			X
BHR Group (2010) [23]	X		X	X	X	X	X			X
Lutze et al. (2010) [24]	X	X	X	X	X		X	X	X	X
PIN (2010) [25]	X		X	X	X	X			X	X

mass efficiency via the reaction conversion and selectivity, which reduces the formation of by-products. Increased energy efficiency and mass efficiency as well as waste reduction also have an impact on the costs of a process.

Safety is another important point of attention in chemical processing. It is related to many other sustainability criteria such as miniaturization (size), residence time, and simplicity. Intensified technologies can both make the process safer (e.g. reduced holdup and less piping through integration of unit operations) or less safe (e.g. more difficult process control in integrated processes). Because of its importance and also its specificity, it will be further dealt with in one of the following sections. The financial **cost** is obviously a very important parameter to compare chemical processes. Both investment and operational costs need to be taken into account. Investment costs are related to equipment size and operational costs are related to energy and mass efficiency. A cost that should also be taken into account is that for the development of a process [23]. In an intensified process, this can either be reduced (e.g. multiplication of modular units) or augmented (e.g. multifunctional reactors can require more development time for complex process control models), compared to a conventional process.

The **quality** of the end product is not always mentioned in the assessment of process intensification. However, this criterion is often essential for the manufacturer. Quality also includes consistency of the product properties over time. A reduction of **residence time** of the reactants in the process also has some advantages in terms of sustainability, such as reduced holdup of potentially hazardous chemicals and the opportunity of just-in-time production, leading to less unused products and to reduced storage of products, thus decreasing the safety implications [22, 25]. One of the most mentioned properties of an intensified process is the reduced **size of the installation**. Reductions in volume with a factor of 100 or more are no exception [14]. Miniaturization is beneficial for sustainability from a number of perspectives. A smaller plant requires less material and small land area. Smaller reaction volumes are also easier to control, leading to easier safety measures. Finally, miniaturization also allows for distributed manufacturing: local production in small (mobile) plants, possibly leading to reduced large-scale transport. The **flexibility** of a process consists of the ability of a process to deal with changes. This is important as chemical industry evolves toward more variability in process conditions, process feed, and product demand. Flexibility therefore includes modularity as this enables easy changes in product properties and volumes. The **simplicity** of a process and its flow sheet relates to the number of process steps and auxiliary devices. Process intensification often shows here an ambiguous picture: integration of unit operations can reduce their number, but it can also increase the level of operational complexity [15, 18].

From the previous paragraph, it is obvious that many of the criteria for process intensification assessment are interrelated. Figure 7.2 illustrates this web of dependency. Cost, safety, flexibility, and product quality are four criteria that are affected by others but do not affect themselves other criteria. This leads authors to classify the criteria on different levels. Hessel et al. [29] distinguished four levels in which the PI criteria can be positioned: reaction/reactor, plant/process, company/economy, and society/environmental. Lutze et al. [24] proposed almost

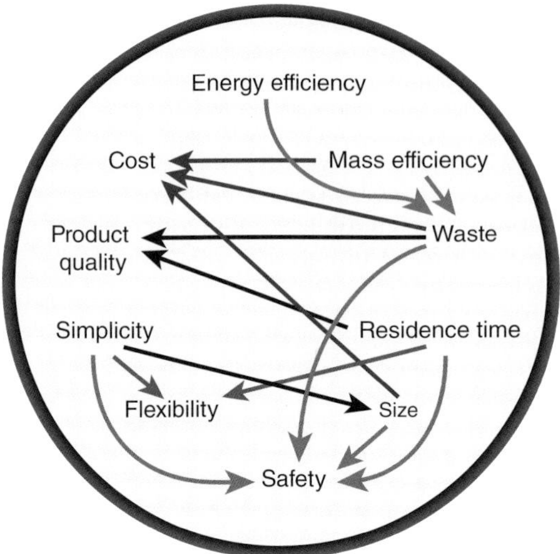

Figure 7.2 Interdependency of the 10 PI criteria.

the same criteria as listed in Table 7.2 (except for the quality criterion) and they classify them into four classes: environmental (the waste, efficiency, and energy criteria), safety (the safety criterion), economic (capital costs and operational costs, Table 7.2, integrated into one cost criterion), and intrinsic intensified (residence time, flexibility, size, and complexity). Figure 7.3 shows another classification, similar to the one of Lutze but now in conformity with the three-dimensional sustainability metrics system covering economic, social, and ecological aspects. An additional class has been added in order to cover the technical criteria.

Now that a classification of PI criteria is in place, one has to look if and how these criteria can be assessed. The most efficient way is to apply the existing assessment tools to do the job. A large number of assessment tools for sustainability evaluation have been developed (e.g. see overview papers of Carvalho et al. [30], Angelakoglou and Gaidajis [31], and Kralisch et al. [32]). These mostly cover one or more of the three dimensions of sustainability (economic,

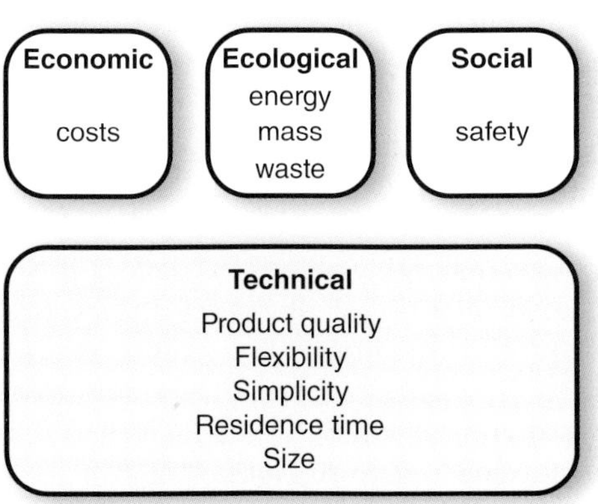

Figure 7.3 Structured classification of PI criteria.

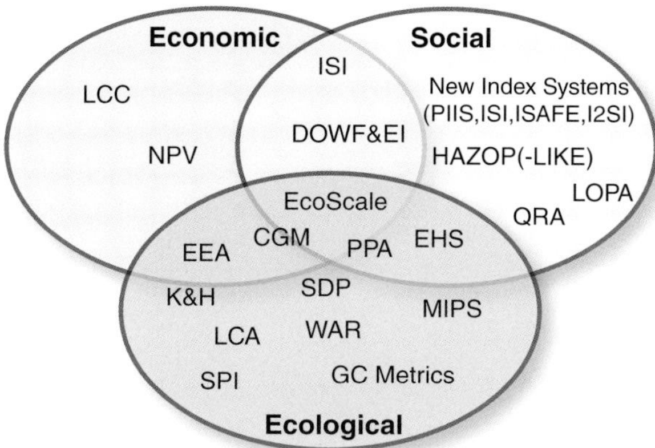

Figure 7.4 Existing tools for sustainability assessment. LCA, life-cycle assessment [33, 34]; WAR, Waste Reduction algorithm [35, 36]; SPI, sustainable process index [37]; GC Metrics, green chemistry metrics [38, 39]; MIPS, material input per unit service [40]; LCC, life-cycle costing [41]; NPV, net present value [42]; EEA, eco-efficiency analysis [43]; K&H, Kheawhom & Hirao [44, 45]; EHS, Environment, Health, and Safety [46]; SDP, sustainable development progress [47]; CGM, Carvalho & Gani & Matos [48], EcoScale [49]; PPA, process profile analysis [50]; HAZOP, hazards and operationability analysis [51]; HAZOP-LIKE, HAZOP analysis dedicated to microstructured devices [52]; DOW F&EI, DOW fire and explosion index [53]; LOPA, layer of protection analysis [54]; PIIS, prototype index of inherent safety [55]; ISI, inherent safety index [56, 57], iSafe [58, 59]; I2SI, integrated inherent safety index [60]; QRA, quantitative risk analysis [61].

ecological, and social). Figure 7.4 shows a selection of these, classified along the sustainable dimensions they cover.

When these existing assessment tools are evaluated on their capacity to be applied to the 10 process intensification criteria, it becomes clear that there is no unique assessment tool available for a complete assessment (Table 7.3). Obviously, these tools are adequate for the aim they were designed for, i.e. the assessment on one, two, or three dimensions of sustainability. However, they are much less suitable for the evaluation on the technical criteria (quality, flexibility, simplicity, residence time, and size). Table 7.3 shows that the eco-efficiency analysis (EEA), Kheawhom & Hirao (K&H), and Carvalho & Gani & Matos (CGM) tools are the most comprehensive at this moment. Combination of any of these tools with the tool(s) to evaluate the technical aspects of a process seems to be the optimal solution.

A word is devoted to the technical criteria for PI evaluation. Quality requirements differ from product to product and it is useless to settle on a general quantitative assessment method on this criterion. However, it should be noted that often a certain purity, which can be lower than 100%, is set as a fixed threshold for a process, in particular for the end products. In that case, it might not be worthwhile to intensify processes to achieve higher purities, but the intensification may allow alleviating process conditions (temperature, residence time, etc.) while still achieving the threshold purity. Obviously, for processes that yield products requiring subsequent purification steps, a higher purity than the one conventionally obtained is advantageous in terms of the full process sustainability score.

Table 7.3 Applicability of existing sustainability assessment tools to the PI criteria.

Dimension	Tool	Ecological			Social	Economic	Technical				
		Energy	Mass	Waste	Safety	Cost	Product quality	Flexibility	Simplicity	Residence time	Size
Ecological	LCA	xx	xx	xx							
	WAR	x	x	xx							
	SPI	xx	xx	xx							xx
	GC Metrics	xx	xx	x							
	MIPS	x	xx								
Social	Safety (many)				xx						
Economic	LCC					xx					
	NPV					xx					
Social–ecological	EHS			xx	xx						
Economic–ecological	EEA	xx	xx	xx	x	xx					xx
	K&H	xx	xx	xx		xx		xx			xx
Economic–social–ecological	SDP	xx	xx	xx	x	x					xx
	CGM	xx	xx	xx	xx	x					xx
	EcoScale	x	x		xx	x	x		xx	x	
	PPA		x	xx	xx	x			xx	xx	

xx, Very suitable; x, suitable.

Simplicity is hard to quantify accurately. In most cases, a simple and rather inaccurate quantification method is applied by assessing the type and number of unit operations. In many reports on intensified technology, this approach is used as well [15, 18, 24].

Flexibility is another criterion that is not easy to quantify. Swaney and Grossman [62] developed the flexibility index, later on adapted by Lai and Hui [63]. In this index, both the variation induced by changing process conditions and the technical limitations of the process are taken into account. Calculation of the index, however, requires extensive knowledge of the process space, which is difficult to achieve when in the process design stage. Criscuoli and Drioli [19] quantified flexibility for intensified membrane processes by defining two values. One considers the extent to which a process can handle changes in process conditions without the need to adapt the equipment, whereas a second metric quantifies the number of processes (e.g. liquid–liquid and gas–vacuum) that the device can handle.

Regarding the energy efficiency assessment of process intensification, exergy analysis may be of value [64]. Exergy is defined as the maximum theoretical useful work obtained if a system is brought into thermodynamic equilibrium with the environment by means of processes in which the system interacts only with that environment. Exergy analysis allows to quantify thermodynamic imperfections of the process and can thus help in understanding where the weak points of the process are. Although a promising approach, exergy analyses are scarce in process intensification studies and in chemical engineering literature at large. Good overviews of exergy analysis of chemical processes are reported by Luis and Van der Bruggen [64, 65]. The same authors have applied exergy analysis to optimize a membrane crystallization process [66]. Because of this scarcity in PI studies, the concept will not be further discussed in this chapter. However, it is believed that exergy analyses can form a complementary assessment tool to what will be discussed further on.

Finally, some comments regarding the safety and cost assessment of process intensification are presented. Although the CGM and, to a lesser extent, EEA tools incorporate safety analysis in their assessment procedure, a dedicated safety assessment is recommended to be performed as well. Safety assessment is discussed in a separate section further on in this chapter. For the assessment of cost, chemical industry typically applies less complex cost calculations than, for example, lifecycle costing, for which a huge amount of data input is required – data that are often not yet available at the stage of process design. When the investment cost of an installation needs to be translated to the required capacity for a certain process, a popular rule of thumb is the so-called six-tenth rule [67]: the ratio of costs (cost of the required capacity over the cost of the known capacity) is equal to the ratio of capacities to the power of 0.6. However, it can be questioned whether this six-tenth rule still applies for intensified technologies as well. Furthermore, the investment cost is only a part of the total cost, which also consists of installation costs and costs for auxiliaries such as piping and control equipment. This difference is accounted for in chemical industry by another rule of thumb in the form of the Lang factor (typically between 4 and 6), which is to be multiplied by the purchase cost of the installation to end up with the total investment cost. Again, it can be questioned whether the Lang factors also apply to intensified technologies. Hessel [68] calculated the total investment cost for a microreactor plant for Kolbe–Schmitt synthesis and ended up at 2.4 times the purchase cost of the microreactor. It needs to be confirmed in the future whether a reduction in Lang factors can be achieved with process intensification.

7.2 Ecological Assessment of Intensified Technologies

Not many studies have been carried out so far to compare the sustainability performance of intensified processes and conventional processes. Mostly, processes that have reached a certain maturity such as reactive distillation, membrane processes, and microreactors have been studied. When comparing intensified technology with the conventional technology, one should, however, be aware that there is still a large difference in their respective development path. Hence, a comparison study may lead to under- or overestimating the (future) sustainable potential of the technologies that are still in development. Immature processes have more room for improvement compared to more mature ones. As an illustration of the different stages in the learning curve between conventional and intensified technology, Figure 7.5 shows the achieved selectivities in the silver-catalyzed ethylene oxidation process, based on values found in scientific literature [69–86]. The conventional process was commercialized by Shell in 1958 and is carried out in a multitubular reactor with ethylene and oxygen as reactants and solid particles of silver-based catalyst immobilized on an inert support. The intensified process is performed in a microreactor where heat removal rate and oxygen concentration can be increased without risking a thermal runaway or an explosion. The figure shows that the increase in selectivity

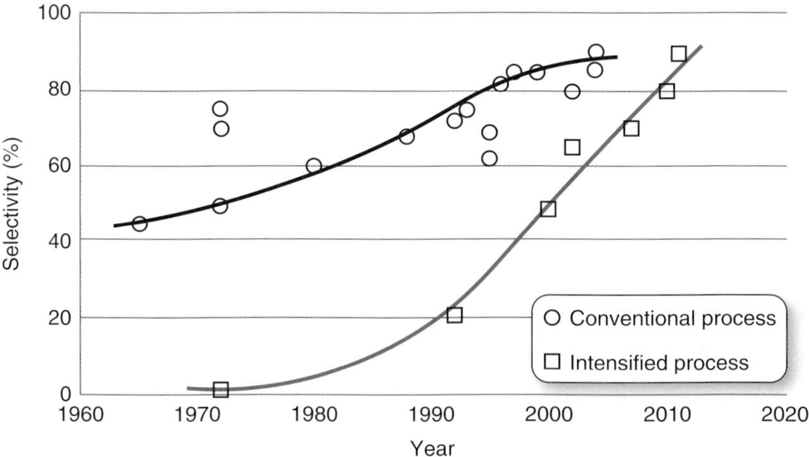

Figure 7.5 Learning curves for the development of a conventional and intensified ethylene oxidation process. Source: Data points extracted from various sources [69–86].

has occurred much quicker in the microreactor technology and that the rate of increase is not yet flattening out. The conventional technology, on the other hand, appears to have reached its optimal performance. It should be noted that the scientific reports, from which the data points in Figure 7.5 are taken, do not necessarily always aim at the highest selectivities as sometimes other improvements are studied.

7.2.1 Microreactor Engineering

A number of life-cycle assessment (LCA) studies comparing microreactor-based and conventional processes have been performed by Kralisch and Kreisel. In the first study [87], the two-step synthesis of *m*-anisaldehyde from *m*-bromoanisole was investigated both at the laboratory (functional unit of 10 kg product) and the industrial scale (1 ton product). In the first step of the reaction, *m*-bromoanisole and *n*-butyllithium are converted into *m*-lithiumanisole and *n*-bromobutane. Then, the reaction mixture is treated with dimethylformamide to produce *m*-anisaldehyde. Both reaction steps apply tetrahydrofuran as the solvent. Afterward, the reaction is quenched using 3 M hydrochloric acid.

On the laboratory scale, the conventional process was carried out in a double-walled batch reactor (the internal reaction volume is 10 l), whereas the intensified process was performed in two continuous microreactors in series (the total internal reaction volume is 2.2 ml). The system boundary spans from the raw material acquisition, through the supply of energy and gas, the performance of the reaction, work-up, and process control to the disposal of wastes. The lifetime of the microreactor was unknown and different scenarios were chosen, from 1 week through 3 weeks and 3 years, up to 10 years of lifetime (the latter equaling the average lifetime of the double-walled glass reactor). Except for the microreactor with a 1-week lifetime, the microreactor process yielded significant reductions in the different impact categories[1] compared to the conventional process, with values between 17% and 63%. The "1-week lifetime" microreactor required frequent replacement, which severely increased the impact of the steel reactor production with the associated chromium and nickel emissions. The microreactors with 3 and 10 years of lifetime almost do not differ in performance and the lower reduction value increases from 17% to 40%. The highest reductions are achieved in the photochemical ozone creation potential, the global warming potential, the abiotic resource depletion potential, the cumulative energy demand, and the nutrification potential. These are mostly related to the savings in energy consumption, the reduction of solvents, and the increase of the reaction yield. For example, a breakup of the different phases in the process showed that the cumulative energy demand was mostly affected by the supply of chemicals and the energy demand during synthesis and work-up. The contribution of the supply of the microreactor itself as well as of the

---

1 The impact categories studied include cumulative energy demand, abiotic resource depletion, global warming potential, ozone depletion potential, photochemical ozone creation potential, acidification potential, nutrification potential, human toxicity potential, fresh aquatic eco-toxicity potential, marine aquatic eco-toxicity potential, and terrestrial eco-toxicity potential.

peripheral equipment was found to be negligible. Because of the immaturity of the technology, not a lot of information is available on microreactor lifetimes. Zschieschang et al. [88, 89] have performed LCA studies on the production phase of microreactors, thus allowing an optimal reactor design and development from the ecological point of view.

On the industrial scale, the conventional process was performed in a 400 l vessel with cryogenic cooling, whereas 10 microreactors (of the same type as in the laboratory scale) in parallel were applied to intensify the process. The study was carried out leaving out the work-up (same in both cases) and the supply of peripheral equipment (shown to be negligible in the laboratory scale). Also, a lifetime of 1 year for the microstructured devices was selected. By replacing the batch installation with a microreactor setup, the impact in the different LCA categories was reduced by 3–38%. The highest reductions were achieved in the marine aquatic eco-toxicity potential and the global warming potential. This result was mostly related to the increase of the reaction temperature and the avoidance of a cryogenic system.

A subsequent study was performed on the Kolbe–Schmidt synthesis of 2,4-dihydroxy benzoic acid (β-resorcylic acid) from resorcinol in an aqueous solution of potassium hydrogen carbonate [90, 91]. The compared technologies were the batch synthesis with oil bath heating and the microreactor-based synthesis with different types of heating (oil and microwaves) and different solvents (ionic liquids and supercritical CO_2). Both were performed on the laboratory scale and the functional unit was 1 kg of β-resorcylic acid. The microwave-heated microreactor with water as solvent yielded the lowest impact with c. 27% reduction in cumulative energy demand compared to the batch process. The higher yield obtained outweighed the higher energy consumption by the microwaves. The oil-heated microreactors with water as solvent yielded a mixed result, sometimes outperforming the batch process, but in other cases (higher temperature combined with shorter residence time) performing worse. The best-performing microreactor with water resulted in a reduction of 13% in the cumulative energy demand and 12% in the global warming potential. The application of ionic liquids as solvent resulted in very high cumulative energy demands because of the energy-intensive supply and the lack of recycling possibility of these chemicals. The use of supercritical CO_2 in this process did not significantly improve the sustainability. At high flow rates, the contribution of the synthesis reaction to the cumulative energy demand becomes negligible (0.1% of the total at a flow rate of 17.2 l/h). The impact in this case is mostly influenced by the supply of the reactants. The conclusion is that via the improvement of the space–time yield, major improvements in environmental efficiency can be achieved by process intensification. Regarding the use of microwave irradiation, further research on the same reaction showed worse results for the microwave case compared to conventional heating (e.g. [92]), which emphasizes that results are highly dependent on the selected process conditions. Furthermore, it has to be taken into account that the energy efficiency of the devices delivering the alternative energies (reported to be 16–20% [92]) can be expected to improve, thus perhaps reversing the results. Also, expanding the system boundaries – for example, including pre- and post-treatment for drying of crystals – might benefit microwave heating and yield different results in comparative studies [32]. Another alternative energy source, ultrasound, was studied in microreactors by Hübner et al. [93] for the hydrolysis of *p*-nitrophenyl acetate in a liquid–liquid process. They report an increase in yield from 11% to 86% compared to the silent flow process. Despite the addition of more energy in the sonicated process compared to the silent one, the global warming potential (GWP) decrease by up to 80% was shown.

Huebschmann also reported on the exothermic homopolymerization of styrene on the laboratory scale, either in a microreaction setup under inert conditions at room temperature or under batch conditions at $-78\,°C$ using an isopropanol/dry ice mixture [91]. The solvents ethanol and *n*-heptane were assumed to be completely recycled. The microreactor allowed for (up to threefold) higher styrene concentration than in the batch system because of its excellent heat removal capacity. At these high styrene concentrations, the microreactor became favorable in view of the cumulative energy demand. However, at styrene concentration equal to that in the batch reactor, the additional pumping energy in the microreactor increased its ecological impact to above that of the batch reactor.

In another study, Huebschmann et al. [94] investigated the LCA of two case studies on the laboratory scale. One was the esterification reaction of phenol with benzoyl chloride giving phenyl benzoate. A different phase transfer catalyst was applied in the batch and microflow systems, which might disturb a fair comparison of the reactor technology. However, results were also reported for the noncatalyzed systems. The functional unit was in all cases set at 1 kg of phenyl benzoate. The microreactor without catalyst performed worse than the batch reactor on the level of all LCA impact categories because of the direct inverse relation between the reaction yield and the environmental impact (high yield results in less process waste and more efficient use of materials and energy). When (different) ionic liquids were used as catalyst, then some microreactors were comparable to the batch system. Microreactors differed on the level of mixing systems (interdigital, Herringbone, emulsification). When comparing the different microreactors, the interdigital mixer outperformed the other from the perspective of product yield and, subsequently, ecological impact. As the energy demand of the plant periphery (mixers) had such a huge influence on the ecological performance of the microreactors, one could conclude that the batch reactor performed more ecologically. However, the authors emphasized that this advantage became less outspoken and even neutralized at the industrial scale, where pumps and controlling systems were also needed for the batch systems. In the second case study, the highly exothermic synthesis of an ionic liquid [BMIM]Cl was investigated. As a functional unit, 1 kg of the product was chosen. Molar ratio, reaction temperature, and residence time were varied. The microflow systems yielded lower ecological impact than the batch system, as long as the ratios close to equimolar were used (the supply of reactants contributes largely to the ecological impact), the residence time was minimized (generated heat can be removed quickly with immediate effects on global warming potential) and the high reaction yields were obtained (direct relation with ecological impact).

In recent years, the groups of Kralisch and Hessel reported on more examples, both in the bulk chemistry (adipic acid production) and in pharmaceutical (nitro-group hydrogenation and rufinamide synthesis) sectors [95–99]. The emerging set of data appears to steer toward three conclusions:

1. The transition from batch to flow processes, in particular applying micro- or milliscale reactors, is able to improve (under certain conditions) the "greenness" of the process.
2. A multicriteria assessment is crucial to elucidate the exact process conditions, under which intensified technologies are indeed better than the traditional processes. These complex answers also do not always lead to one single, the best solution. Simple single-value green metrics such as atom economy, *E*-factor, and carbon efficiency can describe efficiency on a reaction level, but not at process level, let alone at the scaled up and integrated industrial level.
3. Other approaches, such as chemistry intensification (e.g. solvent selection) and flow sheet integration (e.g. recycling), are equally important in the sustainability assessment of the complete process, and optimization of the balance between all approaches may in a particular process result in dropping process intensification as the route of choice, despite its benefits for specific criteria.

7.2.2 Other Intensified Processes

Besides microreactor systems, only a few process intensification technologies have been compared with conventional ones on the ecological performance. Bonet-Ruiz et al. [100] assessed alternative technologies for methyl acetate hydrolysis to methanol and acetic acid, for methyl acetate transesterification to methanol and acetates of higher value than acetic acid, and for *tert*-Amyl methyl ether (TAME) synthesis by etherification of methanol with isoamylenes. For the hydrolysis process, a flow sheet of one reactive distillation column followed by two distillation columns was compared with the industrially applied set of one reactor followed by three columns. Calculations indicated that the energy cost would be 7% higher with reactive distillation. Application of the Waste Reduction (WAR) algorithm showed no significant (c. 1%) difference in the potential environmental impact between the two alternatives. For the transesterification process, multiple flow sheets were assessed, such as extractive distillation followed by transesterification in a reactive distillation system, reactive pressure swing distillation, pervaporation followed by transesterification in a reactive distillation system, and the current industrial process, consisting of traditional hydrolysis followed by reactive distillation and pressure swing distillation. It appeared that the combination of reactive distillation with pervaporation has the lowest energy consumption and the lowest total potential environmental impact: approximately 63% and 26% lower than the industrial process, respectively. However, because of the high number of stages required for the distillation columns, the total economic benefits of the pervaporation-reactive distillation combination are still less than one-third of the benefits achieved with the currently applied process. In the TAME process, a reactive distillation column was compared to a reactor followed by a distillation column. It appeared that in the reactive distillation system, the column was shorter and recycle streams were minimized. The energy consumption was the lowest in the reactive distillation column, providing 10% energy savings in the reboiler.

In the field of membrane processes, Criscuoli and Drioli [19] suggested new metrics to evaluate the performance of membrane operations. They defined the productivity-over-size (PS) ratio, the productivity-over-weight (PW) ratio, two flexibility metrics (one considering the variations for which the technology is able to work and the other considering the number of processes that can be performed with the technology), and a modularity metric. The latter is the ratio of the membrane areas (or volumes in case of conventional technology) minus the ratio of productivities. The authors applied their proposed evaluation criteria to the production of sparkling water comparing hollow fiber membrane contactors with packed columns for the deoxygenation and subsequent carbonation of water. Calculation of the PS, PW, and both flexibility metrics resulted in higher values for the membrane contactor, implying higher productivity and flexibility. With regard to the modularity, the calculated value for the membrane operation was lower than that for the packed column, meaning that membranes are more modular. Finally, mass intensity (i.e. total mass entering the reactor over mass of product) was also calculated, and this gave a lower (i.e. better) value for membranes. This mass intensity was also used to compare the use of a Pd–Ag membrane reactor with a traditional high-temperature reactor for the water–gas shift reaction [101]. Mass intensity and also energy intensity (i.e. ratio between the total energy produced by the reaction in the reactor over the total amount of hydrogen exiting the reactor) were evaluated for a variety of process conditions (temperature, pressure, gas hourly space velocity, and H_2O/CO feed molar ratio). It appeared that the membrane reactor always gave higher values for these two metrics and thus was more intensified than the traditional reactor for comparable process conditions. The membrane reactor can therefore be operated at less severe process conditions to achieve a preset performance in terms of material and energy efficiency. In addition, the membrane reactor allowed to reach mass and energy intensity values that are unattainable with the traditional reactor, in the order of 30% higher than the traditional maximum.

Similar to the ecological impact assessment of the fabrication of a microreactor itself, the processes to fabricate the membranes and the membrane installations need to be taken into account. The review paper of Szekely et al. [102] extensively discusses these issues, taking among other things the mass intensity and solvent intensity into account.

7.3 Process Intensification and Inherent Safety

Since the turn of the century, safety studies of intensified technologies have been published. At first, these were rather general, but nowadays, as more dedicated safety assessment protocols are being reported, quantitative studies are becoming available. In the field of inherently safer process design, intensification has been mostly connected with the strategy of *minimization*, i.e. the use of smaller quantities of hazardous substances [26, 103–105]. In such context, process intensification has usually been limited to continuous reactors with smaller dimensions than the traditional (fed-)batch reactors or continuously stirred tank reactors (CSTRs). Here, intensification is assumed to lead to safety benefits through the reduction of inventory of hazardous substances, the reduction of number of process operations, the ability of better containing thermal runaways and pressure development (e.g. explosion), and the avoidance of process transients, such as start-up and shutdown. However, some potential problems need to be mentioned. These include the possibility to use high-energy inputs (e.g. microwaves, ultrasound, and rotation) that may introduce new hazards and the higher complexity of process and, therefore, process control. The previously mentioned advantages may also bring about disadvantages. A reduced holdup, for example, results in fast dynamics, which leads to less response time for the operators and the control system to stabilize the process [26]. This could push the process into unsafe regions of operation. Larger process variability also adversely affects product quality, thus requiring further work-up processes, which again can induce risks. New, smaller, and faster sensors and actuators are therefore needed to deal safely with fast process systems [106].

An interesting case in that perspective is the perceived inherent safety of microreactors. Controlled oxidation of explosive gases is attractive in microreactor channels. A large amount of literature is available on the maximum dimensions of microscale tubes below which flames cannot propagate (i.e. the quenching distance). It appears that detonations usually propagate in macroscopic tubes if one-third of the so-called detonation cell width λ is lower than the tube diameter [107]. The very first studies on critical tube diameter were reported in 1956 [108], and interest continued on since then (e.g. [109–112]). Recently, the knowledge contained in these publications was picked up by the microreactor community. It was shown that the $\lambda/3$ rule is also valid for microscale tubes [113–115]. Combined with an excellent heat removal, microchannels are perfectly suited to avoid runaway reactions and explosions. Investigated mixtures in microreactors include the ethylene/oxygen mixture for ethylene oxide synthesis [113–115], the hydrogen/oxygen mixture for the highly exothermic production of water and associated heat [116], and the ethylene–nitrous oxide mixture [114, 115]. Figure 7.6 shows the calculated maximum safe capillary diameter for the ethylene/oxygen mixture in various conditions, based on the $\lambda/3$ rule [113]. The safe diameter is a function of mixture stoichiometry, temperature, and pressure: higher temperatures increase the detonation cell width and the safe capillary diameter, whereas higher pressures decrease the safe capillary diameter. Assuming stoichiometric gas composition of ethylene and oxygen and initial temperature and pressure according to the ethylene oxidation process (70% ethylene, $T = 550$, $P = 3$ MPa), the safe capillary diameter for this process can be estimated to be 0.1 mm. This is attainable in microreactors. However, one could argue that the process should be safe throughout the whole range of feed compositions and that the safe capillary diameter, therefore, has to be appropriate for even the most hazardous feed gas mixture. In that case, Figure 7.6 shows that the safe capillary diameter is much lower, namely 0.25 μm, which is not practicable for microreactors. It is therefore

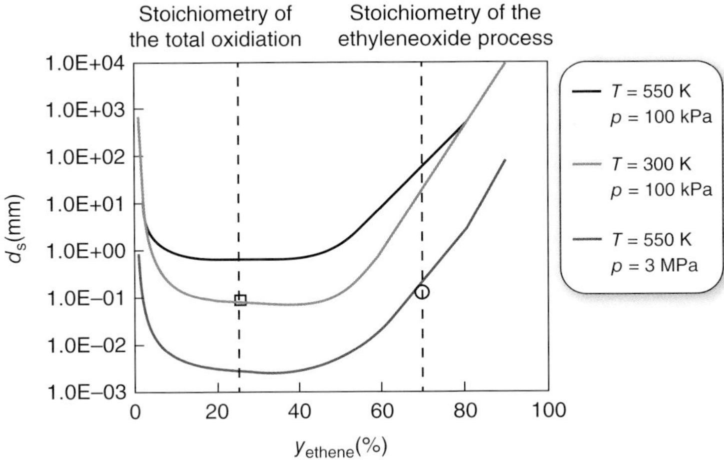

Figure 7.6 Maximum safe capillary diameter versus molar fraction of ethene for different temperatures and pressures. Source: Fischer et al. 2009 [113]. Reproduced with permission of Elsevier.

important to understand that, although the range of safe operation conditions can clearly be extended by performing explosive reactions in microstructured devices, inherent safety is not necessarily achieved [52, 113].

As with the sustainability assessment, it is questionable whether the existing safety assessment schemes are fit for process intensification technologies. A large number of assessment protocols already exist and are applied in chemical industry today to assess process safety. A selection of the well-known assessment tools is listed in Table 7.4, along with their type of analysis, the process development stage, in which a given tool can be applied, and their maturity. For the assessment of process intensification technologies in their design phase, DOW fire and explosion index (DOW F&EI), layer of protection analysis (LOPA), and inherent safety index (ISI) have been used in reports that are discussed below.

Table 7.4 Selection of safety assessment tools.

	Type of analysis	Process development stage	Maturity
HAZOP	Qualitative	Existing installation	High
HAZOP-LIKE	Qualitative	Design	Low
DOW F&EI	Semiquantitative	Design to existing installation	High
LOPA	Semiquantitative	Design to existing installation	High
PIIS	Semiquantitative	Design	Low
ISI	Semiquantitative	Design	Low
iSafe	Semiquantitative	Design	Low
I2SI	Semiquantitative	Design	Low
QRA	Quantitative	Existing installation	High

Kidam et al. [117] studied some of these tools to compare an intensified process with a conventional one. The selected process was the production of hydrogen peroxide, either in a bubble column (160 m^3 effective volume, 1 bar, air as oxidant with 76% oxygen utilization) or in the intensified tubular reactor (7 m^3 effective volume, 2.39 bar, pure (99%) oxygen as oxidant with 93% oxygen utilization). The ISI index decreased from 14 to 12 (reduction of 14%), whereas the DOW F&EI index decreased from 150 to 110 (reduction of 27%). The prototype index of inherent safety (PIIS) index remained at a level of eight in both cases. Ebrahimi et al. [118] performed a LOPA on the same two technologies. They showed that safety was improved by intensification on the basis of three criteria (lower inventory, more regular flow pattern, and better heat exchange) but became worse on the basis of five criteria (e.g. lower ignition energy, no industrial experience, and more complex gas injection), with no effects on the basis of eight criteria. Furthermore, they also evaluated the performance of the same process in a microreactor. It appeared that the microreactor scored better compared to the conventional bubble column on four criteria (less complex, lower holdup, more regular flow pattern, and better heat exchange), but worse on seven criteria (e.g. more explosive, higher pressure, no industrial experience, and more complicated process control), with no difference in five criteria. In particular, in the (first) inherent safety layer, several disadvantages for microreactors were reported. It is difficult to draw conclusions from these numbers because the criteria have different degrees of importance. The most important layer of protection in the conventional LOPA philosophy is the inherent safety layer where no improvements were made by intensification, and for some criteria, there was even a decrease in safety. This is not unexpected because the approach of many PI believers is that due to the improvements in passive protection (the second layer in LOPA) by the reactor design, one can stretch process conditions (e.g. concentrations of chemicals, temperature, and pressure) beyond the inherent safety level. By intensifying the process, there was a worsening in the (first) inherent safety layer, but an improvement in the (second) passive layer. This viewpoint is also illustrated by the hazards and operationability analysis (HAZOP) safety study for the hydrogenation of *o*-cresol in a so-called polyvalent rectilinear stirred minireactor with optimized transfer (RAPTOR) compared to a conventional batch reactor [119].

The research group of Turunen also performed another study where they applied four types of safety assessments on both conventional (stirred tank batch reactor followed by distillation) and intensified technology (continuous microprocess with two mixing steps and a reaction step) for the production of peracetic acid [120]. The DOW F&EI assessment yielded a decrease in index by intensification from 226 to 112 (c. 50%), the ISI index increased from 5 to 6 indicating that the microprocess was less inherently safe, and the Worst Case and Consequence Analysis showed a reduction in worst cases and consequences from 19 to 10 (c. 47% reduction). The ISI index does not take into account the smaller reaction volume and, according to the authors, favors standard and well-known processes and equipment.

Another comparison was performed by Van Calster in the group of Van Gerven [121]. The silver-catalyzed production of ethylene oxidation was compared in the conventional process and in a microstructured process. The conventional process in industry uses a multitubular reactor with packing of silver catalyst on alumina and pure (>99%) oxygen [122–124]. The intensified process is based on a microreactor with the silver catalyst coated on the reactor wall [81]. Both the ISI and the DOW F&EI indices were calculated based on the data found in the literature. By intensification of the process, the ISI index decreased from 42 to 29 (31% reduction), whereas the DOW F&EI index (including credit factors) decreased from 121 to 74 (39% reduction). Both index systems indicate that the microreactor is safer than the conventional system.

The examples reported above show that most comparisons between conventional and intensified technologies have been performed on processes applying microstructured devices. Only one example was found where a reactive separation technology was compared with conventional technology by a safety assessment tool. Ebrahimi et al. [120] compared for hydrogen peroxide synthesis the tubular column as described earlier with a novel design where both reaction and extraction were performed in the same tubular reactor. This novel tubular reactor was filled with static mixer elements and phase separators. The reaction length was the same for both tubular devices because the process rate is mostly limited by the oxidation rate. The LOPA analysis showed that the intensified tube was safer because of a smaller liquid holdup and a larger specific heat exchange surface. However, there are also six criteria for which safety might be reduced, such as the challenge of performing three-phase flow (oxygen gas, organic working solution, and aqueous hydrogen peroxide solution), more complex reactor geometry, and process control. No improvement was made in the (first) inherent safety layer, on the contrary: the intensified systems scored worse on two of the six criteria.

It is clear that all the assessment tools have their strong and weak points, and not necessarily are applicable when comparing a conventional with an intensified technology. Recently, the European IMPULSE project has published a number of papers on the development of HAZOP-LIKE, a HAZOP-based safety assessment tool dedicated to microstructured reactors [52, 125–127]. The authors state that the tool specifically covers characteristic features of microdesigned equipment that are relatively unimportant for conventional equipment, such as short residence times with potential maloperation because of flow and pressure fluctuations, strong dependence of chemical conversion on minor deviations in temperature and flow rate, high surface versus volume ratio because of small diameters with important effects of fouling, structural integrity, and influence of corrosion/erosion on equipment performance. Another attempt to adapt HAZOP to microstructured applications is reported in Kockmann [128] for the lithiation of acetylene, an organolithium reaction, a nitration reaction, and the Grignard reaction. Summarizing, the field of safety metrics for intensified technologies is in the development phase and new, improved assessment tools will see the light in the coming years. For now, it is very difficult to obtain clear and straightforward results, and it may well continue to be so even with improved assessment protocols.

References

1 DOE EIA (2000). Long Term World Oil Supply. AAPG Annual Convention Meeting, April 16–19, New Orleans, Louisiana.

2 Bringezu, S., Ramaswami, A., Schandl, H. et al. (2017). Assessing global resource use: A systems approach to resource efficiency and pollution reduction. A Report of the International Resource Panel. United Nations Environment Programme, Nairobi, Kenya.

3 von Weizsäcker, E., Lovins, A.B., and Lovins, L.L. (1998). *Factor Four: Doubling Wealth, Halving Resource Use*. New York: Earthscan/Taylor & Francis Group.

4 Schmidt-Bleek, F. (1994). How to reach a sustainable economy, Wuppertal Papers Nr.24, Wuppertal Institute.

5 Venselaar J. (2001). The role of chemistry and chemical engineering in an integral approach for transitions towards sustainable development. TOS1 Conference, Budapest.

6 Allen, D.T. and Shonnard, D.R. (2001). Green engineering: environmentally conscious design of chemical processes and products. *AIChE J.* 47: 1906–1910.

7 Bakshi, B.R. and Fiksel, J. (2003). The quest for sustainability: challenges for process systems engineering. *AIChE J.* 49: 1350–1358.

8 Batterham, R.J. (2006). Sustainability – the next chapter. *Chem. Eng. Sci.* 61: 4188 4193.

9 van Berkel, R. (2006). Cleaner production and eco-efficiency. In: *Handbook on Environmental Technology Management* (ed. D. Marinova, D. Annandale and J. Phillimore), 67–92. Cheltenham, UK: Edward Elgar Publishing, Ltd.

10 Anastas, P.T. and Zimmerman, J.B. (2003). Design through the 12 principles of green engineering. *Environm. Sci. Technol.* 37: 94A–101A.

11 Anastas, P.T. and Warner, C.J. (1998). *Green Chemistry: Theory and Practice*. Oxford: Oxford University Press.

12 Tang, S.L.Y., Smith, R.L., and Poliakoff, M. (2005). Principles of green engineering: productively. *Green Chem.* 7: 761–762.

13 Tang, S.L.Y., Bourne, R.A., Poliakoff, M., and Smith, R.L. (2008). The 24 principles of green chemistry and green engineering: improvements productively. *Green Chem.* 10: 268–269.

14 Stankiewicz, A.I. and Moulijn, J.A. (2000). Process intensification: transforming chemical engineering. *Chem. Eng. Progr.* 96: 22–34.

15 Bakker, R.A. (2003). Process intensification in industrial practice: methodology and application. In: *Re-engineering the Chemical Processing Plant: Process Intensification* (ed. A. Stankiewicz and J.A. Moulijn), 419–440. New York: Marcel Dekker.

16 Richardson, J.F., Harker, J.H., and Backhurst, J.R. (2003). *Coulson's and Richardson's Chemical Engineering, Vol. 2: Particle Technology and Separation Processes*. Oxford: Butterworth-Heinemann.

17 Harmsen, G.J., Korevaar, G., and Lemkowitz, S.M. (2003). Process intensification: contributions to sustainable development. In: *Re-engineering the*

Chemical Processing Plant: Process Intensification (ed. A. Stankiewicz and J.A. Moulijn), 464–490. New York: Marcel Dekker.

18 Charpentier, J.C. (2007). In the frame of globalization and sustainability, process intensification, a path to the future of chemical and process engineering (molecules into money). *Chem. Eng. J.* 134: 84–92.

19 Criscuoli, A. and Drioli, E. (2007). New metrics for evaluating the performance of membrane operations in the logic of process intensification. *Ind. Eng. Chem. Res.* 46: 2268–2271.

20 European Roadmap for PI (2008). Creative Energy – Energy Transition. http://www.creative-energy.org (accessed 23 June 2008).

21 Platform Keten Efficiëntie – Energie Transitie (2008). Factsheet KE, Process intensification. http://www.senternovem.nl/energietransitieke/documentatie (accessed 12 September 2008).

22 Reay, D., Ramshaw, C., and Harvey, A. (2008). *Process Intensification: Engineering for Efficiency, Sustainability and Flexibility*. Oxford: Butterworth-Heinemann.

23 BHR Group (2010). Fluid engineering research and consultancy. http://www.bhrgroup.com/process_intensification.aspx (accessed 25 September 2010).

24 Lutze, P., Gani, R., and Woodley, J.M. (2010). Process intensification: a perspective on process synthesis. *Chem. Eng. Process.* 49: 547–558.

25 PIN (2010). Process intensification network http://www.pinetwork.org/about/pi.htm (accessed 30 September 2010).

26 Luyben, W.L. and Hendershot, D.C. (2004). Dynamic disadvantages of intensification in inherently safer process design. *Ind. Eng. Chem. Res.* 43: 384–396.

27 Becht, S., Franke, R., Geisselmann, A., and Hahn, H. (2009). An industrial view of process intensification. *Chem. Eng. Process.* 48: 329–332.

28 Leet, W.A. and Kulprathipanja, S. (2002). *Reactive Separation Processes*. New York: Taylor & Francis.

29 Hessel, V., Cortese, B., and de Croon, M.H.J.M. (2011). Novel process windows – concept, proposition and evaluation methodology, and intensified superheated processing. *Chem. Eng. Sci.* 66: 1426–1448.

30 Carvalho, A., Mimoso, A.F., Mendes, A.N., and Matos, H.A. (2014). From a literature review to a framework for environmental process impact assessment index. *J. Clean. Prod.* 64: 36–62.

31 Angelakoglou, K. and Gaidajis, G. (2015). A review of methods contributing to the assessment of the environmental sustainability of industrial systems. *J. Clean. Prod.* 108: 725–747.

32 Kralisch, D., Ott, D., and Gericke, D. (2015). Rules and benefits of life cycle assessment in green chemical process and synthesis design: a tutorial review. *Green Chem.* 17: 123–145.

33 International Organisation for Standardization – ISO (1997). Environmental management – life cycle assessment – principles and framework: ISO 14040.

34 Scientific Applications International Corporation – SAIC (2006). *Life Cycle Assessment: Principles and Practice*. US Environmental Protection Agency.

35 Hilaly, A.K. and Sikdar, S.K. (1994). Pollution balance: a new method for minimizing waste production in manufacturing processes. *J. Air Waste Manage. Assoc.* 44: 1303–1308.

36 Young, D.M. and Cabezas, H. (1999). Designing sustainable processes with simulation: the waste reduction (WAR) algorithm. *Comp. Chem. Eng.* 23: 1477–1491.

37 Narodolawsky, M. and Krotscheck, C. (1995). The sustainability index (SPI): evaluating processes according to environmental compatibility. *J. Hazard. Mater.* 41: 383–397.

38 Trost, B.M. (1991). The atom economy: a search for synthetic efficiency. *Science* 254: 521–527.

39 Sheldon, R.A. (1992). Organic synthesis: past, present and future. *Chem. Ind.* 23: 903–906.

40 Schmidt-Bleek, F. (1993). MIPS – a universal ecologic measure. *Fresenius Environ. Bull.* 2: 407–412.

41 Dhillon, B.S. (1989). *Life Cycle Costing*. New York: Gordon and Breach Science Publishers.

42 Brealy, R.A. and Myers, S.C. (2000). *Principles of Corporate Finance*, 6e. New York: McGraw-Hill.

43 BASF (2010). Eco-efficiency analysis. http://www.basf.com/group/corporate/en/sustainability/eco-efficiency-analysis/index (accessed 23 September 2010).

44 Kheawhom, S. and Hirao, M. (2002). Decision support tools for process based design and selection. *Comput. Chem. Eng.* 26: 747–755.

45 Kheawhom, S. and Hirao, M. (2004). Decision support tools for environmentally benign process design under uncertainty. *Comput. Chem. Eng.* 28: 1715–1723.

46 Koller, G., Fischer, U., and Hungerbühler, K. (2000). Assessing safety, health and environmental impact early during process development. *Ind. Eng. Chem. Res.* 39: 960–972.

47 Azapagic, A. (2002). *Sustainable Development Progress Metrics*. UK: IChemE Sustainable Development Working Group, IChemE, Rugby.

48 Carvalho, A., Gani, R., and Matos, H. (2008). Design of sustainable chemical processes: systematic retrofit analysis generation and evaluation of options. *Process. Saf. Environ. Prot.* 86: 328–346.

49 Van Aken, K., Strekowski, L., and Patiny, L. (2006). EcoScale, a semi-quantitative tool to select an organic preparation based on economical and ecological parameters. *Belstein J. Org. Chem.* 2: 1–7.

50 Berkoff, C.E., Kamholz, K., Rivard, D.E. et al. (1986). The process profile. *CHEMTECH* 16: 552–559.

51 Kletz, T. (1993). *HAZOP and HAZAN: Identifying and Assessing Process Industry Hazards*, 3e. Institution for Chemical Engineers.

52 Klais, O., Westphal, F., Benaïssa, W. et al. (2010). Guidance on safety/health for process intensification including MS design. Part III: Risk analysis. *Chem. Eng. Technol.* 33: 44–454.

53 American Institute of Chemical Engineers – AIChE (1994). *DOW Fire & Explosion Index Hazard Classification Guide*, 7e. New York: Wiley-Blackwell.

54 Center for Chemical Process Safety – CCPS (2001). *Layer of Protection Analysis, Simplified Process risk assessment*, 1e. AIChE-CCPS.

55 Edwards, D.W. and Lawrence, D. (1993). Assessing the inherent safety of chemical process routes – Is there a relation between plant costs and inherent safety. *Process Safety Environm. Protect.* 71: 252–258.

56 Heikilla, A.M., Hurme, M., and Jarvelainen, M. (1996). Safety considerations in process synthesis. *Comput. Chem. Eng.* 20: 115–120.

57 Heikkila, A.M. (1999). Inherent safety in process plant design. An index-based approach. PhD Dissertation. Helsinki University of Technology, Espoo, Finland.

58 Palaniappan, C., Srinivasan, R., and Tan, R. (2002). Expert system for the design of inherently safer processes. 1. Route selection stage. *Ind. Eng. Chem. Res.* 41: 6698–6710.

59 Palaniappan, C., Srinivasan, R., and Tan, R. (2002). Expert system for the design of inherently safer processes. 2. Flowsheet development stage. *Ind. Eng. Chem. Res.* 41: 6711–6722.

60 Khan, F.I. and Amyotte, P.R. (2004). Integrated inherent safety index (12SI): A tool for inherent safety evaluation. *Process Safety Progress* 23: 136–148.

61 Steinbach, J. (1999). *Safety Assessment for Chemical Processes*. Weinheim, Germany: Wiley-VCH.

62 Swaney, R.E. and Grossman, I.E. (1985). An index for operational flexibility in chemical process design. *AIChE J.* 31: 621–630.

63 Lai, S.M. and Hui, C.W. (2007, 2007). Measurement of plant flexibility. In: *Proceedings of the 17th European Symposium on Computer Aided Process Engineering – ESCAPE17* (ed. V. Plesu and P.S. Agashi), 27–30. Bucharest, Romania: Elsevier.

64 Luis, P. (2013). Exergy as a tool for measuring process intensification in chemical engineering. *J. Chem. Technol. Biotechnol.* 88: 1951–1958.

65 Luis, P. and Van der Bruggen, B. (2014). Exergy analysis of energy-intensive production processes: advancing towards a sustainable chemical industry. *J. Chem. Technol. Biotechnol.* 89: 1288–1303.

66 Luis, P., Van Aubel, D., and Van der Bruggen, B. (2013). Technical viability and exergy analysis of membrane crystallization: closing the loop of CO_2 sequestration. *Int. J. Greenh. Gas Contr.* 12: 450–459.

67 Seider, W.D., Seader, J.D., and Lewin, D.R. (2004). *Product & Process Design Principles – Synthesis, Analysis and Evaluation*, 2e. Wiley.

68 Hessel, V. (2007). Exploring novel processing routes by microstructured reactors. CPAC Satellite Workshop, Micro-reactors and Micro-Analytical Workshop, Rome.

69 Liberti, G., Mattera, A., Soattini, S. et al. (1972). Pulse microreactor study of ethylene oxidation over silver-oxide. *Atti Della Accademia Nazionale deil Lincei Rendiconti – Classe di Scienze Fisiche-Matematiche & Naturali* 52: 392–401.

70 Metcalf, P.L. and Harriott, P. (1972). Kinetics of silver-catalyzed ethylene oxidation. *Ind. Eng. Chem. Process. Des. Dev.* 11: 478–489.

71 Kniel, L., Winter, O., and Stork, K. (1980). *Ethylene: Keystone to the Petrochemical Industry*. New York: Marcel Dekker.

72 Choudhary, V.R. and Rane, V.H. (1992). Pulse microreactor studies on conversion of methane, ethane and ethylene over rare-earth-oxide in the absence and presence of free oxygen. *J. Cat.* 135: 310–316.

73 Gavriilidis, A. and Varma, A. (1992). Optimal catalyst activity profiles in pellets – study of ethylene epoxidation. *AIChE J.* 38: 291–296.

74 Weissermel, K. and Arpe, H.J. (1993). *Industrial Organic Chemistry*, 2e. Weinheim, Germany: VCH.

75 Borman, P.C. and Westerterp, K.R. (1995). An experimental study of the kinetics of the selective oxidation of ethane over a silver on alpha-alumina catalyst. *Ind. Eng. Chem. Res.* 34: 49–58.

76 Goncharova, S.N., Paukshtis, E.A., and Balzhinimaev, B.S. (1995). Size effects in ethylene oxidation on silver catalysts – influence of support and Cs promoter. *Appl. Cat. A – Gen.* 126: 157–171.

77 Wittcaff, H.A. and Renben, B.G. (1996). *Industrial Organic Chemistry*. New York: Wiley.

78 Nakatsuji, H., Nakai, H., Ikeda, K., and Yamamoto, Y. (1997). Mechanism of the partial oxidation of ethylene on an Ag surface: dipped adcluster model study. *Surf. Sci.* 384: 315–333.

79 Löwe, H. and Ehrfeld, W. (1999). State-of-the-art in microreaction technology : concepts, manufacturing and applications. *Electrochim. Acta* 44: 3679–3689.

80 Kestenbaum, H., de Oliveira, A.L., Schmidt, W. et al. (2000). Synthesis of ethylene oxide in a microreaction system. In: *Microreaction Technology: Industrial Prospects* (ed. W. Ehrfeld), 207–212. Springer-Verlag Berlin Heidelberg.

81 Kestenbaum, H., de Oliveira, A.L., Schmidt, W. et al. (2002). Silver-catalyzed oxidation of ethylene to ethylene oxide in a microreaction system. *Ind. Eng. Chem. Res.* 41: 710–719.

82 Kolb, G. and Hessel, V. (2004). Micro-structured reactors for gas phase reactions. *Chem. Eng. J.* 98: 1–38.

83 Stegelmann, C., Schiodt, N.C., Campbell, C.T., and Stoltze, P. (2004). Microkinetic modeling of ethylene oxidation over silver. *J. Cat.* 221: 630–649.

84 Kolb, G., Hessel, V., Cominos, V. et al. (2007). Selective oxidations in micro-structured catalytic reactors – for gas-phase reactions and specifically for fuel processing for fuel cells. *Cat. Today* 120: 2–20.

85 Carucci, J.R.H., Halonen, V., Eranen, K. et al. (2010). Ethylene oxide formation in a microreactor: from qualitative kinetics to detailed modeling. *Ind. Eng. Chem. Res.* 49: 10897–10907.

86 Mazanec, T., Desmukh, S., and Silva, L.J. (2011). Process for making ethylene oxide using microchannel process technology. US Patent 2011/0009653 A1, filed 13 July 2009 and issued 13 January 2011.

87 Kralisch, D. and Kreisel, G. (2007). Assessment of the ecological potential of microreaction technology. *Chem. Eng. Sci.* 62: 1094–1100.

88 Zschieschang, E., Pfeifer, P., and Schebek, L. (2012). Environmentally optimized microreactor design through life cycle assessment. *Green Process. Synth.* (4): 375–384.

89 Zschieschang, E., Pfeifer, P., and Schebek, L. (2013). Life cycle assessment in chemical and micro reaction engineering. *Chem. Eng. Technol.* 36: 911–920.

90 Hessel, V., Kralisch, D., and Krtschil, U. (2008). Sustainability through green processing – novel process windows intensify micro and milli process technologies. *Energy Environ. Sci.* 1: 467–478.

91 Huebschmann, S., Kralisch, D., Hessel, V. et al. (2009). Environmentally benign microreaction process design by accompanying (simplified) life cycle assessment. *Chem. Eng. Technol.* 32: 1757–1765.

92 Kressirer, S., Kralisch, D., Stark, A. et al. (2013). Agile green process design for the intensified Kolbe–Schmidt synthesis by accompanying (simplified) life cycle assessment. *Environ. Sci. Technol.* 47: 5362–5371.

93 Hübner, S., Kressirer, S., Kralisch, D. et al. (2012). Ultrasound and microstructures – a promising combination? *ChemSusChem* 5: 279–288.

94 Huebschmann, S., Kralisch, D., Loewe, H. et al. (2011). Decision support towards agile eco-design of microreaction processes by accompanying (simplified) life cycle assessment. *Green Chem.* 13: 1694–1707.

95 Wang, Q., Gürsel, I.V., Shang, M., and Hessel, V. (2013). Life cycle assessment for the direct synthesis of adipic acid in microreactors and benchmarking to the commercial process. *Chem. Eng. J.* 234: 300–311.

96 Dencic, I., Ott, D., Kralisch, D. et al. (2014). Eco-efficiency analysis for intensified production of an active pharmaceutical ingredient: a case study. *Org. Process Res. Dev.* 18: 1326–1338.

97 Ott, D., Kralisch, D., Dencic, I. et al. (2014). Life cycle analysis within pharmaceutical process optimization and intensification: case study of active pharmaceutical ingredient production. *ChemSusChem* 7: 3521–3533.

98 Sundaram, S., Kralisch, D., Wang, Q., and Hessel, V. (2015). Sustainability lessons from practice: how flow intensification can trigger sustainability and modular plant technology in EU projects. *Asia-Pac. J. Chem. Eng.* 10: 483–500.

99 Ott, D., Borukhova, S., and Hessel, V. (2016). Life cycle assessment of multi-step rufinamide synthesis – from isolated reactions in batch to continuous microreactor networks. *Green Chem.* 18: 1096–1116.

100 Bonet-Ruiz, A.E., Bonet, J., Plesu, V., and Bozga, G. (2010). Environmental performance assessment for reactive distillation process. *Resour. Cons. Recyc.* 54: 315–325.

101 Brunetti, A., Drioli, E., and Barbieri, G. (2014). Energy and mass intensities in hydrogen upgrading by a membrane reactor. *Fuel Process. Technol.* 118: 278–286.

102 Szekely, G., Jimenez-Solomon, M.F., Marchetti, P. et al. (2014). Sustainability assessment of organic solvent nanofiltration: from fabrication to application. *Green Chem.* 16: 4440–4473.

103 Hendershot, D.C. (2004). Process intensification for safety. In: *Re-engineering the Chemical Processing Plant: Process Intensification* (ed. A. Stankiewicz and J.A. Moulijn), 441–463. New York: Marcel Dekker.

104 Etchells, J.C. (2005). Process intensification: safety pros and cons. *TransIChemE B Process Safety Environ. Protect.* 83: 85–89.

105 Srinivasan, R. and Natarajan, S. (2012). Developments in inherent safety: a review of the progress during 2001-2011 and opportunities ahead. *Process Safety Environ. Protect.* 90: 389–403.

106 Nikacevic, N.M., Huesman, A.E.M., Van den Hof, P.M.J., and Stankiewicz, A.I. (2012). Opportunities and challenges for process control in process intensification. *Chem. Eng. Process.* 52: 1–15.

107 Nettleton, M.A. (1987). *Gaseous Detonation: Their Nature, Effects and Control*. London: Chapman & Hall, Ltd.

108 Zeldovich, I.A.B., Kogarko, S.M., and Simonov, M.M. (1956). An experimental investigation of spherical detonation of gases. *Soviet Physics Techn. Physics* 1: 1689–1713.

109 Knystautas, R., Lee, J.H., and Guirao, C.M. (1982). The critical tube diameter for detonation failure in hydrocarbon-air mixtures. *Combust. Flame* 48: 63–83.

110 Munteanu, V. and Jarosinski, J. (2007). Study of propane-air flame propagating in narrow channels under quenching condition. *Analele Universitatii din Bucuresti – Chimie* 16: 71–77.

111 Wu, M., Burke, M.P., Son, S.F., and Yetter, R.A. (2007). Flame acceleration and the transition to detonation of stoichometric ethylene/oxygen in micro scale tubes. *Proc. Combust. Inst.* 31: 2429–2436.

112 Kitano, S., Fukao, M., Tsuboi, N. et al. (2009). Spinning detonation and velocity deficit in small diameter tubes. *Proc. Combust. Inst.* 32: 2355–2362.

113 Fischer, J., Liebner, C., Hieronymus, H., and Klemm, E. (2009). Maximum safe diameters of microcapillaries for a stoichiometric ethene/oxygen mixture. *Chem. Eng. Sci.* 64: 2951–2956.

114 Hieronymus, H., Fischer, J., Heinrich, S. et al. (2011). Safety investigations for operation of microreactors inside the explosion region. *Chem. Ing. Tech.* 83: 1742–1747.

115 Liebner, C., Fischer, J., Heinrich, S. et al. (2012). Are micro reactors inherently safe? An investigation of gas phase explosion propagation limits on ethene mixtures. *Process Safety Environ. Protect.* 90: 77–82.

116 Janicke, M.T., Kestenbaum, H., Hagendorf, U. et al. (2000). The controlled oxidation of hydrogen from an explosive mixture of gases using a microstructured reactor/heat exchanger and Pt/Al_2O_3 catalyst. *J. Cat.* 191: 282–293.

117 Kidam, K., Hassim, M.H., and Hurme, M. (2008). Enhancement of inherent safety in chemical industry. *Chem. Eng. Trans.* 13: 287–294.

118 Ebrahimi, F., Virkki-Hatakka, T., and Turunen, I. (2012). Safety analysis of intensified processes. *Chem. Eng. Process.* 52: 28–33.

119 Machefer, S., Falk, L., and de Panthou, F. (2013). Intensification principle of a new three-phase catalytic slurry reactor. Part II. Eco-efficiency and techno-economic performances. *Chem. Eng. Process.* 70: 267–276.

120 Ebrahimi, F., Kolehmainen, E., and Turunen, I. (2009). Safety advantages of on-site microprocesses. *Org. Process Res. Dev.* 13: 965–969.

121 Van Calster, S. (2010). Safety aspects of process intensification (in Dutch). Master dissertation. MSc Chemical Technology, KU Leuven.

122 Redsbat, S. and Mayer, D. (2000). Ethylene oxide. In: *Ullmann's Encyclopedia of Industrial Chemistry*, 6e. New York: Wiley-VCH.

123 Moulijn, J.A., Makkee, M., and Van Diepen, A. (2001). *Chemical Process Technology*. Chichester: Wiley.

124 Zhou, X.G. and Yuan, W.K. (2005). Optimization of the fixed-bed reactor for ethylene epoxidation. *Chem. Eng. Process.* 44: 1098–1107.

125 Klais, O., Westphal, F., Benaïssa, W., and Carson, D. (2009). Guidance on safety/health for process intensification including MS design. Part I: Reaction hazards. *Chem. Eng. Technol.* 32: 1831–1844.

126 Klais, O., Westphal, F., Benaïssa, W., and Carson, D. (2009). Guidance on safety/health for process intensification including MS design. Part II: Explosion hazards. *Chem. Eng. Technol.* 32: 1966–1973.

127 Klais, O., Albrecht, J., Carson, D. et al. (2010). Guidance on safety/health for process intensification including MS design. Part IV: Case studies. *Chem. Eng. Technol.* 33: 1159–1168.

128 Kockmann, N. (2012). Safety aspects during process development and small scale production with microreactors. *Chem. Ing. Techn.* 84: 715–726.

8

How to Design a Sustainable Intensified Process?

8.1 Conceptual Process Intensification Design

As stated by Gourdon et al. [1], in order for process intensification methods to break through, not only the process needs to be intensified but also the way (i.e. the methodology) for intensifying it. In the next paragraphs, we discuss a possible approach toward process intensification design.

The course "Process Intensification in Chemical Industry" is taught in the Chemical Technology MSc programme, both at Delft University of Technology (TU Delft) in the Netherlands and the University of Leuven (KU Leuven) in Belgium. The course is based on the process intensification concept with principles, domains, and scales, as discussed in this book. The Master students are asked to conceptually redesign a chemical process making use of intensified technologies. The case study processes include the Bhopal carbaryl process (subdivided into three separate subprocesses), the Flixborough Nypro process (subdivided into two separate subprocesses), the methanol carbonylation process, the urea process, and the melamine process. The first process is discussed more in depth in the next section.

The procedure is as follows (Figure 8.1). Students are divided into groups of 3–5. They start with an analysis of the current conventional plant and identify safety and process bottlenecks. Next, they define key assessment criteria, possibly differentiated by weight factors to account for their importance. In the second phase, they generate a long list of intensified alternatives to address the identified bottlenecks, through brainstorm sessions. Then, they converge to a short list by thorough screening, completing the concept selection table (Table 8.1) and selecting 1–3 most promising and viable options. The concept selection table presents a matrix of generated PI options versus the key assessment criteria and is filled in a qualitative way. The selected options are then further developed in a preliminary design study that results in a new process flow sheet and basic dimensioning of the equipment. The students receive as course material a description of the assigned process and, as supplementary material, Appendix 1 of the European Roadmap for Process Intensification report of 2008 [2], which contains a list and brief descriptions of 72 PI technologies.

Recently, Commenge and Falk [3] proposed a useful methodological framework to perform the assignment by applying a three-step approach. In line

The Fundamentals of Process Intensification, First Edition.
Andrzej Stankiewicz, Tom Van Gerven, and Georgios Stefanidis.
© 2019 Wiley-VCH Verlag GmbH & Co. KGaA. Published 2019 by Wiley-VCH Verlag GmbH & Co. KGaA.

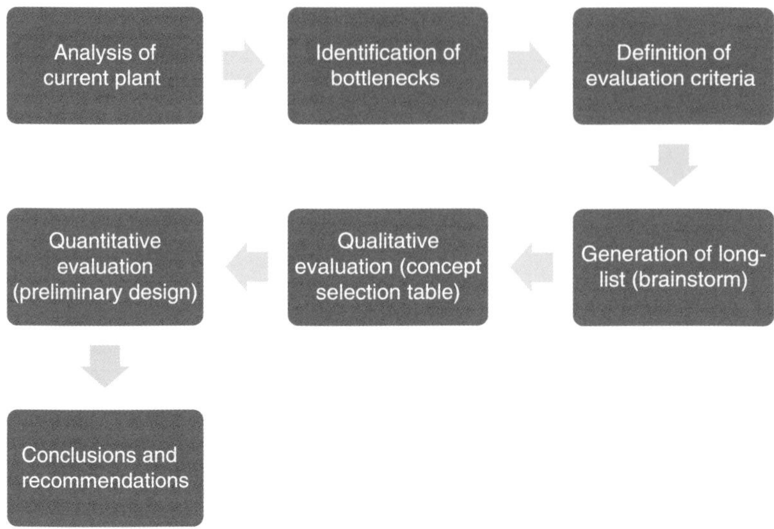

Figure 8.1 Procedure for the conceptual process intensification design assignment.

with the methodology described above, the first step is to identify process limitations. The authors identify 17 possible limitations, classified as elementary limitations (related to, e.g. heat and mass transfer, mixing, and residence time distribution) due to fundamental phenomena, and complex limitations (related to, e.g. equipment volume, safety, pressure drop, and energy consumption), as a result of coupled phenomena. The identification of these limitations can be based on feedback from operators, experimental data, modeling data, and characteristic time analysis. In the second step, these limitations should be matched with useful strategies. In total, 17 strategies classified in three scales (molecular, process, and equipment) are proposed. Examples include change of phases, concentration, temperature, pressure, segmentation, periodic operation, alternative energy sources, geometric structuring, and shear rate. In the third step, the strategies are translated into relevant intensified technologies. These technologies are classified along technologies (reactive and nonreactive) and methods (multifunctional reactors, hybrid separations, alternative energy sources, and other methods), as proposed by Stankiewicz and Moulijn [4]. The whole procedure is illustrated in Figure 8.2. The proposed methodology appears to be very well implementable for the Master student assignment.

Building upon the methodology proposed by Commenge and Falk, a fourth step could be added in the design process by including the sustainability and safety assessment of the selected technology or process compared to the original one. This would entail using the approach illustrated in the previous sections.

8.2 Case Study of Bhopal

The Bhopal case is described in a separate text box below. The process scheme, retrieved from the limited information available in the literature, is divided into

Table 8.1 Example of a concept selection table to reduce possible options from a long list to a short list.

Criterion/reactor type	Mixing degree (toward plug flow)	Mass transfer rates	Investment and operational costs	Reliability and maintenance	Safety of operation	Importance of constructional material	Possibility for pure oxygen feed
Mechanically stirred tanks (in series)	O	+	O	$--$	$--$	+	$-$
Bubble column with motionless mixers	++	++	$-$	++	++	$--$	+
Tubular system with motionless mixers	++	++	$-$	+	+	$--$	+
Membrane reactor	+	?	$-$(?)	?	+	?	+

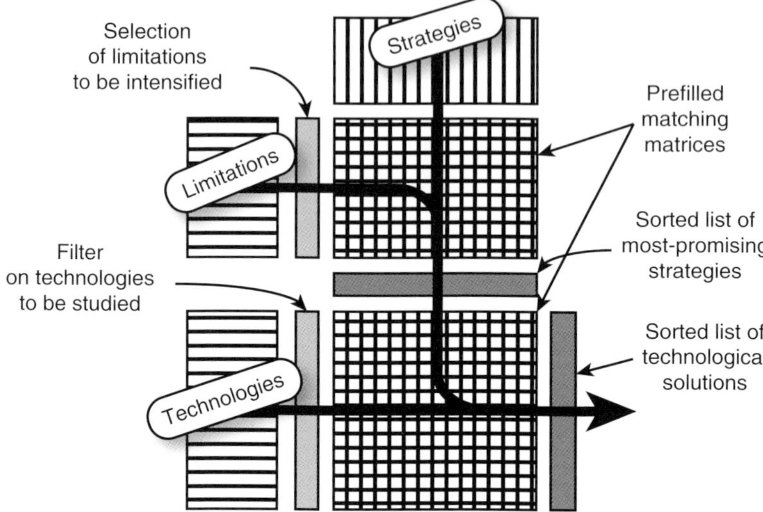

Figure 8.2 A view on the work flow in the selection methodology of Commenge and Falk. Source: Commenge and Falk 2014 [3]. Reproduced with permission of Elsevier.

three sections: the synthesis of methylcarbamoyl chloride (MCC), the MCC pyrolysis, and the carbaryl synthesis. The assignment is to conduct analysis of the assigned section of the Bhopal plant and to develop sustainable alternative(s) of the Union Carbide process assuming the targeted plant capacity as licensed in 1983 (5250 t/yr).

The MCC synthesis project has been assigned to eight groups of students in KU Leuven in the period 2008 till 2014. The highly ranked bottlenecks in the original process were the excess of phosgene and the related need to separate hydrogen chloride from the equilibrium process, the issues with safety in the process (e.g. the presence of chloroform and phosgene and the reactor volume), and the need for better heat removal given the exothermic nature of the reaction. The student groups then developed solutions for these identified bottlenecks. Table 8.2 lists the techniques that were included in their respective long and short lists and that they eventually selected. The final scores on the assignments are also given. These scores are based not only on the selected technique but also on the depth of the study, the arguments used in the process design and the level of interactions within the team. It can be seen that no single solution to the process intensification problem is reached. This is also not expected, as in reality, there is no unique optimal solution. Overall, it seems that most groups rank three of the solutions as viable ones: the microreactor, the heat exchange (HEX) reactor, and the membrane reactor. These solutions solve the bottlenecks by reducing the reactor volume, more intensely removing the produced heat and driving the equilibrium reaction to the reaction product side by removing the hydrogen chloride product, thus alleviating the need of excessive phosgene.

The project assignment helps students to explore novel technologies and relate the process deficiencies with potential improvements. Students also learn to team up and discuss technical aspects among themselves. A further

Table 8.2 Overview of the PI technologies investigated by Chemical Engineering students as process options for the methylcarbamoyl chloride synthesis.

Technique	Group							
	1	2	3	4	5	6	7	8
Microreactor	XX	XXX	X	XX	XXX	XXX	XX	
Static mixer reactor		XX	X	X	X			
Static mixer heat exchanger			X					
Heat exchange (HEX) reactor	XX	X	XXX	XXX	X		XXX	XX
Membrane reactor	XXX		XX		X	XX	X	XXX
Rotating packed bed reactor			X			XX		
Rotor–stator mixer			X					
Spinning disc reactor	X							
Centrifugal absorption			X					
Reactive absorption					X		X	XX
Reactive adsorption							X	
Reactive distillation			X		X			
Membrane-assisted reactive distillation					XX			
Overhead adsorptive distillation			X					
Divided wall column			XXX	XX				
Heat-integrated distillation column				X				
Reverse flow reactor			X		XX			
Oscillatory flow reactor					X			
Pulsed compression reactor			X					
Fluidized bed reactor			XX					
Supercritical reactor	X							
Electric field					X			
Laser field		XX						
Microwave field		X						
Fractal distributors						XX		
Score	**18**	**17**	**15**	**14**	**13**	**12**	**12**	**11**

X, long-listed option; XX, short-listed option; XXX, selected option.

improvement could be to combine process intensification with chemistry intensification (new solvents, new catalysts, etc.) and scale up considerations (e.g. layers of protection analysis and process control), thereby addressing more strategies of the inherently safer process design. A more quantitative approach toward the technology selection would also be beneficial. The approach proposed by Commenge and Falk is a first step in this development. Exergy analysis could also provide useful insights into that respect. However, this would entail much more detailed data to be available regarding the operational conditions and features of the new technologies, which underlines the need for further data generation, both by experimental and modeling studies.

Bhopal: Could We Have Avoided It?

(Case Study for Process Intensification and Inherently Safer Process Design)

Introduction

Shortly after midnight on 3 December 1984, the worst industrial disaster in the history of mankind took place in the Indian town of Bhopal. A leak of c. 40 tons of poisonous gas, methyl isocyanate (MIC), from the pesticide plant of Union Carbide has caused deaths of almost 4000 people and disabled another 3000. The horrifying events of that night have been extensively described in literature, for instance, in *"A Killing Wind"* by D. Kurzman [5] or in *"Bhopal – Anatomy of Crisis"* by P. Shrivastava [6]. A brief technical analysis of Bhopal disaster can be found in volume 3 of the book *"Loss Prevention in the Process Industries"* by F.P. Lees [7].

The Bhopal plant was opened in 1969. At first, it only formulated carbamate pesticides from concentrates imported from the United States, but in 1975, the plant was licensed by the Indian government to produce its own carbaryl (trade name "Sevin"). MIC was a chemical intermediate in the Sevin manufacturing process chosen by Union Carbide. For a time, the Bhopal plant depended on MIC imported from Union Carbide's plant in Institute, West Virginia. However, UC added an MIC production unit to the Bhopal plant in 1979. The unit was approved and designed by UC in the United States. The licensed registered capacity of the plant was 5250 tons in 1983.

In opinion of many experts, the tragedy of Bhopal could have been prevented, or at least minimized, by using the principles of inherently safer process design during the development and design of Union Carbide process. The aim of the present assignment is to analyze the technology used in Bhopal and try to develop a safer alternative, focusing on process intensification issues.

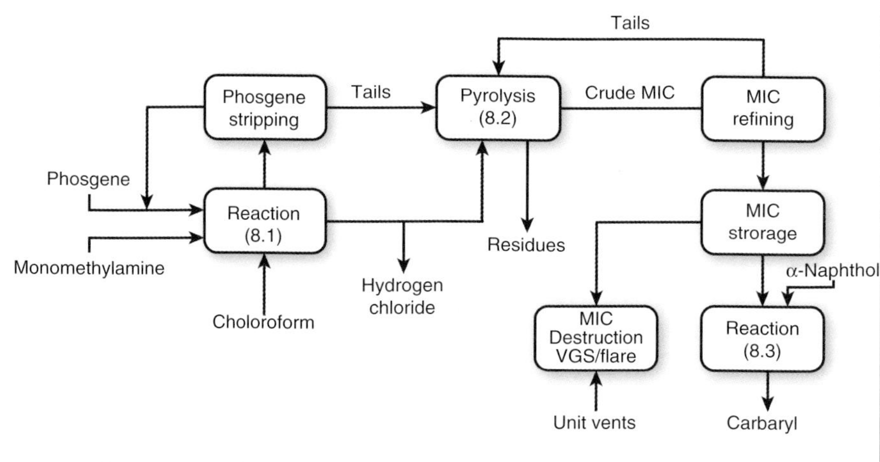

Process Description

The Union Carbide process in Bhopal produced carbaryl (Sevin), a pesticide, via the so-called MIC route, as shown in a simplified scheme below.

The MIC process route begins with the reaction of monomethylamine (MMA) with excess phosgene (used as a lethal gas in World War I and produced in Bhopal on-site from chlorine and carbon monoxide). The reaction is carried out in the vapor phase delivering MCC and hydrogen chloride:

$$COCl_2 \ + \ CH_3NH_2 \ \rightarrow \ CH_3NHCOCl \ + \ HCl \ + \ Heat$$

Phosgene   Monomethyl-   Methylcarbamoyl   Hydrogen

amine   chloride   chloride

(MMA)   (MCC)   (8.1)

Phosgene preheated to 205 °C enters the reactor together with MMA preheated to 240 °C. Molar ration between phosgene and MMA is 1.25 : 1. An excess of phosgene is required to prevent formation of methylamine hydrochloride in the gaseous phase. The noncatalytic gas-phase reaction (8.1) is strongly exothermic, very fast (residence time in the reactor is 1.5 second), and runs at 260 °C, at normal pressure. At the reactor outlet MCC (selectivity 100%) and unreacted phosgene are absorbed and quenched in chloroform (5 °C), down to 40 °C.

The unreacted phosgene is separated by distillation from the quench liquid and recycled to the reactor. The liquid from the still is fed to the pyrolysis section where MIC is formed. The pyrolysis reaction

$$CH_3NHCOCl \ + \ Heat \ \rightarrow \ CH_3NCO \ + \ HCl$$

Methylcarbamoyl   Methyl   Hydrogen

chloride   isocyanate   chloride

(MCC)   (MIC)   (8.2)

is liquid phase and reversible. It is endothermic and heat has to be added in order to shift the equilibrium to the MIC side. Reaction (8.2) is noncatalytic; however, the presence of hydrogen chloride under these conditions catalyzes the side reaction of polymerization of MIC and expensive polymerization inhibitors must be added to the reaction mixture to avoid this unwanted side reaction. The pyrolysis reaction takes place at temperature 90 °C, pressure 9.5–10 bar, and with a residence time of 21 hours. About 80% of the MMC is decomposed to MIC and HCl. The vapor from pyrolysis reactors is cooled to remove most of the HCl and condensed to form a mixture of chloroform, MMC, MIC, and HCl. That mixture is distilled in a 45-plate column, where about 60% of the MIC is recovered as distillate; the remaining 40% is combined with HCl to form MMC and is recycled to pyrolysis section for decomposition. From the bottom of distillation section, chloroform is discharged, evaporated, condensed, and recycled. Heavy ends are discharged for incineration.

The MIC is then run to the storage. It is here, in one of the storage tanks, where the highly exothermic hydrolysis of MIC took place in the night of 3 December 1984, leading to Bhopal disaster.

(Continued)

(Continued)

From the storage tanks, MIC is fed to the third reaction section, where it reacts with α-naphthol to form the final product – carbaryl.

The reaction

$$CH_3NCO + \text{(α-Naphthol)} \longrightarrow \text{(Carbaryl)} \tag{8.3}$$

Methyl isocyanate (MIC)   α-Naphthol   Carbaryl (1-naphthyl methylcarbamate)

utilizes anion exchange resin (e.g. Amberlite) catalyst to convert weak base groups to their catalytically active free amine form, while maintaining any strong-base groups in their catalytically inactive salt form. The process is carried out in a continuous flow system. Carbon tetrachloride, CCl_4, is used as a solvent. Generally, the reaction between hydroxy-substituted organic compounds and compounds containing isocyanate groups are exothermic in nature and, therefore, some means should be provided to maintain the temperature below the selected upper limit. Typical conditions for Reaction (8.3) are given below:

Feed

1-Naphtol	18.0 wt%
MIC	7.2 wt%
CCl_4	74.8 wt%
MIC:1-naphtol mole ratio	1.01

Temperatures

Jacket	83 °C
Reactor bottom	74 °C
Reactor middle	93 °C
Reactor top	88 °C

Products

Product (1-naphtyl methylcarbamate) yield	91.2%
Productivity	3.2 kg product/kg resin/h
Product purity	99.8%

Assignment

Conduct analysis of the Bhopal plant and develop sustainable alternative(s) of the Union Carbide process assuming the targeted plant capacity as licensed in 1983 (5250 t/yr). The analysis should identify the bottlenecks in the process by applying the generic PI principles. In the generation of PI concepts, make use of four fundamental approaches of process intensification in the spatial, thermodynamic, functional, and temporal domains. Consider all scales, from molecular to the scale of processing units. The sustainable alternative(s) should make broad use of process-intensive equipment and methods and should allow to minimize the risk to humans and environment.

Final Remark
Full details of operational and technical data of the Union Carbide plant in Bhopal are not available to the public. Therefore, in the present assignment, data on some parts of the process have been retrieved based on published papers, patent literature, and SRI reports on Union Carbide's carbaryl technology. These data may not reproduce the situation in Bhopal with 100% accuracy. In case of any missing piece(s) of information, the assignees are expected to make reasonable assumptions.

References

1 Gourdon, C., Elgue, S., and Prat, L. (2015). What are the needs for process intensification? *Oil Gas Sci. Technol. – Rev. IFP Energ. Nouv.* 70: 463–473.
2 European Roadmap for Process Intensification (2008). Creative energy – energy transition. http://efce.info/EUROPIN.html (accessed 15 February).
3 Commenge, J.M. and Falk, L. (2014). Methodological framework for choice of intensified equipment and development of innovative technologies. *Chem. Eng. Process.* 84: 109–127.
4 Stankiewicz, A.I. and Moulijn, J.A. (2000). Process intensification: transforming chemical engineering. *Chem. Eng. Progr.* 96: 22–34.
5 Kurzman, D. (1987). *A Killing Wind*. McGraw-Hill.
6 Shrivastava, P. (1992). *Bhopal – Anatomy of Crisis*. Paul Chapman Publishing.
7 Lees, F.P. (1996). *Loss Prevention in the Process Industries*. Butterworth-Heinemann.

Index

The Fundamentals of Process Intensification, First Edition.
Andrzej Stankiewicz, Tom Van Gerven, and Georgios Stefanidis.
© 2019 Wiley-VCH Verlag GmbH & Co. KGaA. Published 2019 by Wiley-VCH Verlag GmbH & Co. KGaA.